D0554800

COLLIDING CONTINENTS

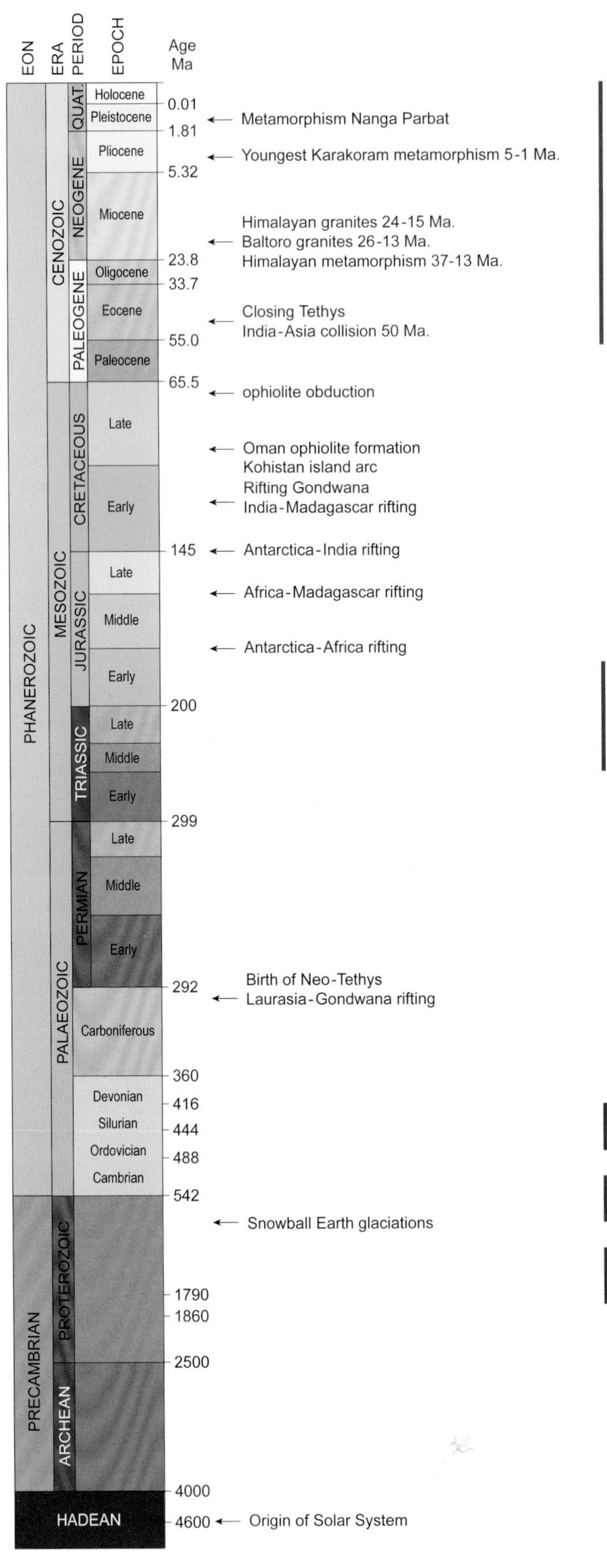

Geological timescale, after the International Commission on Stratigraphy, showing some of the major orogenic mountain-building events and key tectonic events in the Alpine-Himalayan mountain belt. Note the non-linear scale. Age is in million years (Ma).

colliding CONTINENTS

A geological exploration of the Himalaya, Karakoram, & Tibet

MIKE SEARLE

OXFORD
UNIVERSITY PRESS

Great Clarendon Street, Oxford OX2 6DP

Oxford University Press is a department of the University of Oxford.
It furthers the University's objective of excellence in research, scholarship,
and education by publishing worldwide. Oxford is a registered trade mark of
Oxford University Press in the UK and in certain other countries

First Edition published in 2013
Impression: 1

British Library Cataloguing in Publication Data
Data available

Library of Congress Cataloging in Publication Data
Data available

ISBN 978–0–19–965300–3

Typeset by Sparks—www.sparkspublishing.com
Printed in China by C&C Offset Printing Co. Ltd

DEDICATION

This book is dedicated to the memory of my parents, Geoff and Pauline Searle, with my love, my gratitude, and my thanks for a wonderful peripatetic upbringing.

PREFACE

I was lucky enough to have been born and brought up in some of the most wild and beautiful lands of the Far East, in Malaya, Java, and Borneo. As a geology undergraduate at Aberystwyth University in Wales I had been on a three-month-long climbing expedition to Patagonia with four friends, an expedition that evolved into an eight-month trip along the Andes from Patagonia along the length of Chile, through Bolivia and Peru to Ecuador. I had climbed some of the dreaming rock spires of Patagonia, had seen the great North Patagonian Ice Cap with its immense glaciers flowing straight into the Pacific Ocean, the steaming volcanoes and the magnificent ancient Araucania forests of the Chilean Lake District, and the wild, empty stretches of the Atacama Desert. I had travelled across the high Altiplano of Bolivia, trekked across the Peruvian Andes, and had climbed Cotopaxi, one of the great equatorial volcanoes of Ecuador. I had three weeks left in South America, and decided that I had to see the Galapagos Islands before I left. Galapagos was where Charles Darwin had made his studies of the famous finches and began to formulate his ideas on the origin of species, natural selection, and the survival of the fittest. These ideas culminated in the theory of evolution, the greatest scientific breakthrough of the time. Darwin had read Charles Lyell's *Principles of Geology* on the voyage of the Beagle along South America, and this classic work sowed the seeds of his theory of evolution and the book that changed the scientific world, *The Origin of Species*. As Darwin's great champion Thomas Huxley said, 'Biology takes its time from geology.'

I was reading Darwin's *Origin of Species* at the time, in between watching the famous Galapagos finches, the fabulous frigate birds, boobies, and sea eagles, diving with sea lions and marine iguanas, and tramping over the beautifully preserved lava flows. Here, cruising through the Galapagos Islands, Charles Darwin had come up with probably the most important scientific theory since Albert Einstein's theory of relativity. Einstein had linked gravity, space, and time into one all-encompassing theory of general relativity; Charles Darwin related biology and time into the theory of evolution. The fourth dimension, time, is the cornerstone of the science of geology. Time, it seemed to me, was the key to understanding almost everything in the natural world, how the Earth formed in the past, how it evolved, and how it works today.

My first three trips to the Himalaya were on mountaineering expeditions. From vantage points high on the peaks of the Kulu Himalaya and Langtang Lirung in Nepal I looked across vast oceans of mountains rippling away to the far distant horizons in the Karakoram and Tibet. No one can fail to be inspired by the views of these peaks. Soon my passion for climbing was equalled by my passion for the geology, and my mission became to understand what forces of nature were operating to make these spectacular mountains. The collision of India with Asia had created the most awe-inspiring and beautiful mountain ranges on Earth, the Himalaya, the Karakoram, and the mountains of Tibet. In the Himalaya we needed to learn how to unwind time by using every bit of evidence we could find in the rocks, the minerals, and the structures, in order to figure out the processes involved in this continental collision, how mountains are made, and how they evolve through time. After thirty years of my own geological research, spanning four post-doctoral fellowships lasting twenty-five years and now six years as a Lecturer in Earth Sciences at Oxford University, having been to the mountains of the Himalaya, Karakoram, and Tibet every year, I feel that the rocks are at last giving up some of their secrets.

This book traces my journey of discovery from its beginnings working on rocks associated with the Oman ophiolite in Arabia, mapping in the mountains of Ladakh, Zanskar, and Kashmir, through geo-mountaineering expeditions to the Karakoram and Himalaya. It includes expeditions to K2, the Trango Towers, Masherbrum and Biale in the Karakoram, and to Everest, Makalu, Shisha Pangma, Manaslu, Annapurna, and Shivling in the Himalaya. It includes travels across the Tibetan Plateau and the remote Hindu Kush and Pamir to more recent geological work in the countries of Indochina, Burma, Thailand, Vietnam, and Malaysia. The book traces the development of geological ideas on how mountains are made, from the plate tectonic revolution of the 1970s through to more recent geophysical, geochemical, and analytical methods employed to decipher the processes of orogeny or mountain building in space and time. Finally, we attempt to unlock the hidden secrets tied up in the rocks and minerals, to unwind time itself and interpret the geological evolution of the magnificent mountains of the Himalaya, Karakoram, and Tibet.

Mike Searle
Oxford, March 2012

ACKNOWLEDGEMENTS

Most of my Himalayan work was funded through the Natural Environment Research Council (UK) for which I will always be greatly indebted. Additional funds have come from the National Science Foundation (US), the Royal Society, the Royal Geographical Society, the Mount Everest Foundation and the Alpine Club. I am extremely grateful to many collaborators in particular Brian Windley, David Cooper, Tony Rex, Lewis Owen, Kevin Pickering, Dave Waters, Rick Law, Randall Parrish, Steve Noble, Kip Hodges, Roger Bilham, Ernie Rutter, Laurent Godin, Kyle Larson, Dick Brown, Rolf Pedersen, Andy Carter, Jonathan Aitchison, Bruno Scaillet, Steve Reddy, Gerhard Bax, Simon Lamb, Tony Watts, John Elliott, Owen Green, Laurence Robb, Marc St-Onge, Jingsui Yang, Songyong Chen, Sun-Lin Chung, Mary Yeh, the late Charles Hutchison, Azman Ghani, Masa Sone, Andrew Mitchell and Chris Morley.

Our PhD and D.Phil students working throughout the Himalaya, Karakoram and Tibet have of course done most of the hard work and I am grateful to all of them. Many have worked in extremely tough field conditions, and I must apologise to those that have inadvertently been caught up in local skirmishes, Taliban insurgents in the NW frontier of Pakistan, gem miners conflicts in Kohistan, or Maoist revolutionaries in Nepal. Thanks to the late Paul Metcalfe (Garhwal), Mark Crawford, James Fraser, Andy Thow, Richard Phillips (Karakoram), Peter Hildebrand (Hindu Kush), Simon Gough (Kohistan), Jon Burton, Malcolm Dransfield, Ben Stephenson, Mike Ayres, James Walker, Christian Walker, Richard Corfield (Ladakh, Zanskar), Rob Simpson, Ian Brewer, Adam Szulc, Mike Streule, Brendan Dyck (Nepal), Jon Cox, Clare Warren, Tom Jordan, Robbie Cowan (Oman), Micah Jessup, John Cottle, Gavin Chan, Richard Palin, Owen Weller (Tibet), and Samuel Ng (Malaysia). Thanks to David Sansom for drafting, Jenny and Ian Colls for slide scanning, Andrew and Debbie Hall of the Rose and Crown Pub in Oxford, where many an expedition was planned.

In addition to my colleagues and students, I would like to gratefully acknowledge the following for many years of lively discussion and debate: the late Augusto Gansser, the late Doug Nelson, Patrick LeFort, Arnaud Pêcher, Jean-Philippe

Avouac, Laurent Bollinger, Peter Molnar, Paul Tapponnier, Xavier LePichon, John Dewey, Kevin Burke, Bert Bally, Clark Burchfiel, Kristen Cook, Peter Treloar, Nigel Harris, Mike Bickle, Vinod Gaur, Philip England, Talat Ahmad, Asif Khan, Qasim Jan, Bishal Upreti, Brad Hacker, Keith Preistley, Shamita Das, Steve Roeker, Peter Clift, Yani Najman, Igor Villa, Eduardo Garzanti, Maurizio Gaetani, Bernhard Grasemann, Bob Spicer, Alastair Robertson, Mike Johnson and many other participants of our annual Himalaya-Karakoram-Tibet Workshop meetings. Special thanks also Leo Dickinson, Paul Harris, Nick Groves, Steve Razzetti, Ben Stephenson and David Willis for use of some of their photos. All photographs in this book were taken by the author except those that are acknowledged separately.

Most of my early expeditions to the Himalaya and Karakoram have been joint geological and mountaineering expeditions. I am extremely grateful to my trekking and climbing companions especially Martin Stephens, Neil Hopkinson, David Cooper, Tony Rex, Dana Coffield, Nick Groves, Maryrose Fowlie, Steve Razzetti, Sean Smith, Simon Yates, Mark Miller, Jon Tinker, Paul Harris, Tony Howard and Di Taylor. In particular I thank Nick Groves who taught me to climb on the crags of North Wales and has been my longest-lasting rope-buddy in the UK and on several expeditions to the Himalaya and Karakoram. Thanks also to Jon Tinker, David Hamilton and Kenton Cool for collecting rock samples from the summit of Everest down to the South Col and also to Henry Todd, Victor Saunders and Russell Brice for hospitability at their base camps on both the Nepal and Tibet sides of Everest.

One of the most delightful aspects of my fieldwork has been interacting with the local people. It seems to be a fact that the mountain environment spawns the most friendly and hospitable people anywhere and this has most certainly been true in my experience. In Ladakh Fida Hussein Mitoo from Leh has been a friend since my first visit in 1981 and has helped me on every trip with trekking logistics, cooking and pony and yak loading. Many Ladakhis and Zanskaris have also been great companions, Lobsang and Namgyal Tsering, Rintzing Tanta, Tsering Norboo lama, Urgen Dorjay from Padam, Sonam Targis from Marka, Tshering Tukten from Sani, and Sonam Norboo and his family from Jildo. Special thanks also to our intrepid porters on our Winter *Chaddur* expedition along the frozen Zanskar River, Lobsang, Mutup, Tashi, Tondup, Tenzing and Dorjey all from the village of Pishu, Zanskar. Ghulam Murtaza in Padam was a great host in Padam. Thanks to Punchuk Dawa, the *Giapo* (King) of Zangla for great hospitality and unlimited supplies of *chang* (barley beer). In Kashmir, thanks to Mohammed and Ayub Shalla, owners of the magnificent Houseboat Rolex on Dal Lake, our haven of rest and recuperation after strenuous treks in Ladakh and Zanskar. Ali Hussein Shillikchey and Mohammed Amin from Kargil were extremely helpful on several occasions.

In Manali thanks to John Banon whose delightful guesthouse in the wonderful apple orchards of the Kulu valley was a haven after long treks in Himachal Pradesh.

In Pakistan many thanks to my good friends from the University of Peshawar, especially Rashid Khan Tahirkheli, Asif Khan and Qasim Jan for looking after our jeep and for years of friendship and collaboration. Hakim, Anwar Ali and Abdullah Javed from Hushe were all excellent sirdars on various Baltoro and Hushe expeditions. Mohammed Ali Changazi, Ghulam Ahmed and Jaffar Ali from Skardu were also very helpful. Our four intrepid porters on our crossing of Snow Lake from Hunza to Askole, Shambi Khan, Dolat Kazi, Tafas Shah and Faisal Baig, all from Shimshal, were the most efficient and best porters in all Pakistan. In the Hindu Kush thanks to Shakirullah and Zafar Ali Khan from Chitral for their help and the jeep ride up to see the classic polo tournament on the Shandur pass.

In Nepal, Shiva Dhakal of Royal Mountain Travel has been the most efficient and helpful agent for many years. Tashi Sherpa, Manoj Kumar Rai, Dorje Sherpa, all from Sola Khumbu and Mingu Lama from Langtang were all extremely pleasant and efficient sirdar - porters and great companions. Thanks to Rinzin Dorjey Lama from Langtang village for great hospitality. In western Nepal, many thanks to super-porter and friend Suka Ghale from Kashigaon, Gorkha who carried my rucksack over many high passes in the Annapurna, Machapuchhare and Manaslu Himalaya. Thanks also to Saroj Lama from Pokhara, and the late Pasang Tamang from Kathmandu for their help. In Darjeeling and Sikkim, thanks to Tashi Sherpa, Gyalzen, Sonam and Doma Sherpa for their wonderful hospitality. In Bhutan thanks to Dhodo Tshering from Thimpu.

In Tibet, special thanks to Réne Schrama of Shigatse Travels for organising the endless permissions, jeeps, guides and also to Kris and Réne Schrama and Fred Madern for some great discussions in the infamous Dunya restaurant in the Yak Hotel, the best watering hole in Lhasa. Lan Hui also did a great job getting permissions for our work around Kailas and SW Tibet. Thingley and Sonam Wangdu from Tingri have been excellent guides and good friends on several expeditions to the Tibetan side of Everest and Cho Oyu, and to SE Tibet. Special thanks go to Songyong Chen for diplomatically extricating us from a Chinese police station at Bayi.

In Burma thanks to Andrew Mitchell, Tin Hlaing, Myint Naing Win, Myint Thein Htay and Koko Than from Ivanhoe Mining Myanmar Ltd. in Rangoon and Meiktila. In the Andaman Islands thanks to Nimal Sunanda and Titus Bwa (Mayabunder) for *doongie* boat rides around the islands and to Benny Jacob for a wonderful trip to some remote islands and reefs around Havelock Island. In Laos special thanks to Boonruang Kotchompu and Sithun Tammawong from Vientienne for great boat trips along the Mekong River and some excellent jungle treks around Vong Vien and Luang Prabang, hunting for those elusive sapphires.

In Thailand many thanks to Chris Morley (PTT Exploration) and Punya Charusiri (Chulangkon University, Bangkok) for trips to the Three Pagodas and Mae Ping fault zones. Special thanks to Joseph and Sally Poon for great hospitality in Phuket and to Joseph for being my dive buddy on our various Scuba diving trips around the Andaman Sea, Burma banks and Indonesia. Thanks to Taen Sakra and Tawan Noiha for some wonderful treks along the Thai-Burma border regions and in Isaan. In Vietnam special thanks to my great friends Mai Anh Tuan, Thoan and their children, Tuan Minh and Nam for wonderful hospitality in Hanoi and Ha Long Bay, and to Phan Ngoc Hai from Sapa for guiding me on our climb of Mount FanSiPan, the highest peak in Vietnam.

This book would never have seen the light of day had it not been for the sterling efforts of my editor at Oxford University Press, Latha Menon. Many thanks to her and also to Emma Marchant, Jenny Lunsford and Phil Henderson at OUP for editorial and production matters. Thanks also to Debbie Sutcliffe for proof reading.

Finally, my lasting love and thanks to my parents, Geoff and Pauline Searle, now both sadly passed away, to whom this book is dedicated.

Mike Searle
Oxford, March 2012

CONTENTS

FIGURES

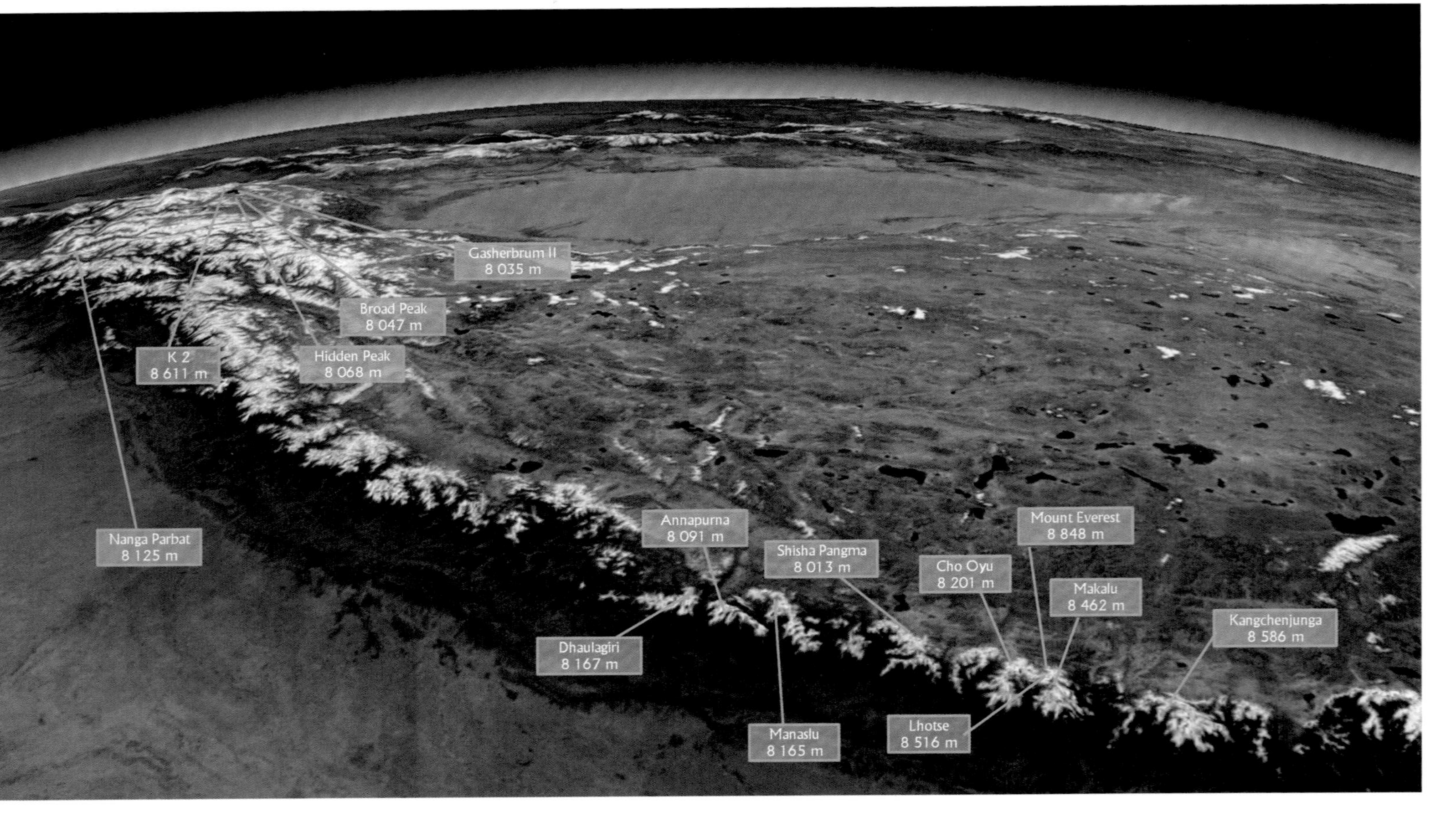

Digital Elevation Model showing the Himalaya, Karakoram, and Tibetan Plateau with locations of the major peaks.

CHAPTER 1

Oceans and Continents

We find no vestige of a beginning, no prospect of an end.

James Hutton, 1788, pondering on the geological timescale

Sitting atop a rickety Indian bus trundling up the Beas Valley to the delightful hill station of Manali in the Kulu Valley of Himachal Pradesh, peering between swaying bodies and piles of luggage, I caught my first glimpse of the snowy peaks of the Himalaya. Yes, those white streaks way up in the sky were not, after all, clouds: they were glints of sunlight on impossibly high and steep ice-fields plastered onto the sides of mountains that tore up into the sky. It was a sight to take one's breath away and I knew instantly that this was going to be the start of a great adventure. We were a typical shoestring British student expedition of five friends who could fit easily into two overloaded rickshaws, heading for the mountains around the Tos Glacier. Mountaineers dream about climbing in the Himalaya. Since my earliest days of climbing the hills and crags of Snowdonia and northern Scotland, I had yearned to see and climb those magical Himalayan Mountains. Now here I was, and the reality of the Himalaya was even better than I imagined.

I had taken three months off from my PhD studies on the geology of the Oman Mountains to go on this expedition. We had driven a Land Rover out from England to Muscat through a snowy Europe and across the Empty Quarter of Arabia from Syria and Jordan to the United Arab Emirates and Oman. After three months' fieldwork in Oman I caught a passenger ship, the MV *Dwarka,* last of the British East India Company merchant vessels that plied the Gulf route from Basra via Kuwait,

Bahrain, Dubai, and Muscat to Karachi, and then travelled through Pakistan by train into India. That first expedition to Kulu was a revelation. We camped on the Tos Glacier, four days' walk above the village of Manikarin in the Parbati Valley of eastern Kulu, for about four weeks. During that time the weather was perfect almost every day. We climbed a total of twelve mountains, including five first ascents, and scaled the most beautiful peak of the Kulu Range, White Sail, also called Dharmsura (6,446 m)—a magnificent Matterhorn-like peak with spectacular views in every direction (Figure 1.1). The climbing was superb and the entire expedition was thoroughly enjoyable (Figure 1.2). This Himalayan climbing business is easy I thought: no need for a big Raj-style expedition with hundreds of porters. We had hired five ponies and two local porters for the last two days on the Tos Glacier to get all our food and gear to base camp. We had bought a month's supply of staple foods—rice, *dhal* (lentils), *atta* (chapatti flour), and basic supplies from the local bazaar in Manali, and the cost was less than living at home in England. On top of that we had climbed lots of wonderful new routes and stood on unclimbed Himalayan summits. This was the life!

Figure 1.1 White Sail (6,446 m) in the eastern Kulu Himalaya, India.

Figure 1.2 Nick Groves climbing on Conival's west ridge, Kulu Himalaya, India.

* * * * *

The Himalaya is the greatest mountain range on Earth: the highest, longest, youngest, the most tectonically active, and the most spectacular of all. The Himalaya formed when the Indian plate collided with Asia approximately 50 million years ago, closing a great ocean called Tethys that, for over 100 million years, had lain between them. For climbers and explorers, Himalayan peaks are the ultimate goal. For geologists, they hold the secrets of how tectonic plates collide, how continents amalgamate, and how mountains are formed. Critical clues are locked up in the rocks of the Himalaya, and the great ranges to the north along the Karakoram, Hindu Kush, Pamirs, Tien Shan, Kun Lun, and the great plateau of Tibet. To understand fully the geological processes involved in the collision of continental plates we need to look elsewhere along the great Alpine-Himalayan Ranges. The geology that we see and map along the Himalaya represents a snapshot in time of a process that has lasted more than 70 or 80 million years, but we can use geology to decipher the past, much as a detective uses clues to solve a mystery.

The abrupt rise of the great Himalayan Mountains along the northern margin of India was one of the major pieces of evidence used to propose the theory of Plate Tectonics (Figures 1.3 and 1.4). At first the theory of Plate Tectonics was widely and flatly rejected, but later, more and more evidence from geology, geophysics, seismology, and geochemistry appeared to agree with the radical concept that parts

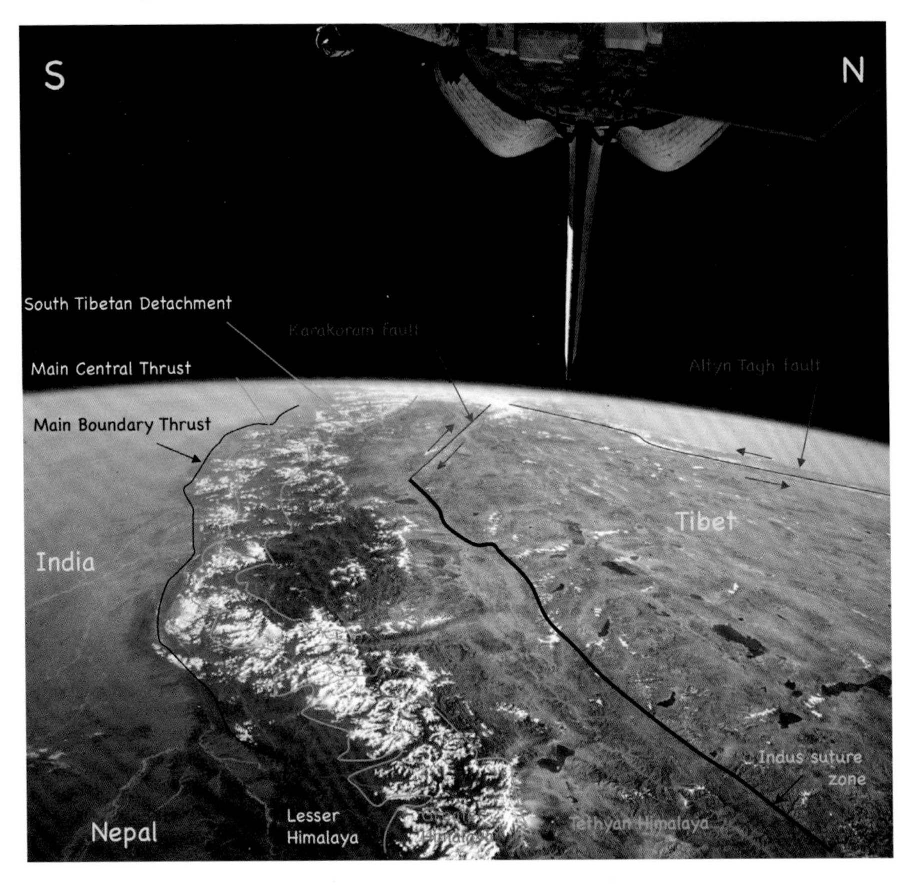

Figure 1.3 The Himalaya and Tibetan Plateau looking west, taken from the Space Shuttle (photo courtesy of NASA). The thick black line marks the zone of collision (Indus suture zone) between the Indian plate to the south and Asian plates to the north.

of the Earth's crust and upper mantle were being pushed, pulled, and moved hundreds, even thousands of kilometres across the globe. Plates spread apart to create oceans, and plates collided to create mountain belts. Although geochronologists have dated the Earth's oldest rocks using radioactive isotopes of uranium, thorium, and lead to at least 4.5 billion years old, there is good evidence from the geological record that plate tectonic processes have been operating at least for the last 2.5 billion years. In the 1970s these relatively rigid plates were thought to extend throughout the crust into the lithospheric mantle. Earthquakes were indicative of brittle fracture and, although most earthquakes are restricted to the upper 20 kilometres or so of the Earth's crust in some places, such as along subduction zones, earthquakes extended deep down into mantle depths.

During the break-up of the southern hemisphere super-continent Gondwana, India broke away from southern Africa and Madagascar and rifted apart from

Antarctica about 120 million years ago. About 50 million years ago India collided with Asia somewhere near equatorial latitude and since then has continued pushing north into central Asia. Before India collided with Asia, closing the ocean, huge sheets of oceanic crust and upper mantle rocks called *ophiolites* were thrust onto the continental margins of India and Arabia. Along the Himalaya the evidence for this earliest phase of the mountain-building or *orogenic* process has largely been subsequently removed by erosion, but in Oman in eastern Arabia the Tethyan Ocean has not yet closed, continental collision has yet to occur, and the ophiolite thrust sheet is wonderfully exposed in these desert mountains. Along the Himalaya a few remnant ophiolites are still preserved in the highest peaks of the Ladakh-Zanskar Himalaya and in a few localities in southern Tibet, but nowhere are these ophiolites so perfectly preserved as they are in Oman. An entire orogenic cycle lasting approximately 20 million years is known to have involved the emplacement of these ophiolite thrust sheets onto the continental margins of Arabia and India prior to the collision with Asia.

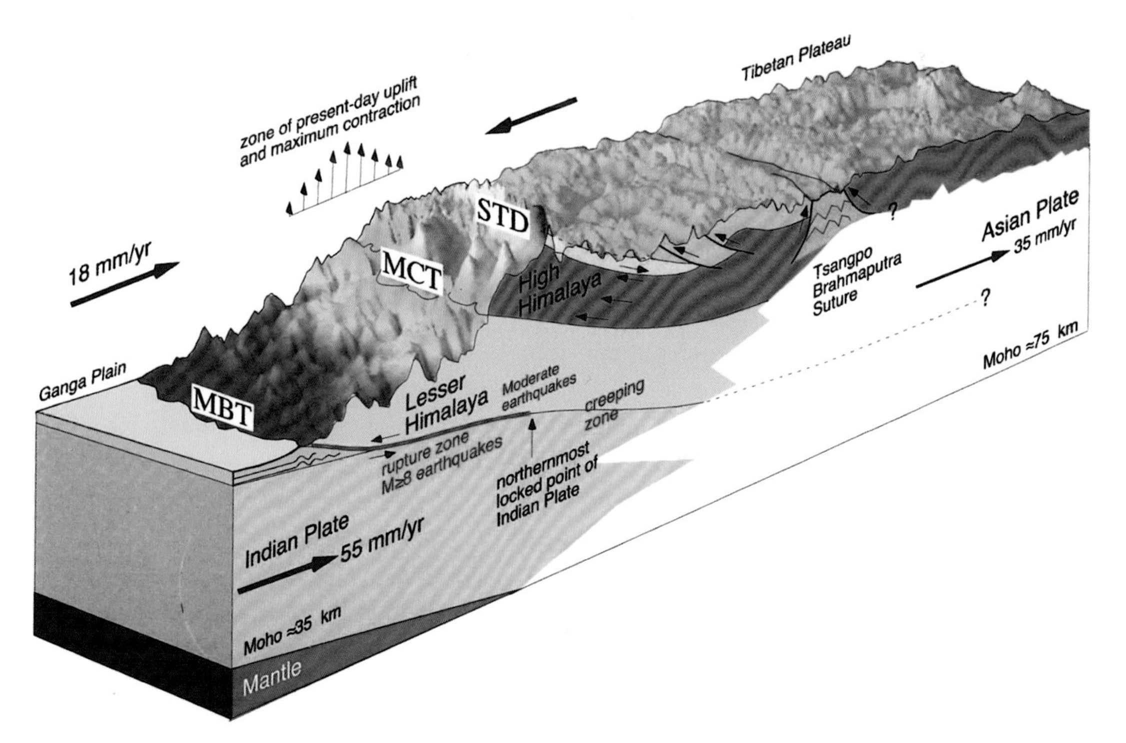

Figure 1.4 Block diagram illustrating the structure of the Himalaya. STD is the South Tibetan Detachment—the low-angle normal fault that bounds the northern margin of the Greater Himalayan metamorphic rocks and granites (magenta colour); MCT is the Main Central Thrust that places the metamorphic rocks south over unmetamorphosed rocks of the Lesser Himalaya. MBT is the Main Boundary Thrust, the active southern margin of the Himalaya and the fault zone along which earthquakes are triggered.

(a)

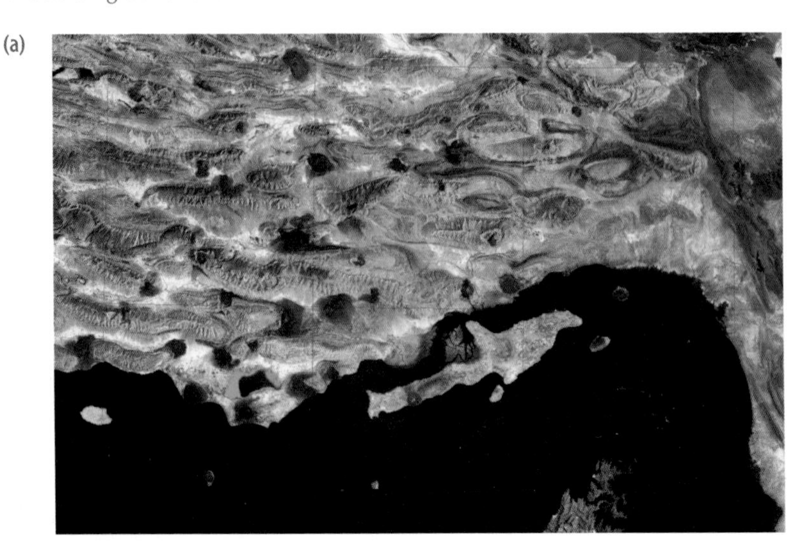

Figure 1.5 (a) (*above*) Landsat satellite photograph of southern Iran showing the impressive large-scale folds and intrusive salt domes (circular dark masses) in the Zagros Mountains. Dark-green-coloured rocks in the top right are ophiolites that mark the suture zone between the Arabian plate (including the Zagros Mountains) and the central Iran plate. The drowned coastline of the Musandam peninsula is in the bottom right and the Straits of Hormuz separates Oman from Qishm Island and Iran to the north. (b) (*opposite*) Large anticlinal folds of the Zagros Mountains seen from the air.

The earliest stages of the continental collision process are best studied in the Zagros Mountains of Iran (Figure 1.5a, b), where the initial collision occurred only around 15 million years ago when the intervening Tethys Ocean closed and the two continental plates of Asia and India met. The line along which the plates collided, the suture zone, is marked by a narrow zone of smashed-up ophiolites and deep-sea sedimentary rocks—the sole remnants of the once wide Tethyan Ocean—that stretches from the Mediterranean along the Bitlis Mountains of south-east Turkey and along the Zagros Mountains of Iran. The crust has buckled, forming giant folds called 'anticlines', many of which contain huge oil and gas reserves. Almost two-thirds of the world's oil reserves are held in the huge folds along the Zagros Fold belt and along the Arabian foreland. The oil and gas reserves in Iraq, Iran, Kuwait, Saudi Arabia, Qatar, and the United Arab Emirates are almost exclusively preserved in this zone along the north-eastern margin of the Arabian plate. The folds that preserve these oil and gas fields were formed as a direct result of the collision of Arabia with central Iran.

Following the initial continental collision, the Indian plate ploughed on northwards, rucking up mountains as it penetrated further, indenting into Asia. The crust to the north of India, in Tibet, grew thicker, doubling in thickness to over 70 or 80 kilometres in places, but to the west, in Afghanistan and Iran, and to the east in Indochina, the effects of lateral extrusion, in which the continental crust has

(b)

been shunted sideways out of the way of India, can best be seen in South East Asia along the mountains of Burma, Thailand, Laos, and Vietnam. In this region, large tracts of crust slid past one another along enormous San Andreas-type transform faults—fault lines that extend from the high plateau of Tibet all the way to Vietnam and the South China Sea and south to Burma and the Andaman Sea.

The effects of the continental collision along the Asian plate are seen in the Tibetan Plateau, the largest area of high elevation, averaging 5 kilometres above sea level, and thick crust anywhere on Earth. However, as erosion rates are very small in the high, arid plateau of Tibet, the best place to study the deep crustal rocks is in the mighty Karakoram Ranges where thick crust, extreme topography, and erosion reveal rocks that crystallized deep in the crust and now crop out along the highest peaks. In this book we shall visit all of these mountain ranges, following the story of how we can use the rock record to interpret the processes involved in the greatest continental collision the Earth has seen, that of India and Asia some 50 million years ago.

The Great Trigonometrical Survey

The Himalaya and Tibet have always been at the forefront of geological thought simply because they are the largest, most spectacularly impressive, and youngest of all mountain ranges, but the story has to start with their initial explorations, geographical discoveries, mapping, and the determination of the height of the mountains. During the British Raj, one of the earliest priorities was to map out the entire Indian sub-continent and particularly the great mountain ranges along its northern borders with both Tibet and the Russian tsarist empire to the north-west. The Great Trigonometrical Survey of India was an incredibly ambitious project destined to last fifty years and map out India using simple triangulation surveys.[1] It was started by a genius of a geodesist named William Lambton early in 1802 when he laid out the first seven and a half miles of base line near Madras (nowadays called Chennai). The triangles gradually and painstakingly spread along the length and breadth of India. Lambton used a massive theodolite which had to be carried everywhere by teams of porters, elephants, even camels, and hoisted on top of specially built survey towers. In 1815 the closed Kingdom of Nepal ceded the central Himalayan districts of Garhwal and Kumaon to British India, enabling the geodesists to gain access to the Himalaya.

A young George Everest joined the team in 1818, and after Lambton's death in 1823 Everest completed the triangulation all the way north to the Himalaya. Having established that the Himalaya was higher than the Andes, the survey measured the peak of Nanda Devi in the Kumaon Himalaya at 25,479 feet, and for twenty-five years this was thought to be the highest mountain in the world. Later, Andrew Waugh, Everest's successor as Surveyor-General thought that maybe Dhaulagiri, measured at 26,862 feet, was the highest peak, and then in 1847 the western peak of Kangchenjunga was measured even higher at 28,176 feet. Even as Kangchenjunga's height was being calculated, Waugh suspected that another mountain 120 miles away along the Tibetan border, simply called peak 'b' would exceed this height. Peak 'b' was shyly hiding behind a massive rampart of a mountain, peak 'a', now known as Makalu, the world's fourth highest mountain. Indeed, a preliminary height was measured at 28,799 feet, but it was not until two seasons later in 1849 that peak 'b', now officially called Peak XV, was found to be the highest mountain in the world. In March 1856 Andrew Waugh was finally satisfied with all the measurements and calculations. He wrote that the highest peak was located at latitude north 27° 59' 16.7", longitude east of Greenwich 86° 58' 5.9" and height 29,002 feet above sea level on the Nepal–Tibet border. Waugh said that 'to perpetuate the memory of that illustrious master of accurate geographical research I have determined to name this noble peak of the Himalaya Mont [*sic*] Everest'. So it was that the world's highest mountain came to be named in honour of George Everest, the Surveyor-General who finally completed the Great Trigonometrical Survey across the length of the Indian continent.

The next major advance arose from Everest's computation of the geodetic position at several localities along the Himalayan foothills where he had already measured the astronomic positions. He found a constant difference in the latitudinal position, which he attributed to small errors in the triangulation survey. In 1854 a Cambridge-trained mathematician, J. H. Pratt, who happened also to be Archdeacon of Calcutta, proposed that the discrepancy was caused by the deflection of the vertical as a result of the gravitational pull of the nearby Himalaya. He thought that low-density regions underlay mountains whilst high-density regions underlay oceans. Shortly afterwards, in 1855, George Airy, the Professor of Astronomy at Cambridge University, provided a neat and simple explanation: that areas of high topography like Tibet and the Himalaya were underlain by a deep 'root' rather like an iceberg floating on water. The higher the iceberg, or the mountain, the deeper the crustal 'root', following Archimedes' Principle. The term 'isostasy' was first used by C. E. Dutton in 1882 to explain the progressive uplift of mountains and the progressive subsidence of sedimentary basins. Geological forces were constantly striving to maintain isostatic equilibrium, in other words erosion was constantly trying to weather mountains down to sea level, and basins were constantly filling up to achieve stability or equilibrium.[2]

Richard Oldham, following his father, joined the Geological Survey of India in 1879 and became Superintendent. He witnessed the great Assam earthquake of 12 June 1897, which caused immense devastation across north-east India, reduced all buildings to rubble in an area the size of England, and uplifted the northern part of the Shillong Plateau by 11 metres instantly. Oldham made the most detailed observations ever recorded in a large earthquake, and first identified on seismograms the arrivals of primary (P), secondary (S), and tertiary (surface) seismic waves, previously predicted by mathematical theory.

Despite the Himalaya being the youngest and highest mountain range on Earth, very little was known about its geology. Burrard and Hayden published the first main work in 1907, but at that time so few geologists had actually trekked into the great Himalayan Mountains that real geological data was scarce.[3] Their four volumes included final results of the great trigonometrical and geographical surveys on the mountain ranges, rivers, and physical features of the Himalaya and Tibet. Nowadays satellites give us instant access to precise topography, and seismic and gravity surveys give us clues as to the underlying properties of the crust. The 1907 work represented the culmination of a heroic effort by the early surveyors who painstaking mapped the Himalaya and Tibet using plane tables and theodolites in the early part of the 20th century.

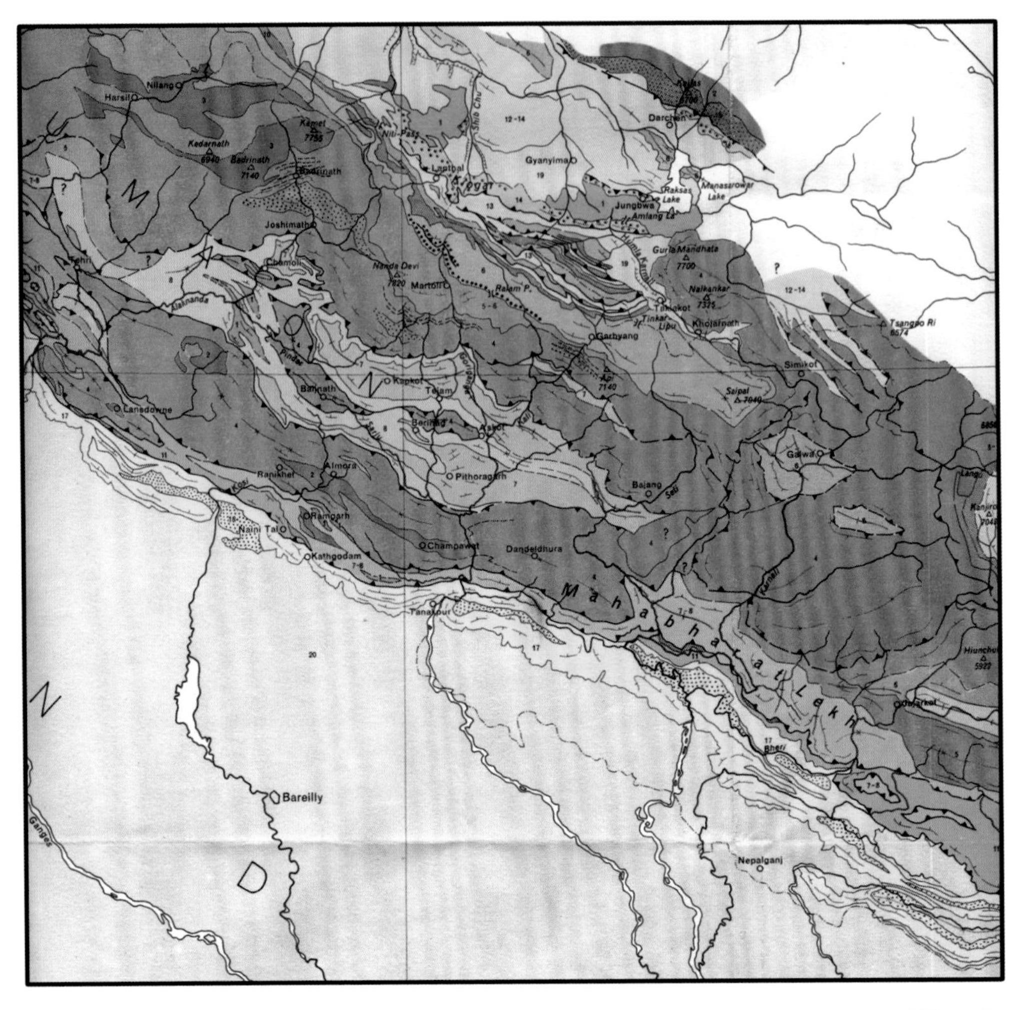

Figure 1.6 A portion of the geological map published by Augusto Gansser in 1964 in the Garhwal–south Tibet region where Heim and Gansser worked in 1936. The green-coloured rocks in the north-east are the ophiolites around the Amlang-la that mark the zone of India–Asia collision. Mount Kailas is in the far north-east.

Following World War Two a period of chaos descended on the Indian sub-continent with the fall of British India and the three Indo-Pakistan wars that followed the partition of Pakistan and India in 1947. Travel into the Himalaya was extremely difficult, and the remote Kingdoms of Nepal, Sikkim, and Bhutan were completely closed. In 1936 the first major geological expedition was organized by a Swiss team including Arnold Heim and a young Augusto Gansser. Heim had mapped large tracts of the Swiss Alps, and had been on a field excursion to north-west Scotland with the great geological mappers from the British Geological Survey, Ben Peach and John Horne. He had seen the newly discovered thrust faults of the Moine

Thrust zone, a series of compressional faults that represent the eroded remnants of a Himalayan-type mountain belt, and had already mapped enormous fold structures in the Alps. Heim and Gansser made the first critically important observations in the central Himalaya, including the stratigraphy of the Lesser Himalaya, the inverted metamorphism along the Main Central Thrust, and the large-scale structure of the Himalaya.[4] In 1964 Gansser published his classic book *Geology of the Himalaya*, the first all-encompassing account of the geology and structure of the entire mountain range (Figure 1.6).[5]

Plate Tectonics

Two-thirds of the surface of the world is covered by sea, much of which to this day remains unexplored. It is relatively easy for geologists to walk around most of the continents, to map the rocks, and explore for mineral, oil, and gas deposits. The ocean, however, is a realm unto itself. We can drill wells into the upper crust or drive submersibles through the depths, but this is only scratching at the surface. It is extremely difficult to actually see and touch and collect rock samples from the ocean floor, let alone the deeper levels of the lithosphere. Geophysical methods such as seismic reflection or refraction profiling, and magnetic and gravity surveys can give us important information on the structure of the lithosphere. The ideas that sparked Plate Tectonics—continental drift and sea-floor spreading, can be traced back to the early 1900s.[6] Alfred Wegener, a German meteorologist, is usually credited with being the first to propose that the continents were all originally joined together in one super-continent called Pangaea, which later split apart as a result of ocean-floor spreading. Wegener did not provide a mechanism for continental drift, but Arthur Holmes really laid the foundations for plate tectonic ideas when he proposed in 1928 that continents were moved by convention currents in the mantle, powered by heat from radioactive decay.[7] In 1924 Emile Argand came up with the insightful idea that the great Tibetan Plateau had been completely underthrust by India, in order to explain the high topography and thick crust, an idea that pre-dated the plate tectonic revolution by forty years.[8]

Before the plate tectonics revolution, geologists had a very 'fixist' view of the Earth, in which basins were sinking and mountains rising somewhat statically. In the 1950s the Cold War was the driving force behind the scientific exploration of the world's oceans. The US Navy needed accurate maps of the ocean floor so that their long-range nuclear submarines would not bump into hidden shoals, so numerous expeditions were sent out to map the ocean bathymetry using echo-sounding devices and seismic experiments. As the oceans were mapped out, some amazing new discoveries came to light. Long, linear mountain ranges ran the length of the oceans that appeared to coincide with the youngest basaltic eruptions along mid-ocean ridges. Fault lines thousands of kilometres long appeared to offset geological features on the ocean floor. Occasionally these transform faults or fracture zones would hit the continent and could be traced into strike-slip faults such as the San Andreas fault in California. Isolated seamounts in the middle of oceans like Hawaii reached heights over 10 kilometres, greater than the height of Mount Everest. Linear volcanic island chains were mapped out alongside deep trenches. These linear ocean trenches reached amazing depths, as much as 12 kilometres in the western Pacific off the Mariana Islands, and were exactly coincident with the deep earthquakes along subduction zones. Indeed, it was geophysicists who supplied the initial data that gave rise to the greatest revolution in the Earth Sciences that occurred during the late 1960s and early 1970s, the theory of Plate Tectonics.

The theory of Plate Tectonics states that the lithosphere, comprising the Earth's crust and upper mantle, is divided up into several 'rigid' plates that are bounded by either constructive plate margins, such as mid-ocean ridges, where new oceanic crust is formed, or destructive plate margins. Destructive plate margins can be Andean-type margins, where an oceanic plate is subducting beneath a continental plate, or Himalayan-type margins where two continental plates collide, or along island arcs (a line of oceanic volcanoes) where one oceanic plate subducts beneath another, such as the western Pacific island arcs of the Marianas or Philippines. Continental crust is composed of granites, which are slow-cooled igneous rocks, together with sedimentary and metamorphic rocks. By contrast, oceanic crust is composed almost entirely of the igneous rocks basalt and gabbro. The lighter continental crust is dominated by the presence of the minerals quartz and feldspar, whereas denser oceanic crust is dominated by olivine and pyroxene. The continents are therefore more buoyant than the oceans, and 'float' on the denser mantle rocks, according to the principle of isostasy, an example of Archimedes Principle, rather as a buoyant iceberg floats on the denser ocean water. Both continental and oceanic crust lie above the mantle, which is composed almost entirely of rocks called peridotites, made up of olivine and pyroxene. The boundary between the crust and the mantle is the Mohorovičić discontinuity, or Moho for short, a geophysical boundary named after the Yugoslavian mathematician who first discovered it.

During the 1960s all these discoveries began falling nicely into place.[9] It was clear that new ocean crust was formed along the mid-ocean ridges and spread apart sideways to accommodate new basaltic eruptions. Sea-floor spreading created new oceans like the Red Sea, and pushed the continents apart. In places where the continents are rifting apart, great rift valleys such as those along the East African rift system extend off into the ocean. The Red Sea is a young ocean that is presently splitting the Arabian plate off from its African plate neighbour. We can actually witness the birth of a new ocean in the Afar region of Ethiopia, where the East African rift system meets the southern Red Sea. In Afar, the land is below sea level and volcanic eruptions have split the continental crust asunder. The continental margins of Arabia and Africa can be restored and fitted together simply by winding back the clock and closing the Red Sea.

Oceanographers discovered whole mountain ranges extending along the length of the world's great oceans from bathymetric studies. Geophysicists proved that the oceans were actually spreading apart by mapping and matching magnetic anomalies in the oceans. Seismologists proved that crustal plates were actually descending deep into the mantle along subduction zones. Geologists mapping the continents found narrow zones of oceanic rocks in the middle of continents. These 'suture zones' actually mapped out ancient plate boundaries. Huge 'nappes' or folds were discovered in many mountain ranges, and structural geologists could unwind these structures to work out the amounts of crustal shortening across mountain belts.

One of the major factors that kicked off the plate tectonic revolution of the 1960s and 1970s was the new marine magnetic anomaly maps of the world's oceans. Using magnetometers towed behind ships the magnetic anomalies throughout the oceans were systematically mapped out. The resulting magnetic 'stripes' of normal and reversed polarity were found to parallel the mid-ocean ridges and recorded these magnetic reversals back in time.[10] As lavas were erupted along mid-ocean ridges and then cooled through a critical temperature known as the Curie temperature (about 500°C) the rocks were magnetized in the Earth's magnetic field of that time. Tiny magnetite and ilmenite crystals in the lavas became orientated, pointing towards the Earth's magnetic north pole. It is a curious fact that the Earth's magnetic field has flipped numerous times in the past as a consequence of the nature of the flow of liquid iron in the Earth's outer core. These magnetic reversals have occurred at least as far back as the Jurassic, the age of the oldest known ocean crust in the world's oceans today. By dating the ocean floor basalts it is possible to reconstruct both the history of the Earth's magnetic field over time and the age of the sea floor in the oceans. Geologists can use this palaeomagnetic data to reconstruct the movement of plates back in time.

Oceanic crust is generated along mid-oceanic ridges, mountain ranges in the middle of the great oceans, along which upwelling magma erupts along deep fissures, crystallizes, and is pushed sideways by later volcanic eruptions. Iceland is one of the few places on Earth where a mid-oceanic ridge, in this case the Mid-Atlantic Ridge, is actually exposed on land. Approximately 3.5 km^2 of new ocean crust is added to the Earth's surface every year along the spreading centres of the world's oceans. This is approximately balanced by the amount of oceanic crust that is consumed into the mantle along destructive plate margins, subduction zones. During major earthquakes such as the 26 December 2004 Sumatra earthquake, several metres of ocean crust were subducted almost instantaneously. When the Atlantic Ocean was born, a great rift in the Earth's crust split the continents of America and Africa. Magma generated from the mantle welled up and was erupted as basalt along a chain of mountains, forming the Mid-Atlantic Ridge. America and Africa were slowly pushed apart as the sea floor began to spread. Today the oldest oceanic crust in the Atlantic Ocean occurs along the continental margins of the Americas and Africa, while the youngest basalts are being erupted along the spreading centres of the mid-ocean ridges. Unwinding the geological clock and restoring the plates to their position before the Atlantic opened, the eastern continental margin of South America can be geometrically fitted very closely with the south-west African margin.

Some plate boundaries are giant transform faults, along which plates slide horizontally past one another. The best example of this tectonic setting is southern California where the San Andreas fault comes on land. Other plate boundaries are zones of compression and collision, where two plates converge. Along the western margin of South America, the Pacific Ocean plate is converging with the South American continent. The oceanic plate is denser, and sinks beneath the continental margin along a deep subduction zone. These subduction zones are marked by numerous earthquakes set off when the plates slide past one another, and a line of active volcanoes such as that along the length of the Andes today. In the oceans, where two oceanic plates converge, a similar deep subduction zone, marked by intense seismic activity, is also evident from an island arc. These island arcs are especially common in the western Pacific Ocean (for example the Philippine, Mariana, New Hebrides, or Tonga island arcs) and also in the Caribbean (Lesser Antilles arc). In southern Alaska, the Aleutian island arc continues along strike into the North American continent.

Along Andean-type plate margins, the ocean crust is being destroyed, plunging back down into the mantle along deep subduction zones beneath the continents. Directly above these subduction zones is a line of highly explosive volcanoes, like Mount St. Helens in the western US or the great volcanoes of Chimborazo and Cotopaxi in Ecuador. These inclined subduction zones are marked by a narrow zone of deep earthquakes called Benioff zones. In the Western Pacific numerous subduction zones are associated with the destruction of oceanic crust and lines of highly explosive island arc volcanoes, like Krakatau in Java, Pinatubo in the Philippines, or Montserrat in the West Indies.

The greatest mountain ranges occur where two continental plates collide. Pushed by ocean-floor spreading and pulled by subduction zones, continental plates can be forced together. The results are spectacular. Along the Alpine-Himalayan belt two continental plates collided, the Gondwana super-continent in the south (Africa, Arabia, India) with the Laurasian (Europe and Asia) landmass to the north, closing the great Mesozoic Tethys Ocean. Tethys has now almost completely vanished, and hundreds of kilometres of crustal shortening and thickening have resulted in impressive mountain belts like the Alps, the Zagros, and of course the Himalaya.

Break-up of Gondwana and the Northward Drift of India

After the break-up of Gondwana, India became an island drifting across the Indian Ocean, pushed northwards by mid-ocean spreading ridges.[11] From the timing of initial continental break-up, oceanic spreading centres formed to the south, pushing India towards the north and Antarctica to the south as the Indian Ocean widened. In the Indian Ocean magnetic anomalies are laid out in stripes parallel to the major crustal spreading ridges, the Central Indian Ocean Ridge and the South-East Indian Ridge. These were active plate boundaries where basaltic rocks were erupted from long linear magma chambers in the same way that the Mid-Atlantic Ridge formed, pushing apart Africa and South America. By unwinding the marine magnetic anomaly stripes in the Indian Ocean it is possible to reconstruct the path of India's northward drift since it detached from its neighbours on Gondwana, Madagascar, and Antarctica, some 120 million years ago (Figure 1.7).

The northward drift of India can also be traced using the Réunion 'hot spot'. Major 'hot spots' or 'super-plumes' like Hawaii and Réunion are generally thought to be long-lived and static regions of high thermal anomalies, arising from deep in the mantle. Réunion Island is situated east of Madagascar and is composed of a large alkali basaltic volcano that is presently quite active. The lavas erupted from Réunion are similar to those found in other ocean islands such as Hawaii, the Comores Islands, Tristan da Cunha, and the Canary Islands, for example, some distance away from mid-ocean ridges. These volcanic rocks are very different from mid-ocean ridge basalts in that they are far richer in alkali elements and hotter, indicating that they have been generated from melting far deeper in the mantle. Fragments of the mantle were ripped off from the deep melting source region and became entrained in the lavas.

As the Indian oceanic lithospheric plate drifted across the fixed Réunion 'hot spot', it left a continuous line of volcanoes in its wake, each volcano becoming successively older away from the fixed 'hot spot'. In the Indian Ocean a continuous line of volcanic sea-mounts is beautifully preserved along the Maldive–Laccadive–Chagos Ridge, a 2,500 km-long chain of ocean islands with coral atoll caps that

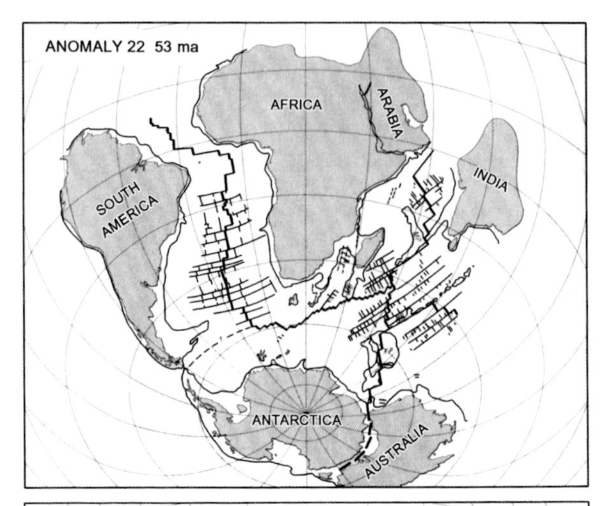

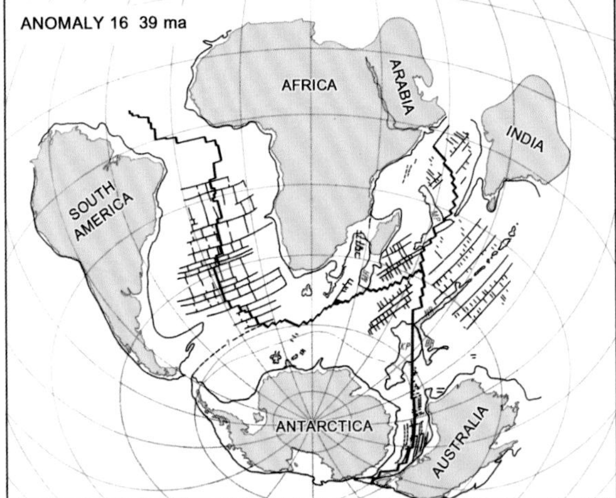

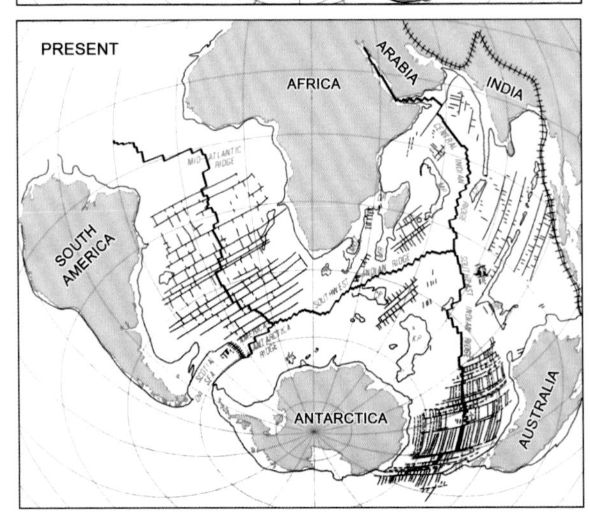

Figure 1.7 The northward drift of India as deduced from palaeomagnetic reconstructions of the Indian Ocean. Marine magnetic anomaly 22 corresponds to 53 million years ago and anomaly 16 corresponds to 39 million years ago.

become progressively younger to the south. The northern part of the chain extends right into the vast Deccan Traps region of the Western Ghats, India. The huge outpourings of basaltic lavas in the Deccan Traps, which erupted between 67.5 and 64.5 million years ago and originated from deep in the mantle are thought to have been partially responsible for severe environmental change and the mass extinctions, including that of the dinosaurs, that occurred around the Cretaceous–Tertiary boundary. Although somewhere around 75 per cent of faunas including the dinosaurs became extinct at the 'K–T' boundary 65 million years ago, not all were affected. Many molluscs, sharks, boney fish, placental animals, and all amphibians remained unscathed. The extinctions are now thought to have resulted from a combination of factors including a giant meteorite impact (at Chixculub in the Gulf of Mexico), the enormous volcanic outpourings of the Deccan Traps, and a combination of environmental factors that triggered extreme biological stress.[12]

Fifty million years ago the Indian continent had reached equatorial latitudes as a result of its rapid northward drift. Tethys was gradually closing. Along its western boundary a giant transform fault, known as the Owen Fracture Zone, separated the Indian plate from Arabia. Along its eastern boundary the Indian plate was subducting towards the north beneath the island arcs of the Andaman and Nicobar Islands, and the huge volcanic chain of Sumatra, Java, and the Indonesian islands to the east. This plate boundary remains one of the most active in the world, regularly experiencing large earthquakes and spectacular volcanic eruptions.

Great slabs of oceanic crust and upper mantle, known as ophiolite complexes, had earlier been thrust onto the northern margin of the Gondwana continents, Arabia and India. These ophiolites, now beautifully preserved in the mountains of Oman, sporadically occur all around the Indian plate margin from Baluchistan and Waziristan across Ladakh and southern Tibet, as far east as the Andaman Islands and Indonesia. The closing of vast tracts of ocean and the emplacement of ophiolites onto continental margins was a long-lasting process and resulted in a major mountain-building episode even before India collided with Asia. Along the Himalaya this early phase of mountain building would now be hard to unravel, given the later folding, thrusting, uplift, and erosion that occurred during the subsequent collision with Asia.

India–Asia Collision

The collision of the Indian plate with Asia and the closing of that part of the Tethyan Ocean occurred at equatorial latitudes approximately 50 million years ago, and since then India has continued to move northwards relative to stable Siberia, indenting into Asia and converging at a rate of around 55 mm/year at the present time. The Himalaya are the folded and thickened leading edge of the Indian plate, where the crust has been crumpled and compressed, shortening by over 800 kilometres and thickening to double its normal thickness (from about 35 to 70 kilometres thick), elevating the highest peaks to nearly 9 kilometres above mean sea level.

How can we know when India and Asia first collided? We know the timing from the precise dating of stratigraphic sections across the suture zone from north-western Pakistan, Ladakh, and parts of southern Tibet.[13] Marine sediments dominated the stratigraphic record of the Indian plate for over 200 million years from at least Permian times (approximately 250 million years ago) to Early Eocene time (approximately 50 million years ago). These shallow-water limestones, full of marine fossils, corals, bivalves, coiled gastropods, and millions of tiny foraminifera, abruptly gave way to continental clastic sediments that were deposited in lakes and rivers on continental crust. The youngest marine sediments known from several locations along the zone of collision have been very precisely dated at 50.5 million years old (Early Eocene), implying that after that time there was no ocean left between India and Asia.

Ophiolites and Oceanic Crust

Ophiolites comprise rocks formed beneath ancient oceans, and in a few places of the world are preserved to such an extent that they can be used as a proxy for the structure and composition of the ocean crust (Figure 1.8).[13] Since oceanic crust and mantle are mainly dense basalts and peridotites composed dominantly of the minerals olivine and pyroxene, the oceanic crust normally sinks beneath the continents along plate margins, such as the Andes, where the Pacific and Nazca plates subduct towards the east, beneath South America. Only in very rare cases, such as in Oman, the Troodos Mountains of Cyprus, and in the Bay of Islands, western Newfoundland, do these oceanic rocks sit on top of the continental crust.[14] Along the Himalaya the ophiolites that record the earliest orogenic history have mostly been eroded away, so geologists have used the Oman ophiolite as an example of what the Indian plate margin probably looked like before the collision with Asia 50 million years ago.

Geologists have long puzzled over how these large thrust slices of dense oceanic crust and upper mantle were emplaced onto the more buoyant continental

Figure 1.8 (*opposite*) The rock succession or stratigraphic column through the Oman ophiolite sequence, with photos illustrating each rock type on the right. The crustal section shows from the top downwards: pillow lavas, sheeted dykes, and layered gabbros. The mantle section comprises almost entirely peridotites (darker-coloured harzburgites and pale-coloured dunites). The crust and mantle are separated by the 'Moho' discontinuity which can be a distinct boundary but is more often a zone of mixed peridotites and gabbros. The 'metamorphic sole' consists of a thin thrust slice of amphibolites and greenschists along the base of the mantle sequence. The Oman ophiolite is interpreted as an intact thrust slice of Tethyan oceanic crust and upper mantle formed 95 million years ago and emplaced onto the northern continental margin of Arabia during the Late Cretaceous.

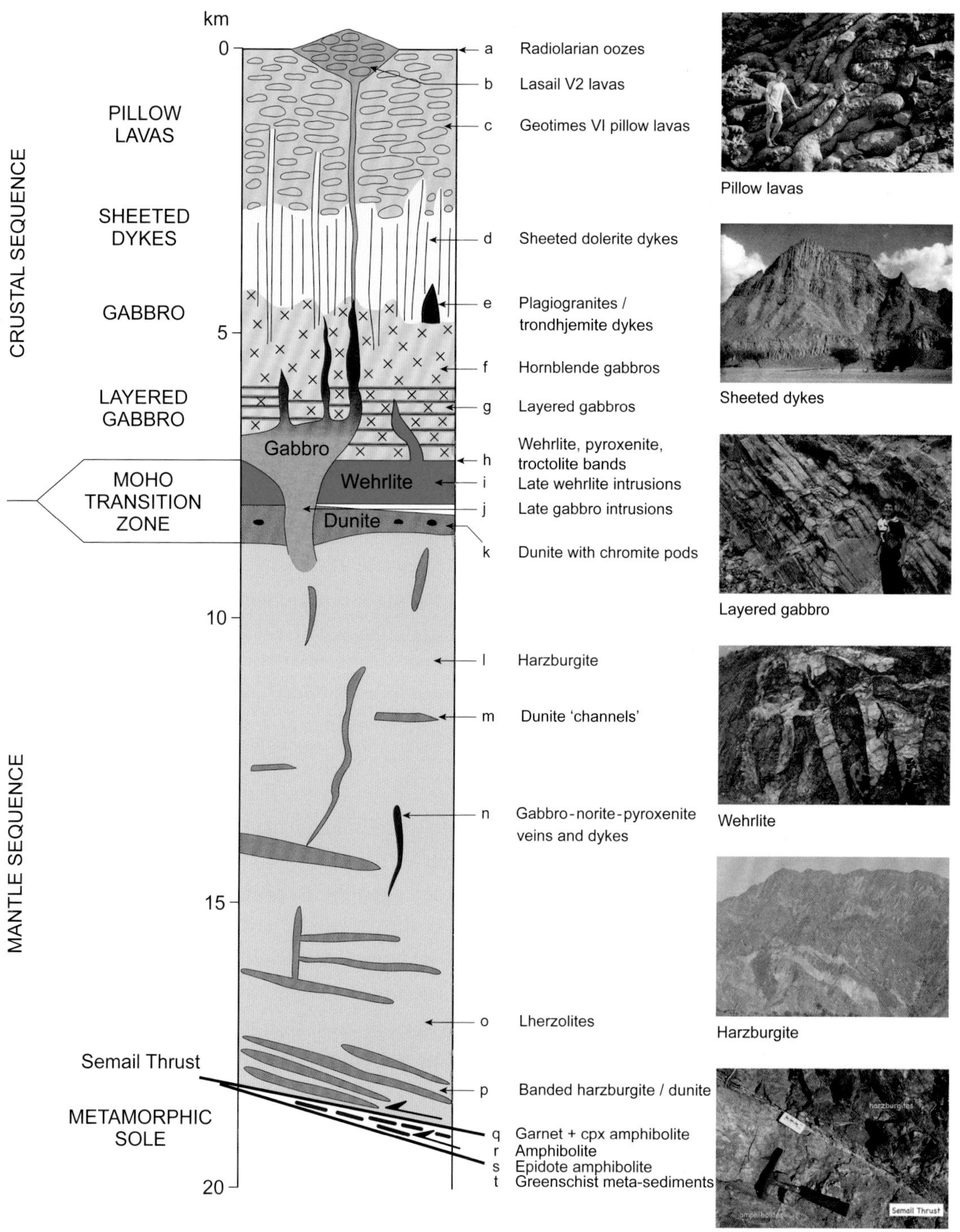
km
0
5
10
15
20
CRUSTAL SEQUENCE
MANTLE SEQUENCE
PILLOW LAVAS
SHEETED DYKES
GABBRO
LAYERED GABBRO
MOHO TRANSITION ZONE
Semail Thrust
METAMORPHIC SOLE
Gabbro
Wehrlite
Dunite
a Radiolarian oozes
b Lasail V2 lavas
c Geotimes VI pillow lavas
d Sheeted dolerite dykes
e Plagiogranites / trondhjemite dykes
f Hornblende gabbros
g Layered gabbros
h Wehrlite, pyroxenite, troctolite bands
i Late wehrlite intrusions
j Late gabbro intrusions
k Dunite with chromite pods
l Harzburgite
m Dunite 'channels'
n Gabbro-norite-pyroxenite veins and dykes
o Lherzolites
p Banded harzburgite / dunite
q Garnet + cpx amphibolite
r Amphibolite
s Epidote amphibolite
t Greenschist meta-sediments
Pillow lavas
Sheeted dykes
Layered gabbro
Wehrlite
Harzburgite
Semail Thrust

margins, apparently defying the laws of gravity. The process of ophiolite emplacement was termed 'obduction' as opposed to the normal process of subduction that operated along destructive plate margins. At the base of the ophiolite, narrow bands of metamorphic rocks occur that record the process by which these massive sheets of oceanic rocks were pushed up on top of continents. These metamorphic rocks are called 'amphibolites', and contain the minerals hornblende, plagioclase feldspar, garnet, and pyroxene. They are basalts that were metamorphosed at high temperatures and pressures during subduction of ocean-floor basalts deep into the mantle. The Oman ophiolite shows some of the best examples of these metamorphic rocks anywhere in the world and it is here that we need to begin our quest to understand the geological evolution of the Himalaya.

The Mountains of Oman

The northern margin of the Arabian plate lies along the rugged mountains of the Bitlis and Zagros ranges of eastern Turkey and south-west Iran (Figure 1.9). Here a narrow suture zone, containing highly crushed oceanic rocks separates the Arabian plate in the Zagros Mountains from the central Iran plate. Further east, in Oman, this great clash of continents has not yet occurred. The Gulf of Oman is a relic piece of the Tethyan Ocean, caught between the great continents of Arabia and Asia. It is for this reason that the great ophiolite thrust sheet is so beautifully preserved in the Oman Mountains (Figure 1.10). To the west in the Zagros, and to the east along the Himalaya, these great sheets of ophiolites were thrust high upon the continents. After the continuing collision, they were the first to be eroded and removed from the rock record, only the few highly deformed and smashed-up pieces remaining along the actual zone of plate collision.

The Oman Mountains lie along the eastern margin of the Arabian plate. Arabia used to be a part of the giant African plate until about 15 million years ago, when it split apart along the fracture now occupied by the Red Sea. As the continental crust stretched and eventually fractured, the void was filled by upwelling basaltic magma, and a juvenile ocean was born. In the southern part of the Red Sea, the spreading centre continues east to connect with the Central Indian Ocean Ridge. In the north, the Red Sea splits into two arms, one continues north to the Mediterranean along the now-failed Suez rift, whilst the main active plate boundary branches off along the Gulf of Aqaba into the Dead Sea fault. The Dead Sea is 200 metres below sea level, sinking fast and rifting Palestine and Jordan apart. Eventually, in a few millions years' time, Lebanon, Palestine, and the West Bank will become separated by a juvenile ocean from Jordan, Syria, and the rest of Arabia.

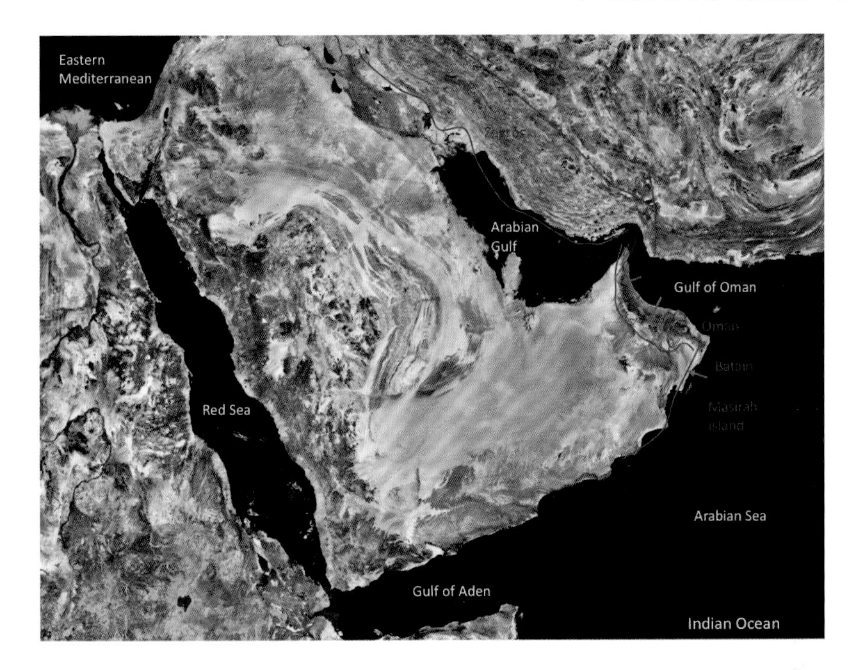

Figure 1.9 Landsat satellite photo of the Arabian plate showing the major structures in the Arabian peninsula and the Zagros Mountains of Iran. The Oman ophiolite comprises the dark-coloured rocks exposed along the mountains of northern Oman.

Figure 1.10 Aerial view of the Oman ophiolite. Pale-coloured rocks in foreground are gabbros of the lower crust separated from dark-coloured peridotites of the mantle sequence by the Moho. The ophiolite has been thrust over the Mesozoic limestones of the Arabian plate margin seen along Jebel Nakhl in the distance.

The mountains of Oman in eastern Arabia were known to contain the largest and best-preserved ophiolite complex in the world,[15] and it was here that I headed in 1977 to work for my PhD. It was an exciting time to start out on a research career in geology. After the plate tectonic revolution and the development of new geochemical and isotopic dating techniques, geologists were looking at mountain belts in a new light. My supervisor at the Open University, Professor Ian Gass, had given me a free hand in deciding exactly what topic to work on, and so for the first few weeks, I drove around the northern mountains with John Smewing and Adrian Lewis, two post-doctoral researchers, also at the Open University. Eventually I decided that I was going to work on the structures and the metamorphic rocks along the base of the ophiolite sequence, rocks that would give us an indication of the processes operating during the emplacement of this great slab of oceanic rocks onto the Arabian continental margin.

In the 1970s the geological secrets of Oman were only just being discovered. Here in a remote mountain range in eastern Arabia was an entire range made up of rocks that should have only been exposed in the depths of the oceans. The Oman mountain range is one giant geological museum. The mountains rise to nearly 3,000 metres above sea level. A great network of wadis, dried river courses, have carved deep canyons into the heart of the mountains so that standing on the rim of one of these great gorges cut into the Jebel Akhdar, one can literally see right into the heart of the mountain range. The limestones are also full of fossils from a warm tropical ocean, corals, echinoid sea-urchins, bivalves, gastropods, and all sorts of tiny foraminifera shells. The dark rocks around Muscat and along much of the mountain range are extremely unusual rocks to find on continents. These are basalts, dolerite dykes and gabbros of the ocean crust and peridotites, comprising the Earth's deep mantle. The basalts contain important copper deposits and the peridotites contain small deposits of chromium, vanadium, and platinum. It is these ophiolites that give Oman its unique and spectacular, almost lunar-like landscape.

Unlike the continental crust, the oceanic crust was rarely, if ever, seen, and geologists only knew of its composition and structure from remote means, like geophysical experiments or deep-sea drilling. Yet, here in Oman there is an entire ocean laid out on land. One can walk all over the ocean floor, see perfectly shaped pillow lavas spewing out of vertical sheeted dykes, find fossil black smoker vents, collect iron and manganese nodules, even pick out tiny radiolaria fossils embedded in red cherts between the lavas. The entire ophiolite complex is laid out bare in the rocky mountains. One can walk down deep into the crust, follow the lavas and dykes downwards into the magma chamber, even take a stroll along the Moho, the invisible geophysical line that divides the Earth's crust from the mantle. How did this vast sheet of oceanic rocks become trapped and emplaced onto the Arabian continent?

Mapping the Oman Ophiolite

Altogether I spent three winter seasons in Oman mapping the rocks and structures in several key areas along the base of the great ophiolite thrust slice. I was beginning to formulate my own ideas about how the ophiolite became emplaced onto the Arabian continental margin, and it was exciting to see how all the bits and pieces of the puzzle suddenly started to come together. After many weeks out in the field I drove back through the mountains to our house near Sohar, spread out my field maps and started to fit them all together. Then I began to construct geological cross-sections across the mountain range, and even tried to restore them by unwinding all the folds and thrusts that cut through the rocks. I was amazed to discover that the structures of the rocks beneath the ophiolite suggested that hundreds of kilometres of ocean crust had been shortened and compressed by the folding and stacking of one unit on top of another. There was no doubt that the rocks in Oman were originally formed at the bottom of an ocean and that they had been pushed onto the depressed continental margin during the emplacement of the ophiolite. If a relatively small mountain range like that in Oman could be the result of closing an ocean hundreds of kilometres wide yet only a small portion of the great Tethys Ocean, what would a great mountain range like the Himalaya record?

Back at the Open University in England I analysed the rock samples I had collected from the metamorphic sole of the ophiolite. These rocks, as we saw earlier, were originally deep-sea sediments and volcanic rocks that had been metamorphosed to marbles and amphibolites by the increase in temperature and pressure as a giant slab of ocean crust and mantle was thrust over them. By analysing the chemical composition of the garnet, hornblende, feldspar, and micas in these rocks using an electron microprobe, I tried to determine the precise temperature and pressure at which the rocks must have formed. The science of thermobarometry, determining temperatures and pressures from the metamorphic mineral assemblage in a rock sample, was at a very early stage of development at that time, but it was obvious that this was going to become a very powerful tool for geologists working on exhumed deeply buried rocks that form in the deepest levels of mountain belts.

Although certain metamorphic minerals are indicative of specific fields of pressure and temperature, precise conditions could now be determined. Temperature and pressure conditions of formation of individual minerals and combinations of minerals had been experimentally calibrated using chemical composition. The development of electron microprobe instruments also meant that we could obtain precise chemical compositions from tiny grains of minerals, less than a few microns in diameter. For example, the elements iron (Fe) and magnesium (Mg) in the coexisting minerals garnet and biotite mica are known to be temperature-dependent, so analysing the Fe and Mg contents across garnet-biotite pairs in an electron microprobe is a powerful tool to determine very precise temperatures of formation of that rock. The results from my electron microprobe work on the Oman rocks were intriguing. I found that these rocks could only have formed at extremely high temperatures, up to 850°C and pressures of nearly 10 kilobars, equivalent to depths of nearly 40 kilometres beneath the ocean floor. I also found that the entire metamorphic sole to the ophiolite was upside down and extremely highly strained: the hotter rocks were structurally above the colder rocks. The only source of heat was from the overlying mantle peridotites. How could these rocks have formed at these great depths in the first place, and how could they subsequently have found their way back to the surface?

We already knew that the ophiolite in Oman comprises a slice of rock equivalent to 4- or 5-kilometres thickness of the ocean crust (the deep-level gabbros and higher-level sheeted dykes and pillow lavas of the ocean floor), as well as at least 10 kilometres thickness of upper mantle, the dense peridotites beneath the crust. Now I had found that my garnet- and pyroxene-bearing amphibolite rocks along the actual basal contact of the ophiolite were formed at even greater depth and at extremely high temperatures. I figured that the only environment that these rocks could possibly have formed was in a narrow subduction zone, a zone where a colder oceanic plate was descending beneath another hotter oceanic plate.[16] Maybe this was the key to the puzzle of how the ophiolite was emplaced onto the Arabian margin.

Using my field maps and cross-sections as well as my new pressure-temperature data from the metamorphic sole, I began to formulate a model for the formation and emplacement of the Oman ophiolite. The model involved formation of the ophiolite above a subduction zone rather like the marginal basins of the western Pacific, and it was this subduction zone that was responsible for the actual emplacement of the ophiolite (Figure 1.11). In effect the Arabian continental margin was dragged down beneath the ocean as the ophiolite was thrust from the Gulf of Oman onto Arabia. Simultaneously, my colleagues from the Open University, Julian Pearce and Tony Alabaster, who were working on the volcanic rocks of the ophiolite sequence, were formulating their own model for the ophiolite. They found that the geochemistry of the upper pillow lavas of the ophiolite were very similar to volcanic rocks formed at modern-day island arcs, strings of volcanoes formed above subduction zones in the oceans.[17] Analyses of volcanic lavas from islands such as the Marianas, the Bonin Islands, and the Philippines in the western Pacific all formed above a subduction zone, showed a geochemistry very distinct from the pillow lavas dredged from active mid-ocean ridge spreading centres such as the Mid-Atlantic Ridge. Maybe the volcanic lavas so beautifully exposed in Oman were a clue that the ophiolite actually formed above one of these recently developing subduction zones.

Meeting back at our base in Sohar we discussed our results with growing excitement. I discovered that Julian and Tony, working on the pillow lavas along the eastern margin of the mountains, had arrived at very similar conclusions about the origin of the ophiolite as I had, working at much deeper structural levels along the base of the ophiolite. These 'eureka' moments, when several different scientific angles seem to converge towards formulation of a common hypothesis, are the essence of the excitement of scientific research. The publication of our first papers in scientific journals on our new discoveries in the mountains of Oman was, for me, the beginning of a lifetime's quest. Geology was the key to understanding how mountains formed, and the more I got into it the more fascinated I became. Perhaps the most remarkable aspect of geology is the enormous diversity of the subject, from the study of fossils or climate right through the whole spectrum of sedimentary, igneous, and metamorphic rocks to geophysics and seismicity of the

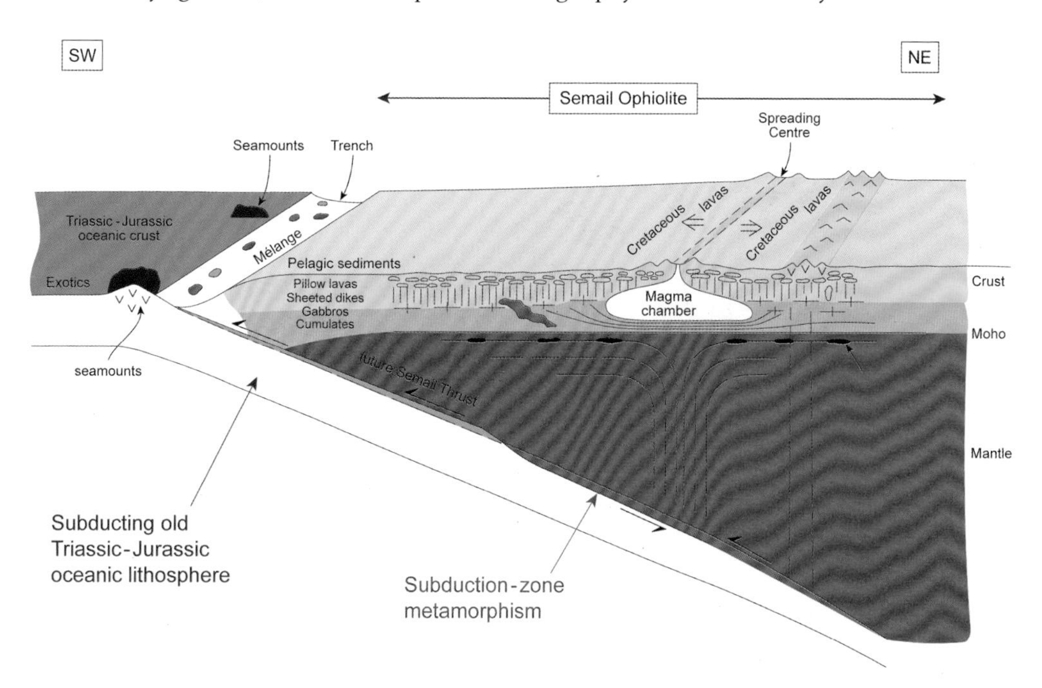

Figure 1.11 Model to explain the formation of the Oman ophiolite above an active subduction zone along which the metamorphic sole rocks are formed simultaneously with gabbros and basalts of the crustal sequence.

deep Earth. I had used large-scale structural mapping of the Oman Mountains at a scale of tens to hundreds of kilometres right down to micron-scale chemical differences in garnet and biotite mineral compositions to formulate my hypothesis. This was exciting!

Climbing Langtang Lirung

After I obtained my PhD in 1980, I decided once again to take time off and set about organizing a second Himalayan expedition, this time to the mountain of Langtang Lirung, a 7,245-metre-high Himalayan giant in central Nepal (Figure 1.12). After our success of peak bagging in Kulu in 1978 I was hopeful that we would be able to climb a new route up the south face. Together with my old climbing buddy Nick Groves we recruited a team of four other friends and set off for Kathmandu in the spring of 1980. We trekked in from Trisuli bazaar, a small village at the road head north of Kathmandu, over seven days, through magnificent rhododendron forests ablaze in pinks, reds, and white, through patches of bamboo jungle with hanging mosses, and up to the tree line, where we set up a comfortable base camp a little distance outside Langtang village.

Figure 1.12 The south face of Langtang Lirung (7,245 m) in the Nepal Himalaya.

The south face of Langtang Lirung soared up over four vertical kilometres straight above us. However, Himalayan reality soon set in, and we realized that we had bitten off more than we could chew. After three weeks of climbing steep and dangerously loose rock, winding a labyrinthine route through hanging glaciers and steep black crags, we were still only two-thirds of the way to the summit, and we found our supply lines were over-extended. The views looking east across the Jugal Himalaya and over to the beautiful fluted ice peak of Gangchempo wetted my appetite for further Himalayan exporation (Figure 1.13). From my viewpoint high on Langtang Lirung wave upon wave of incredible mountains spread across the horizon. Avalanches raked down the nearby face (Figure 1.14). These big Himalayan peaks it seemed really did require major expedition tactics, and well-stocked high camps that could form the base for a summit push. We reached an altitude of 6,500 metres on the south face and were thoroughly exhausted from difficult climbing and spells of bad weather (Figure 1.15). Retreat was called for, but before we did I collected some samples of extraordinary looking granites that contained large black crystals of a mineral called tourmaline, small red garnets, and lots of shiny white muscovite micas.

Figure 1.13 The beautiful peak, Gangchempo (Fluted Peak), in the Jugal Himalaya of central Nepal, seen from high on Langtang Lirung.

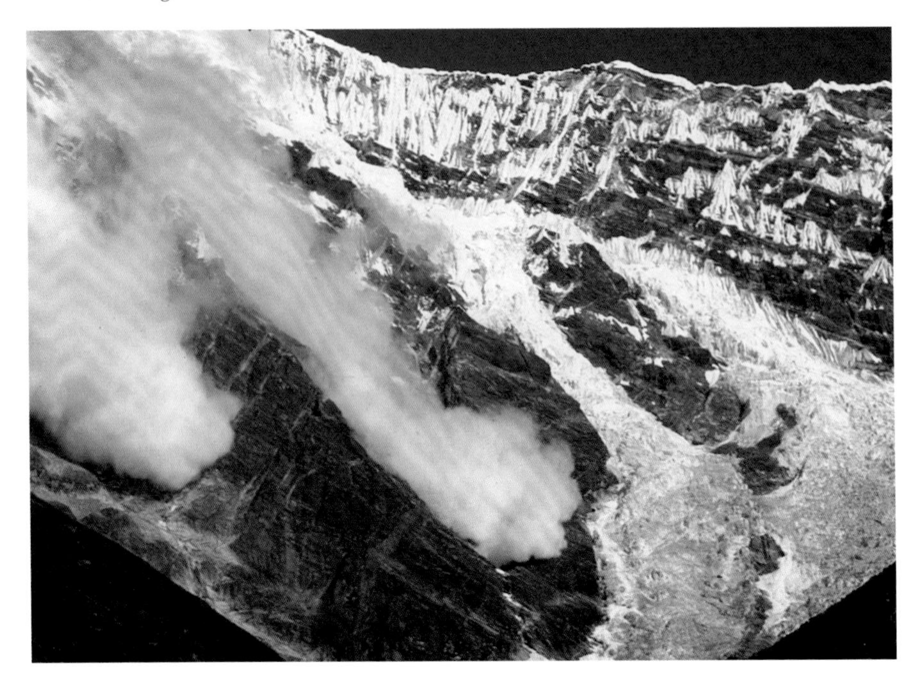

Figure 1.14 Avalanches sweeping down the south face of Langtang Lirung.

The Langtang expedition had opened my eyes to the possibilities of starting a major geological project in the Himalaya. It seemed amazing to me that so few people worked here along the largest and most impressive mountain belt on Earth. The Alps were crawling with geologists but here in Nepal there were only a few and then they were only looking at the most accessible sections along some of the major trekking routes. I decided to try my best to get grants and jobs to enable me to unravel this most spectacular of all geological wonders, the Himalaya.

Figure 1.15 The author abseiling off the south face of Langtang Lirung, Nepal.

CHAPTER 2

Continents in Collision

Kashmir, Ladakh, Zanskar

All India is full of holy men stammering gospels in strange tongues, shaken and consumed in the fires of their own zeal; dreamers, babblers and visionaries, as it has been from the beginning and will continue to the end.

Rudyard Kipling

To understand how the Himalaya were formed it seemed logical to start at the actual zone of plate collision, the Indus suture zone (Figure 2.1). Most of this collision zone runs across southern Tibet, which in the 1970s was almost impossible to travel through. Following Mao Tse-tung's Red Army's invasion and occupation of Tibet in October 1950, that region had remained firmly closed to all foreigners. In the western Himalaya the Indus suture zone runs right across the northernmost province of Ladakh. Ladakh used to be a part of south-western Tibet before the British annexed it during the Raj. Leh, the ancient capital of Ladakh at 3,500 metres in the Indus Valley, was the final outpost of British India before the great trans-Himalayan barrier of the Karakoram Range. Only the Nubra Valley and the Tangtse Valley north of Leh were beyond the Indus, and these valleys led directly up to the desolate high plateau of Tibet. Leh was a major caravan route and a crossroads of high Asia, with double-humped dromedary camel caravans coming south from the Silk Route towns of Yarkhand and Khotan; Kashmiris and Baltis came from the west and Indian traders from the Hindu regions of Himachal and Chamba to the south. Ladakh, Zanskar, and Zangla were three ancient Himalayan kingdoms ruled by a *Giapo*, or King, each from a palace that resembled a small version of the Potala Palace in Lhasa.[1]

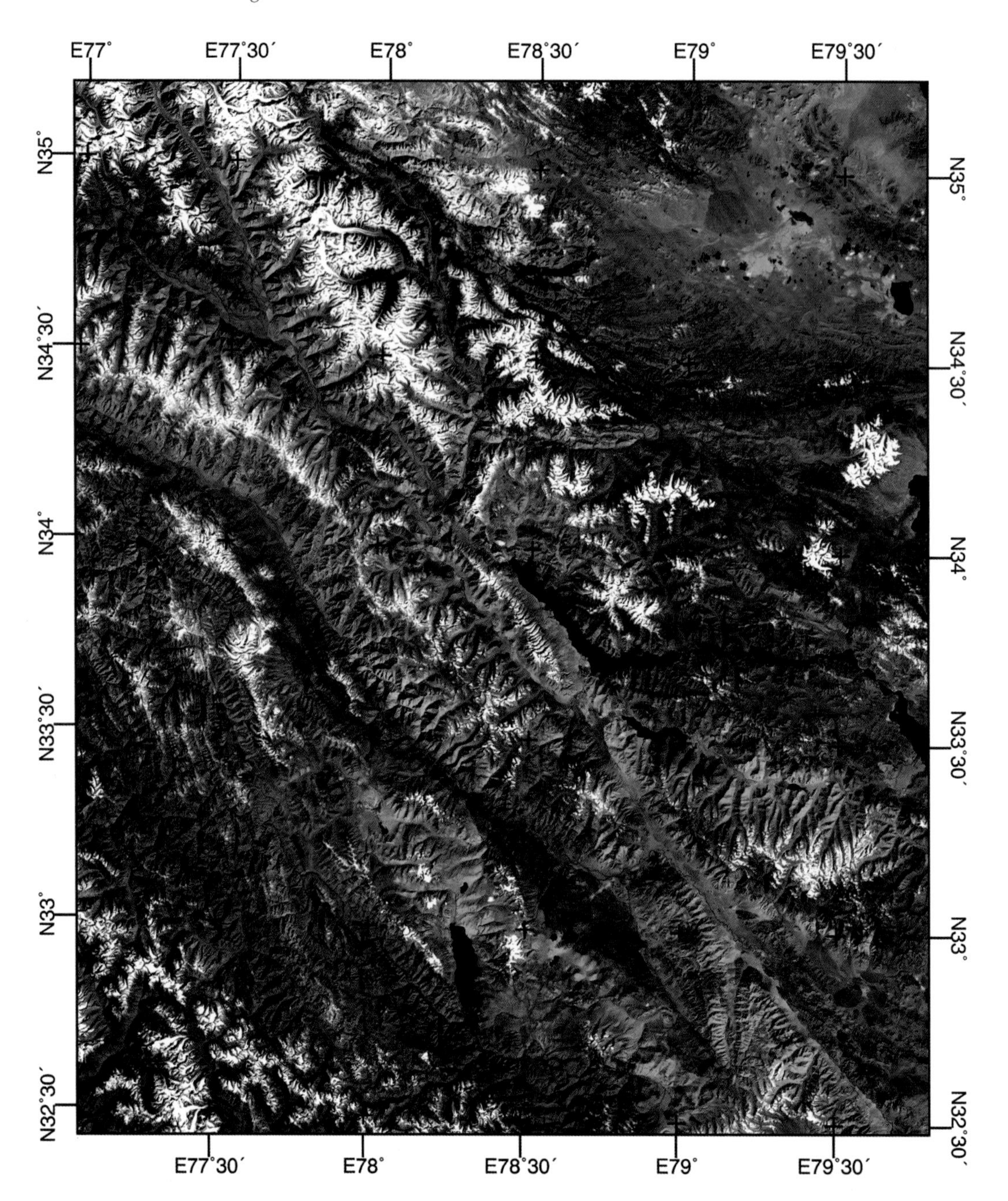

Figure 2.1 Landsat satellite photo of the Zanskar and Ladakh Himalaya, the eastern Karakoram and western Tibet. The Karakoram fault follows the Nubra Valley and Pangong Lake from top left to bottom right. The dark band in the middle is the Indus suture zone, the collision zone separating the trans-Himalayan Ladakh granites to the north from the Indian plate sedimentary rocks in the Zanskar Range to the south. The large lake in the south is Tso Morari which sits in a north–south-aligned rift valley in Rupshu, northernmost India.

In 1978, when we were climbing in the mountains of Kulu, I had looked from our high summits across to the desert mountains of Lahoul and Zanskar, north of the main Himalayan watershed. Here, in the ancient Buddhist kingdoms of Zanskar and Ladakh lay wave upon wave of unexplored and unclimbed mountains (Figure 2.2). They lay north of the monsoon limits and in the rain shadow of the main Himalaya, so the vegetation was sparse, and the geology was laid bare. Flying north

Figure 2.2 Aerial photo of the Zanskar Valley (*above*) with the snowy peaks of the Greater Himalaya in the foreground. The peaks of Nun (7,135 m) and Kun (7,077 m) in Zanskar are the highest mountains, with the Kishtwar Himalaya and Kashmir Valley in the distance.

from Delhi, or east from Kashmir into Leh, the views were simply mesmerizing. Enormous folds the size of British mountains were laid out below tiny patches of green barley fields next to Ladakhi villages revealing their true scale. Huge, winding glaciers carved great chasms between the mountain ranges. In Ladakh the rocks were of every hue and colour, from bright purple, red, ochre, and green, to the browns and greys of the limestones. The air was crystal clear up at these altitudes and the local people were extremely friendly and hospitable, in true Tibetan fashion. Ladakhi women wore the *perak*, heavy embroidered headgear laden with inset stones, blue lapis lazuli and turquoise, red coral, and cowrie shells from some far-away ocean. The men wore a dark red *chuba* or cloak and characteristic upturned hats. Ladakhi houses were spectacular constructions too in the old Tibetan style, with yak stables on the ground floor and living space above, with a central fireplace with mounds of yak dung for fuel and juniper incense. No Ladakhi kitchen was complete without a barrel of *chang*, the potent barley beer, and the circular wooden barrels for churning the ubiquitous yak butter tea.[2]

The Indian state of Jammu and Kashmir is a complete mixture of lands and peoples. It comprises at least four completely distinct parts: Hindu Jammu south of the main Himalayan Range, the predominantly Sunni Muslim Kashmir Valley, and the mainly Buddhist ancient kingdoms of Ladakh, Zanskar, and Zangla, north of the main Himalayan Range. In the western part of Ladakh the town of Kargil is predominantly Shi'a Muslim, cut off from its neighbouring Shi'a towns of Skardu, Khapalu, and Machalu in Pakistani-held Kashmir by the artificial border of the UN ceasefire line, termed the 'Line of Control'. Every few years vicious fighting would break out between the Indian and Pakistani armies, with the local Ladakhis and Kashmiris caught in the middle of this political turmoil.[3]

Ladakh was under restricted access by the Indian government, being along the disputed borders with Pakistan and China, but in 1978 the government lifted the ban and allowed Western tourists in for the first time. The first trip I made to Ladakh in the summer of 1981 I flew up to Srinagar in Kashmir and travelled by bus to Kargil. The winding road had recently been built by the Indian army and on every corner there were wonderful signs erected by the road-builders: 'Be gentle on my curves', 'No hurry, no worry', 'Better late than never', 'Wood is good; tree is better'. There were also a few ominous signs like 'Welcome to Kargil' and right next to it 'Shoot to kill', leaving one with no doubt that up here the Army was in control. The road was only opened for the summer months and buried under metres of snow all winter.

Kargil is a frontier town right on the ceasefire line, and has suffered from numerous border skirmishes by Pakistani tribesmen since the 1947 partition of India and Pakistan. The town is surrounded by spectacular granite mountains, part of the great trans-Himalayan granites that run along the Ladakh Range north of the Indus River and continue east into the Gangdese Ranges of southern Tibet. These rocks were geologically part of Asia, north of the great suture zone. The Ladakh granites are Andean-type granites and contain the minerals hornblende and biotite, remarkably similar to the granites exposed along the length of the Andes in Peru, Bolivia, and Chile. Occasional remnant volcanic rocks, andesites and ignimbrites of almost identical geochemical and isotopic composition to the andesitic lavas that erupted from the giant volcanoes along the length of the Andes also occur in the Ladakh Range. The granites in Ladakh spanned a long period of time, at least from about 110 million years ago to about 50 million years ago. This suggests that the southern margin of Asia during that time must have resembled the Andean margin of Peru and Chile today. Volcanoes such as Cotopaxi and Chimborazo in Ecuador or Mount St. Helens in Washington State are erupting andesite lavas of very similar composition to the remnant lavas seen in parts of the Ladakh Range today. The Ladakh granites also look almost identical to the great granite batholiths in Peru and Bolivia.

Interestingly, the timing of the granite magmatism in Ladakh shows that it appeared to end about the same time as the Indian plate first collided with Asia, about 50 million years ago, when oceanic subduction northwards beneath the Asian continental margin also ended. These granites resulted from melting of the crust above a subducting ocean plate. When the continental collision occurred, the ocean finally closed, ending both subduction beneath and granite intrusion above (Figure 2.3). As soon as the ocean closed, the zone of collision then buckled downwards and subsided, forming a long narrow sedimentary basin that filled up over time with all the detritus eroded from the nearby mountain ranges (Figure 2.4a–d). Sedimentary rocks along this basin were all of continental derivation, great river floods depositing fluvial conglomerates and fine-grained silts and muds that slowly settled along the floors of lakes. The clasts in the conglomerate beds could all be forensically traced to their source regions. Most were granites, with occasional andesites derived from the high Ladakh Range to the north, but a few were serpentinites, basalts, and red cherts—oceanic rocks derived from the great ocean that had closed along the Indus suture zone.

The Hidden Zanskar Valley

I wanted to carry out a large-scale reconnaissance on my first trip to try to get a feeling for the problems ahead, both geological and logistical. After a few weeks of geologizing all around Kargil and along the Suru Valley, I caught a local bus to Rangdum Gompa, the first Buddhist monastic settlement of the Zanskar region. It was an extraordinary

Figure 2.3 Large-scale folds formed by the collision of India with Asia, south of the Spiti Valley in Lahoul.

Figure 2.4 (*opposite*) (a) Stakhna gompa in Ladakh above the Indus Valley. Mountains in the distance show folded sedimentary rocks of the Indus molasse.
(b) The Hemis conglomerate, showing boulders derived from the suture zone (red cherts, green serpentinite), from the Asian margin (pale-coloured granites and andesite lavas), and from the Indian plate margin (white limestones).
(c) Syncline fold in the Hemis conglomerates, part of the Indus molasse deposited along the zone of collision during Tertiary time.
(d) Sketch section across the Indus molasse basin in Ladakh. The blue horizons are marine incursions prior to the collision containing fossils with age ranges spanning the Palaeocene and Lower Eocene. Brown beds are shales and yellow conglomerates, the erosional products of the rising Himalaya. The red rocks are the granites along the South Asian plate margin with their intrusive ages. U_1–U_8 refers to prominent unconformities in the sedimentary strata of the molasse basin with their ages in millions of years.

(a)

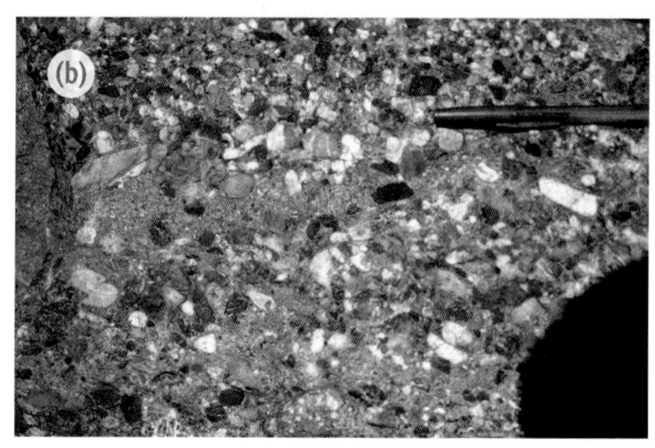
(b)

(c)

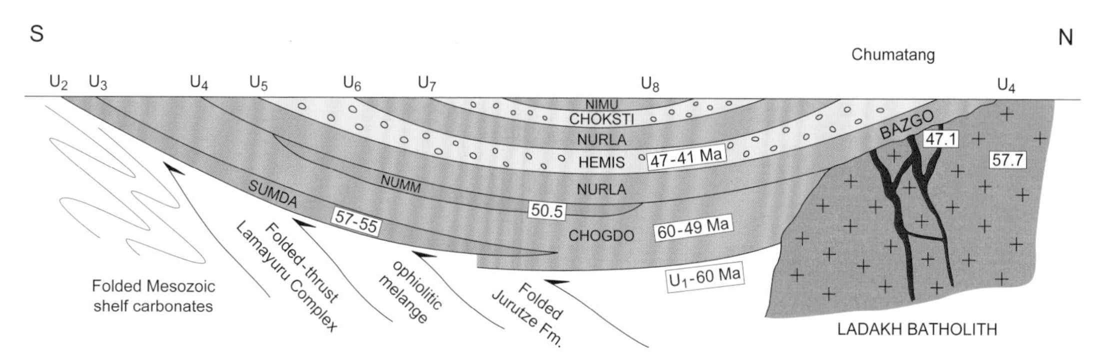
S
N
Chumatang
U_2 U_3 U_4 U_5 U_6 U_7 U_8 U_4
NIMU
CHOKSTI
NURLA
BAZGO
47.1
57.7
HEMIS
47-41 Ma
NUMM
NURLA
SUMDA
50.5
57-55
CHOGDO
60-49 Ma
U_1-60 Ma
Folded Mesozoic shelf carbonates
Folded-thrust Lamayuru Complex
ophiolitic melange
Folded Jurutze Fm.
LADAKH BATHOLITH

(d)

transition from the ultra-orthodox Shia'a Muslims of the villages around Kargil to the smiling, jovial Buddhist monks of Rangdum. In 1981 the road ended at Jildo, the first Buddhist village just below the monastery at Rangdum. Jildo was in a spectacular setting, with the massive snowy mountains of the great Himalaya behind and the barren brown rocky mountains of the Zanskar Range to the north. I stayed with a delightful local family, and became good friends with Sonam Norboo. Sonam had several ponies and agreed to come with me as far as Padam. With a couple of horses as pack animals we set off for the three-day trek over the Pensi-la and along the main Zanskar Valley. It was an enchanting journey, through rustic little villages, houses built in the traditional Ladakhi style, of mud brick with Tibetan-style windows, roofs buckled under piles of birch branches and yak dung, both used for winter fuel and cooking. The trail passed through fields of swaying barley and apricot orchards.

Padam, the main village of Zanskar, is located at the junction of the two main rivers of Zanskar that meet and then flow north through steep gorges to join the Indus River in Ladakh. The mountains of the Zanskar Range to the north-east of the Zanskar Valley are all limestones and other sedimentary rocks with fantastic folds and thrust structures, magnificently exposed in three dimensions along the deeply incised river gorges. To the south-west of the Zanskar Valley the mountains all comprise metamorphic rocks and some interesting tourmaline-bearing granites, similar to the rocks I had collected from high on Langtang Lirung. I mapped out much of the Zanskar Valley as I was going along, and decided that there had to be a major fault running along the valley right through Padam. This fault was aligned NW–SE, parallel to the foliation of the metamorphic rocks below and the bedding planes of the sedimentary rocks above, and dipping at about 40 degrees to the north. It struck me as being very unusual that a normal fault, which results from extension rather than compression, occurred right in the middle of the Himalaya, a mountain belt formed by compressional tectonics as a result of the collision of two very large continental plates. The geometry of the rocks suggests that the metamorphic rocks of the Greater Himalaya were being extruded out from beneath the Zanskar Range and the contact was this very large low-angle normal fault. This was a structure that as far as I knew had not been described before from a compressional mountain range like the Himalaya, and it fired my imagination to make a more detailed geological map of the entire Zanskar Range.

Normal faults occur in the brittle upper crust in regions of the world which are undergoing crustal extension, such as the Dead Sea rift in Jordan, and the East African rift or the North Sea. These normal faults tend to be rather steep and bound the deep rift valleys. Other areas of previously thickened continental crust which are now undergoing extension and 'collapse', such as the Basin and Range Province of continental North America, east of the Sierra Nevada, are characterized by low-angle normal faults which extend into more ductile shear zones at depth. Many of these extensional areas on the continents are commonly associated with volcanic cones and outpouring of viscous lavas along large cracks in the Earth's crust. Some of the larger rifts, such as in the Tanzania–Kenya–Uganda region are associated with huge continental volcanoes like Kilimanjaro, which are sourced from deep in the upwelling mantle beneath the crust. The normal fault along the Zanskar Valley was a very different structure, being at a low-angle and in the middle of a compressional mountain belt.

In Padam I found two local Zanskari traders who were about to set off for Leh, a two-week walk across the Zanskar Mountains north of Padam and I arranged to tag along with them, as the route was very complicated and crossed three high passes. I adapted my geological mapping to the pace of trekking, and soon learnt that I had to be very quick and observant as we covered a lot of mountainous ground in a day's march. After leaving Padam, the path crossed the great flood plains where the two major rivers of Zanskar merge to flow north, cutting a spectacular canyon right across the mountain range. The river winds through the limestone mountains as far as the Kingdom of Zangla and north until it suddenly thunders into an impossibly deep cleft. No trails entered the gorge so the path was forced out of the main valley to climb up steep-sided valleys towards the monastic settlement of Lingshed beneath the great limestones that guard access to the Spontang Mountains. This high country is the site of the largest and best preserved of all the Himalayan ophiolites that crop out (Figures 2.5a, b). The ophiolite was actually quite accessible despite being very high, so I spent a few days looking at several of the main valley sections, whilst my two trekking companions whiled away their time, talking, drinking *chang*, and generally chilling out in the village of Photoksar. The final few days' trekking north followed a spectacular river canyon winding down steeply all the way to the Indus. The trail was in places carved out of the vertical cliffs lining the sides of the river. Eventually we emerged from the canyon and, after crossing one final pass, there in front of us was the dramatically situated fortress of Lamayuru Gompa, a 14th century monastery perched on top of teetering, crumbing cliffs of barely consolidated lake sediments.

Lamayuru is one of the most famous monasteries in Ladakh, a land of several hundred *gompas*, lamaseries, and nunneries. It was common practice for the younger sons of Ladakhi families to become monks, but it seemed like a very free and easy system. Some monks would travel for months or years all along the Himalaya, even down as far as southern India, others would go back to their family farms and villages, and some would stay and learn the Buddhist scrolls and keep the monastic practices going. Inside, cushioned seats would line the main chamber, which would be adorned with many '*thankas*' or paintings, depicting the life of the Buddha, the wheel of life, the Mandala, or any of the hundreds of deities. One *thanka* inside Lamayuru Gompa had a painting of the monastery, but outside the walls was

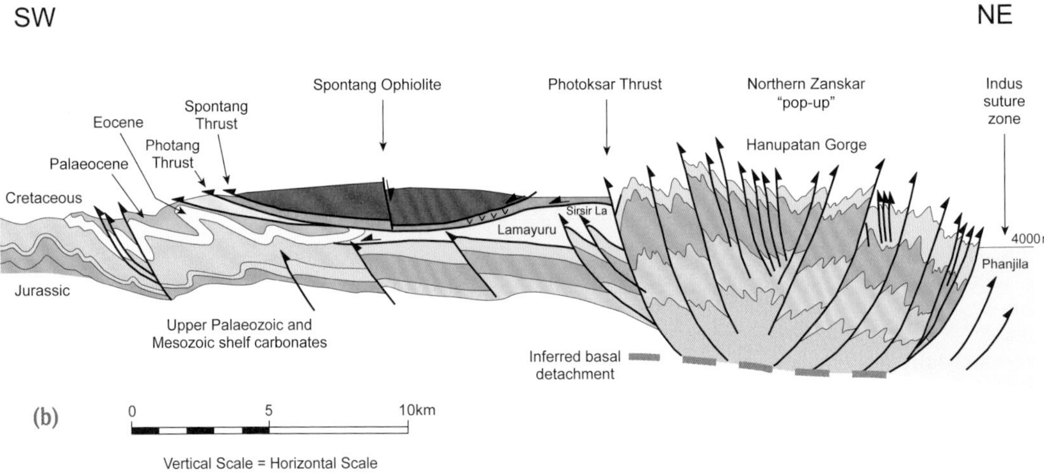

Figure 2.5 (a) Panorama of the Spontang Range in Ladakh showing the ophiolite (dark rocks beneath the clouds), the cliffs of Tertiary limestones above the thrust sheets of deep marine shales (high levels) emplaced onto the Indian margin sedimentary rocks during ophiolite obduction. Note the recumbent folding of Mesozoic limestones in the large cliffs above the Zanskar River indicating major crustal shortening and thickening during the obduction and collision process. (b) Simplified geological section across the Zanskar and Ladakh Himalaya.

a beautiful blue lake. Today, Lamayuru is surrounded by horizontally bedded pale yellow sediments that appear to be relics of the lake shore, the rocks that lined the shores of this now-disappeared ancient lake (Figure 2.6). The monastic village had an incredible assortment of mud-brick buildings, some perched into vertical cliff faces, and an array of tunnels carved into the crumbling, barely consolidated lake sediment. One had the impression that all it needed was one big rainstorm and the entire structure would melt into the valley. At some time in the historic past the lake must have drained out catastrophically along the Yappola Stream, flooding its contents into the Indus River far below.

Figure 2.6 Ancient lake sediments near Lamayuru in Ladakh. The lake waters drained catastrophically into the Indus River after an earthquake in the historic past.

We hitched a ride on a passing truck, and tossing our rucksacks onto the roof clambered up for a winding, scenic ride all the way into Leh. The ancient capital of Ladakh has been a trading post of central Asia for centuries (Figure 2.7). Ancient caravans of horses, ponies, and camels regularly traded along routes northwards towards China, crossing the Ladakh Range and the mighty Karakoram. Approaching Leh from the west along the Indus was a magical experience. Tiers of mountains spread away to the north as the Indus Valley widened out with its green fields of swaying, ripening barley, and little orchards of apricot trees and birch groves. To the south the serrated peaks of the Zanskar Range rose steeply, with the rocks folded and contorted into fantastic shapes. The great palace of Leh Gompa, a small version of the Potala Palace in Lhasa, formed a welcoming beacon with a backdrop of the Ladakh granite hills in every shade of brown, grey, and salmon pink. The truck dropped us off below the bazaar and I strolled up through the narrow streets to the old fort. Behind the fort a maze of tiny narrow alleys led

Figure 2.7 Leh, the ancient capital of Ladakh and an important trading route in central Asia. The old palace was built in the 17th century during the reign of King Senyye Namgyal and is a replica of the Potala Palace in Lhasa. The photo was taken in 1981 before restoration. The mountains behind are composed of the Ladakh granites, the southern margin of the Asian plate.

into the old quarter, full of incredible smells, wafts of burning juniper wood and incense, fluttering prayer flags, and the beautiful chimes of windbells. I found a guest house above the palace on the outskirts of town, almost the last house before a caravanserai where yaks and ponies and even a few double-humped camels, dromedaries from the Nubra valley over in the Karakoram, were grazing. The manager of the guest house was a friendly Ladakhi Muslim by the name of Fida Hussein, with whom I became great friends and still am to this day.

I was extremely pleased with my first trek across Ladakh. I had fallen in love with the wilds of Ladakh and Zanskar right from the beginning. The mountain scenery was spectacular in the extreme, the local people were amongst the pleasantest and most friendly and hospitable people I had met anywhere in the world. The geology of Ladakh and Zanskar was very complicated but extremely well exposed in three dimensions and completely fascinating. I knew that if I was going to try and understand the Himalaya and how it was formed that this was going to be a lifetime's work, but after that first trek, I decided that this was a task that I would absolutely relish.

Indus Suture—The Zone of Collision

It always seems remarkable that an ocean such as Tethys, once thousands of kilometres wide, could have completely vanished, with only a trace remaining along the ancient suture zone. Yet this is what the rocks were telling us. Deep-sea sediments containing fossils ranging in age from Permian times more than 250 million years ago to the Palaeocene time about 60 million years ago are preserved, together with highly sheared pieces of oceanic crust and mantle rocks. The first thing needed was a thorough investigation of the suture zone across Ladakh, and in 1982 I travelled once again through Kashmir, crossing the Zoji-la pass over the Himalaya to the border town of Kargil. It was very pleasant to return to my old friend Haji Ali's guest house and I was given a great welcome. From Kargil, I travelled slowly east along the main road connecting Leh with the outside world. I stayed several days in small villages along the way whilst I studied the geology and tried putting all my observations on a map. After several weeks I had practically walked the entire road section. Most travellers either flew directly into Leh or over-nighted in Kargil on a two-day bus journey from Srinagar, but the small villages and farms I encountered along the way were a real treat to experience. Mulbeck was a delightful village, the first Buddhist settlement coming in from the west with a fascinating little *gompa*, built around a six-metre high statue of the Maitreya, the future Buddha, carved out of solid rock. Once again I stayed at Lamayuru Gompa, which I had visited on my first trip, the largest and most spectacular of the monasteries. The mountains around Lamayuru show one of the best sections across the Indus suture zone, the actual zone of collision of the Indian and Asian plates. Clambering over the hills and descending the huge cliff section north to the Indus River near Khalsi, I discovered green ophiolitic rocks sheared and smeared along giant faults, remnants of the ocean crust that once separated the great continents.

The road descended in a never-ending series of hairpin bends from Lamayuru down to Khalsi. From here towards Leh it followed the Indus River upstream, winding around crags, with stark mud-brick villages clinging to the sides. Ladakhi architecture was impressive, with massive houses built of mud with ramparts all around, their rooftops adorned with piles of yak-dung patties drying for winter fuel, and always strings of colourful prayer flags flapping in the wind. Monastic villages like Saspol and the red-earth village of Bazgo formed perfect stops on the route to Leh.

Back in Leh I returned to Fida Hussein's guest house and spent several days drafting up my geological maps, eating in some of the newly opened trekkers' restaurants, and making plans. This year I decided to trek the whole way across the

Figure 2.8 The Indus suture zone, the site of the India–Asia collision zone in Ladakh. The bedded rocks on the left are Indus molasse sediments overlying Ladakh granites of the Asian plate. The suture zone rocks include green serpentinites and remnant ophiolites above the Indus River and thin-bedded shales originally deposited in the Tethyan Ocean between India and Asia.

Himalaya, starting where I was in Ladakh on the actual zone of the plate collision and walking south all the way to the southern margin of the Himalaya, the Main Boundary Thrust. I planned out a circuitous route starting from Leh, walking through a series of canyons and narrow river valleys in eastern Zanskar and Rupschu to get across the Zanskar Range to the Kingdom of Zangla. This section traversed the continental margin sedimentary rocks that spanned the entire geological history of the Tethyan Ocean from Late Carboniferous–Permian rifting of Gondwana 300 million years ago up to the India–Asia collision 50 million years ago (Figures 2.8 and 2.9). From Zangla it was a short day's walk to Padam in the main Zanskar Valley. There I would have to change Ladakhi porters or horses for Himachal ponies or porters for the route across the main Himalayan Ranges to Darcha and Manali. From Manali I planned to walk south along the main Kulu Valley, where we had climbed during our first Himalayan expedition in the Tos Glacier region to the east. It was an ambitious plan and I reckoned the whole trek, mapping the geology en route, would take me probably seven or eight weeks.

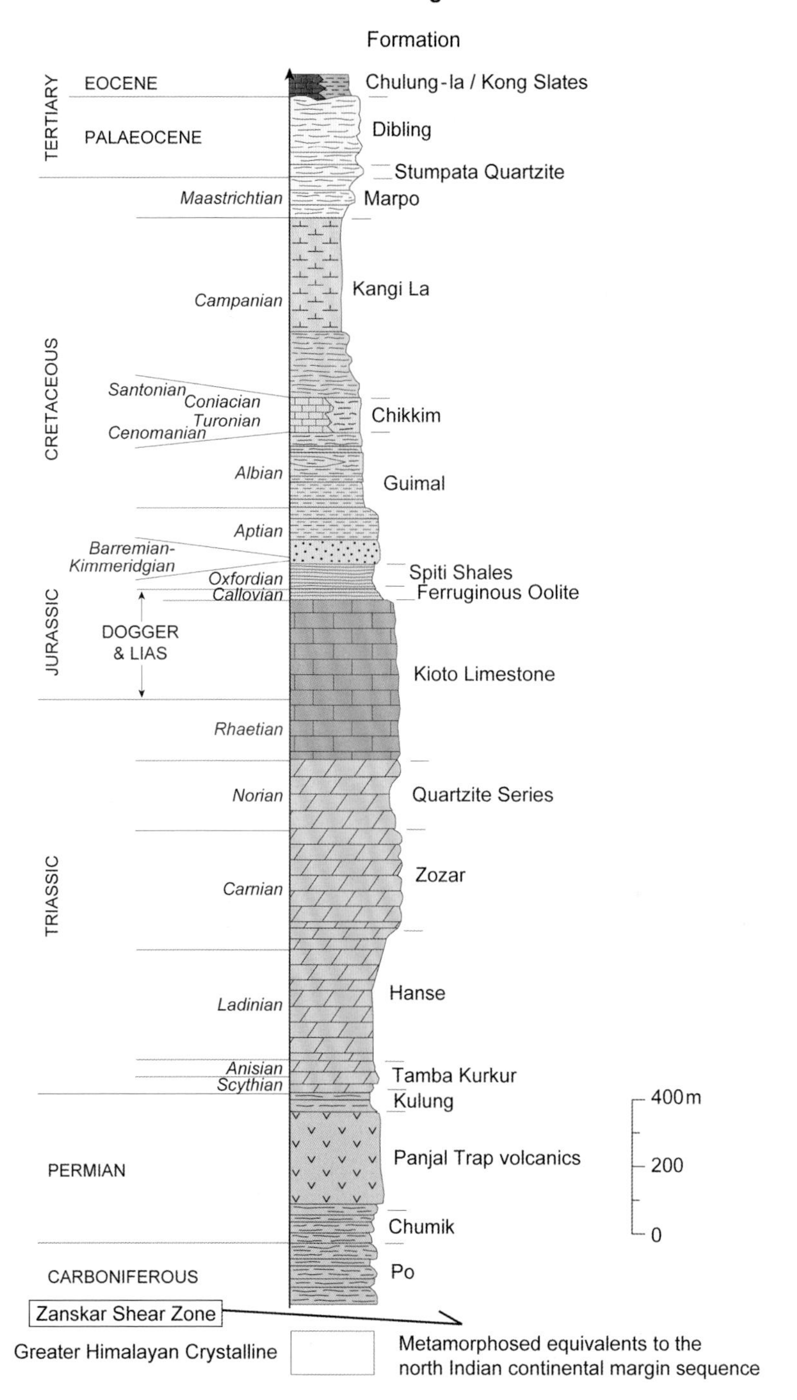

Figure 2.9 Stratigraphic column showing age ranges of rocks exposed along the northern Indian Himalaya in Zanskar.

Geo-Trekking across the Himalaya—the Zhung Lam

I left Leh in May 1982 with Fida Hussein and two ponies laden with several weeks' supplies of rice, *dhal*, vegetables, and fodder for the ponies. We crossed the Indus River at Choglamsar, a Tibetan refugee camp outside Leh, and trekked over the 4,920-metre-high Ganda-la Pass to the Marka Valley. In Marka we stayed with an old friend of Fida's, Sonam Targis. Targis knew the remote mountains of eastern Zanskar and Rupshu like no one else, having traded salt and sheep between Leh and the Zanskar Valley for thirty years. It was complicated country and easy to get very lost among a maze of river valleys, narrow canyons, and steep passes, so we were relieved when Sonam Targis agreed to come with us as far as Zangla.

Sonam Targis wanted to take his yak instead of Fida's ponies, as the route we were taking, the *Zhung Lam*, or 'Middle Way', was a difficult trek involving over fifty river crossings, crossing two 5,000-metre-high passes, the Rubering-la and the Chacha-la, with lots of ascending and descending (Figure 2.10). For ten days we trekked along the wonderful canyon country of Zanskar, traversing ever more remote mountain ranges without seeing any villages, or any other people at all. The gorges led up to high passes, and the final section involved walking up a gorge so narrow that the yak's horns would not fit through. Targis decided that the only solution was to unload the yak and leave it there to graze whilst we humans carried the final loads up the pass. From here on the Chacha-la, we could look way down

Figure 2.10 Sonam Targis leading the horse and Fida Hussein trekking with the author in Ladakh.

onto the green barley fields of Zangla to the west. We bade farewell to Targis who returned to pick up his yak and descend back through the gorges to Marka, whilst Fida and I carried our heavy packs down into the Kingdom of Zangla. In Zangla we stayed with the *Giapo* or King of Zangla in his old Ladakhi house piled high with birch twigs and crusty yak dung. He plied us with yak butter tea, food, and copious glasses of *chang*, and told us stories of how remote and cut off Zanskar used to be when he was a child. During the five or six winter months when all the passes were snowed in and all trekking routes closed, the only way out was along the frozen Zanskar River, the *Chaddur*, a perilous eight-day trek walking on the thin ice north to Chilling and the Indus Valley.

The country east of Zangla is extremely inaccessible and the few trekking paths wind along narrow deep gorges through the jagged limestone mountains (Figure 2.11). The high peaks abound with wildlife, herds of ibex and *bharal* (blue

Figure 2.11 The trekking trail along the Hanupattan gorge in Ladakh.

sheep) leaping with ease across the impossibly steep hillsides. Snow leopards are fairly common but rarely seen, except during harsh winter months when they sneak down from their mountain lairs to raid chicken coups and even take young lambs. There were numerous river crossings every day and the occasional Zanskari rope bridge to negotiate. These structures comprise three great strands of rope, one along the bottom for feet, and two along the top for each hand, with thin strands attached to each. One has to tread carefully, tightrope style, along the single rope strand below, juggling with the hand ropes and trying to avoid upending and falling into the thundering river below (Figure 2.12). After several days of trekking along these great limestone gorges, we emerged into the Tsarap Lingti *chu* and the monastic settlement of Phugtal, tucked into the recesses of a large cave.

Figure 2.12 One of the classic rope bridges high above the Chenab River gorge in Zanskar.

During the trek, I was mapping the superb three-dimensional exposures through the fold and thrust structures all the way, and had attempted to draw a geological cross-section from Leh to Padam across the sedimentary rocks of the Zanskar Range, to work out the three-dimensional layout of the beds (Figure 2.13). In Padam we had a few days' rest and I drew up all my rough sketches into a comprehensive geological profile. The folds and thrust structures were truly spectacular all the way across and I had taken plenty of photographs during our trek. I was amazed at the cross-section that emerged from my pencil. The structures looked wildly implausible, but I had actually seen them all and photographed many of them. I knew that the paper I was hoping to write on this section would have to be accompanied by photographic evidence, otherwise the sceptics would easily believe my cross-section to be largely imaginary.[4]

Figure 2.13 Folds in Mesozoic limestones of the north Indian plate margin beneath the Spontang ophiolite in Ladakh.

The Zanskar Fault and High Himalaya

In Padam, I had to once again change porters and trekking companions. Fida returned to his wife and young son in Leh, and I found a Zanskari from the village of Sani, Tschering Tukten, who was willing to carry my large rucksack over the Shingo-la, the main pass that crossed the Himalayan Range between Zanskar and Lahoul, to Darcha and Manali. This next stretch was about ten days' walk, and once again I was mapping along the way. I had become used to mapping at the pace of a yak, taking long stops at interesting outcrops and rushing fast along river valleys and terraces where there were no rocks to see. The route east from Padam followed the trace of the very large low-angle normal fault I had found the previous year and I took special care mapping along this route. At the time I realized that this was an important structure, but I still could not quite understand how you could get a low-angle normal fault indicating crustal extension here along some of the highest mountains in the world caused by the compression during a continental collision. How could the Himalaya be both compressing and extending in a north–south direction at the same time? How could the mountains maintain their high elevation and steep relief if the crust was extending? I needed an explanation. The answer would require a comprehensive study of the deep crustal metamorphic rocks of the high Himalaya beneath and to the south of the fault.

By the time Tschering and I had crossed the Shingo-la, I thought I was beginning to understand this normal fault, the Zanskar shear zone. As the Indian plate collided with Asia and the crust of India started buckling, folding, and thickening, the upper crust composed of the sedimentary rocks of the Zanskar Range had become decoupled or detached from the structurally deeper metamorphic rocks of the high Himalaya. This decoupling occurred on massive thrust faults, low-angle 'décollement horizons' that detached vast slabs of rocks and separated them into distinct thrust sheets. Later, as India continued to penetrate north pushing under the southern margin of Asia, the basal thrust fault reactivated as a normal fault detachment as the upper plate locked up and deformation propagated downwards and southwards across into the Greater Himalayan metamorphic terrane (Figure 2.14). These rocks, initially deeply buried and metamorphosed into schists and gneisses beneath the Zanskar Range, were later expelled out and extruded southwards along these giant shear zones and fault systems. The middle crust of the Himalaya was literally being squeezed out from beneath the Zanskar fault like the jam from the middle of a jam sandwich that is being squashed flat. The Zanskar normal fault was the upper boundary or detachment plane of the extruding jam-like middle layer of the Himalayan crustal sandwich (Figures 2.15 and 2.16).[5]

Figure 2.14 A side valley of the Zanskar River in Ladakh showing the style of crustal shortening and thickening in rocks of the north Indian continental margin. Yellow lines pick out bedding planes showing the folds, and the red lines are thrust faults.

After two more weeks' trekking over the high Himalaya from Zanskar to Lahoul, Tschering and I crossed the final pass of the main Himalayan Range, the Rhotang-la, and descended to the lush Alpine meadows and beautiful pine forests of the Kulu Valley. After the barren desert mountains of Ladakh it was a delight to see the forests of green pines and deodars and meadows bursting with a profusion of colourful flowers once again. Tschering returned to his village in Zanskar and I had a few days' rest in the bustling hill station of Manali, staying in the delightful apple orchards of John Banon's guest house. I set off on the final gentle stretch along the Kulu Valley, trekking south past the great castle fortress of Nagar, and the town of Kulu, the jumping-off point for our expedition to the Tos Glacier in 1978. Walking southwards down the Kulu Valley we saw that the rock layers mainly dip to the north, so we should have been encountering deeper-level rocks in lower structural positions. Instead I realized that the metamorphic rocks were becoming lower grade—subjected to lower temperatures and pressures. The metamorphism was in fact upside down. This was the famous inverted metamorphic sequence initially discovered almost a hundred years ago. The Greater Himalayan metamorphic slab is nearly 20 kilometres thick and dipping to the north. Along the top is the

Stage 1. ORIGINS

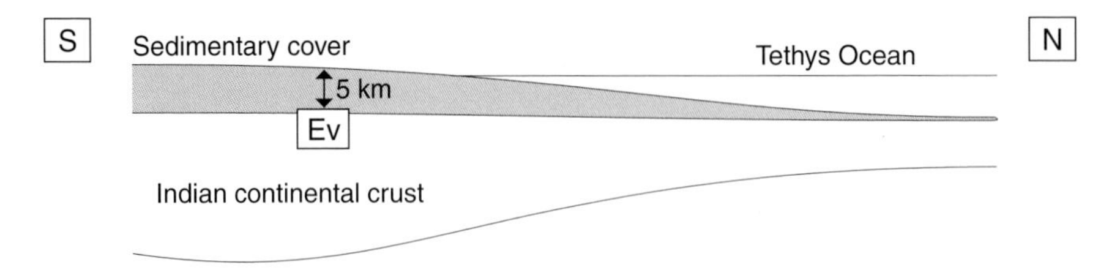

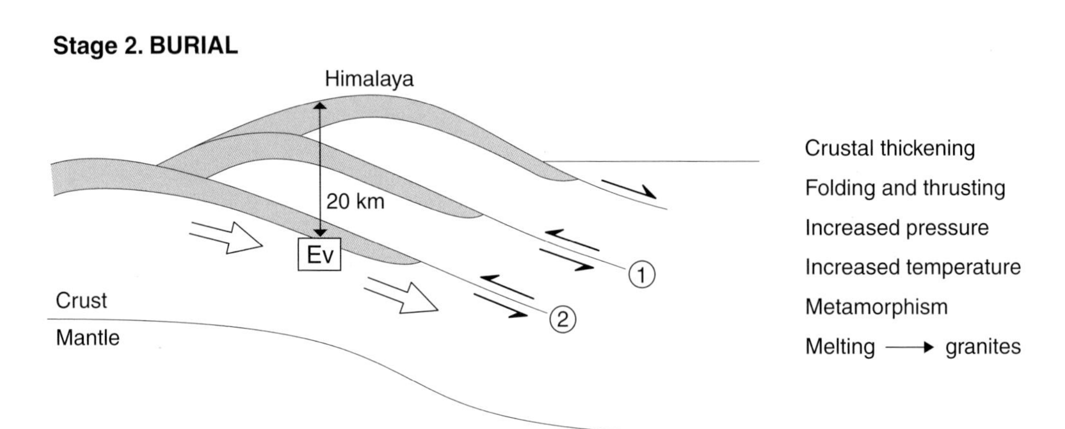

Stage 3. EXHUMATION

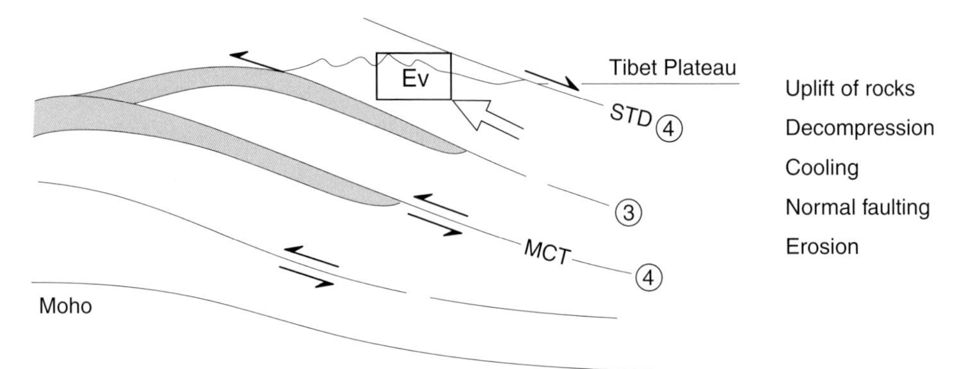

Figure 2.15 (*above*) Sequence of thrusting illustrating the crustal thickening process during plate collision. A thrust fault can become a later passive normal fault with time as deformation propagates down-section, and higher thrust sheets ride 'piggy-back' on younger, deeper thrusts. The box Ev shows how a rock now exposed at the summit of Everest became buried (stage 2) and exhumed (stage 3) by the thrust faulting process combined with erosion.

Figure 2.16 (*opposite*) Cross-sections of the Zanskar Himalaya showing the southward-extruding, partially molten middle crust between the Main Central Thrust below and the Zanskar shear zone low-angle normal fault above.

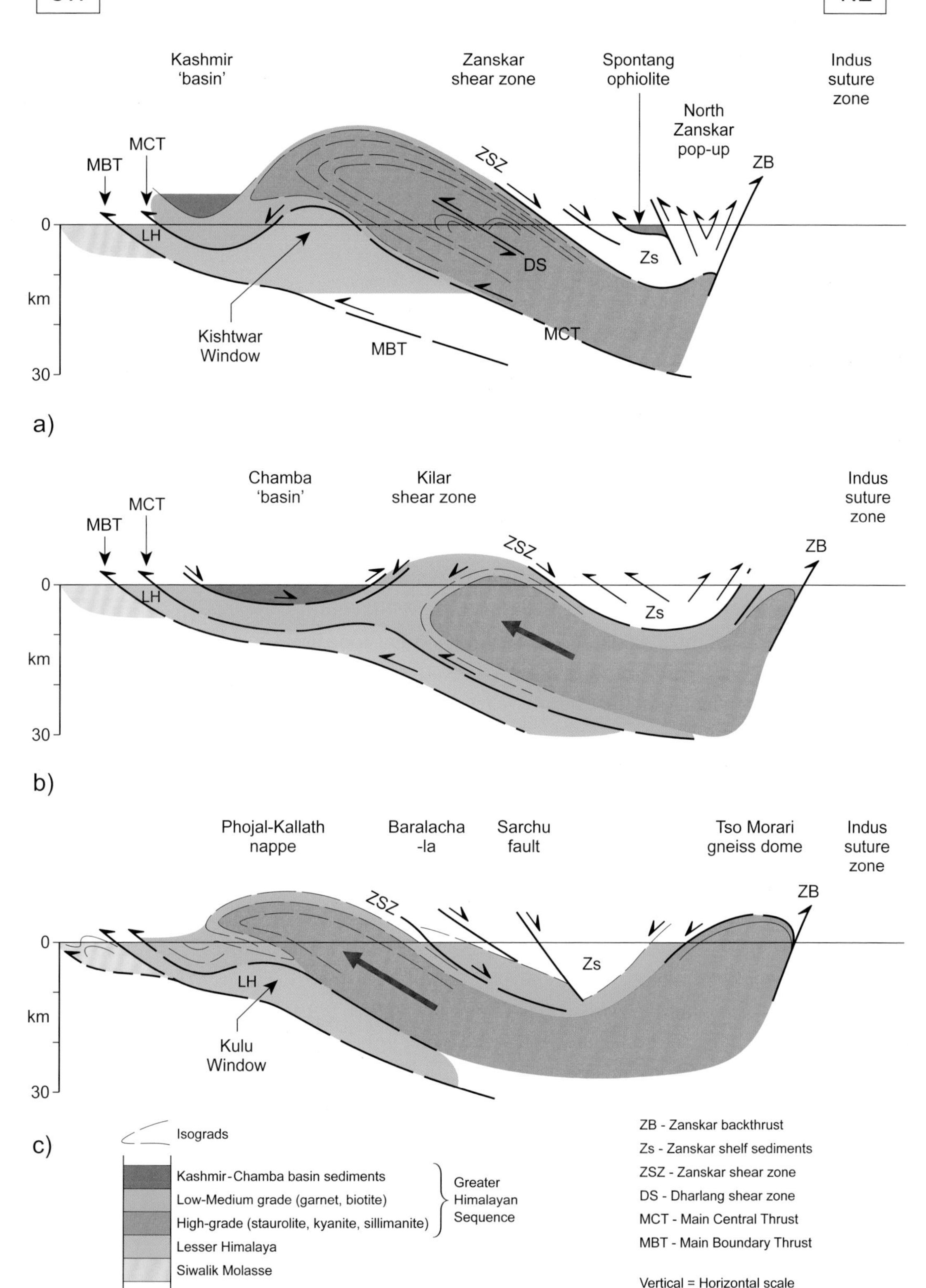
SW
NE
Kashmir 'basin'
Zanskar shear zone
Spontang ophiolite
Indus suture zone
North Zanskar pop-up
MCT
MBT
ZSZ
ZB
LH
DS
Zs
Kishtwar Window
MBT
MCT
0
km
30
a)
Chamba 'basin'
Kilar shear zone
Indus suture zone
MCT
MBT
ZSZ
ZB
LH
Zs
0
km
30
b)
Phojal-Kallath nappe
Baralacha -la
Sarchu fault
Tso Morari gneiss dome
Indus suture zone
ZSZ
ZB
Zs
LH
Kulu Window
0
km
30
c)
Isograds
Kashmir-Chamba basin sediments
Low-Medium grade (garnet, biotite)
High-grade (staurolite, kyanite, sillimanite)
Lesser Himalaya
Siwalik Molasse
Greater Himalayan Sequence
ZB - Zanskar backthrust
Zs - Zanskar shelf sediments
ZSZ - Zanskar shear zone
DS - Dharlang shear zone
MCT - Main Central Thrust
MBT - Main Boundary Thrust
Vertical = Horizontal scale

Zanskar shear zone—the low-angle normal fault I had found the previous year. Along the base of the metamorphic pile another narrow zone of *mylonites*, ribbon-like stretched and sheared rocks, occurs along a structure called the Main Central Thrust, a zone of extreme deformation where the entire Greater Himalayan Range has been thrust towards the south. The cross-sections I drew in my field notebook looked extraordinary, with the entire Greater Himalayan slab extruding to the south, bounded by the two giant ductile shear zones—the Zanskar shear zone above and the Main Central Thrust below (Figure 2.17).

After eight weeks of almost continuous trekking, geological mapping, and nomadic travel I finally reached the town of Mandi, and the southern boundary of the Greater Himalayan Range. The hot and humid lowlands of the southern Himalayan Ranges were, however, far more populated and forested. The geology was much more difficult to make out, covered as it was by forests, and everywhere I stopped it seemed that within minutes I was surrounded by crowds of small children. It was not long before I yearned for the remote desert highlands of Zanskar and Ladakh. The final leg of my great trek that year was to re-cross the Rhotang-la from Manali to Lahoul and then turn west, winding along the spectacular deep gorge of the Chenab Valley to Kishtwar and thence eventually back to the Kashmir Valley. The Chenab Valley had some of the most beautiful and spectacular trekking paths of all, with wildly high, swinging decrepit rope bridges and narrow trails clinging to the sides of vertical canyons.

Mapping along the Chenab Valley in the middle of the Greater Himalayan metamorphic sequence revealed more extremely complicated structures. In the middle part a spectacular vertical cliff face showed layer upon layer of white sills of leucogranite intruding the metamorphic rocks (Figure 2.18). This was the deepest part of the Greater Himalayan slab and it seemed that the rocks reached temperatures high enough to have undergone partial melting. Migmatites and leucogranites, containing crystals of red garnet, black tourmaline, and white micas abound here. Occasionally the granite sills have amalgamated to form impressive mountains like Ghumbur Ranjung, a gleaming white peak that lies at the south-eastern end of the Zanskar Valley (Figure 2.19). It looked like the central part of my southward-extruding slab was formed of partially molten crustal rocks. Approaching Kishtwar to the west I found that the rocks rapidly became overturned and showed lower metamorphic grade. Here was another excellent profile through the classic inverted metamorphic sequence associated with the Main Central Thrust.

Figure 2.17 (*opposite*) Sketch showing metamorphic zones superimposed onto structurally thickening crust. The metamorphic isograds are initially right way up but have been folded during subsequent deformation. Shearing along the Main Central Thrust zone along the base of the Greater Himalaya resulted in inverted metamorphism, and low-angle normal faulting along the top allowed for southward extrusion of the partially molten middle crust.

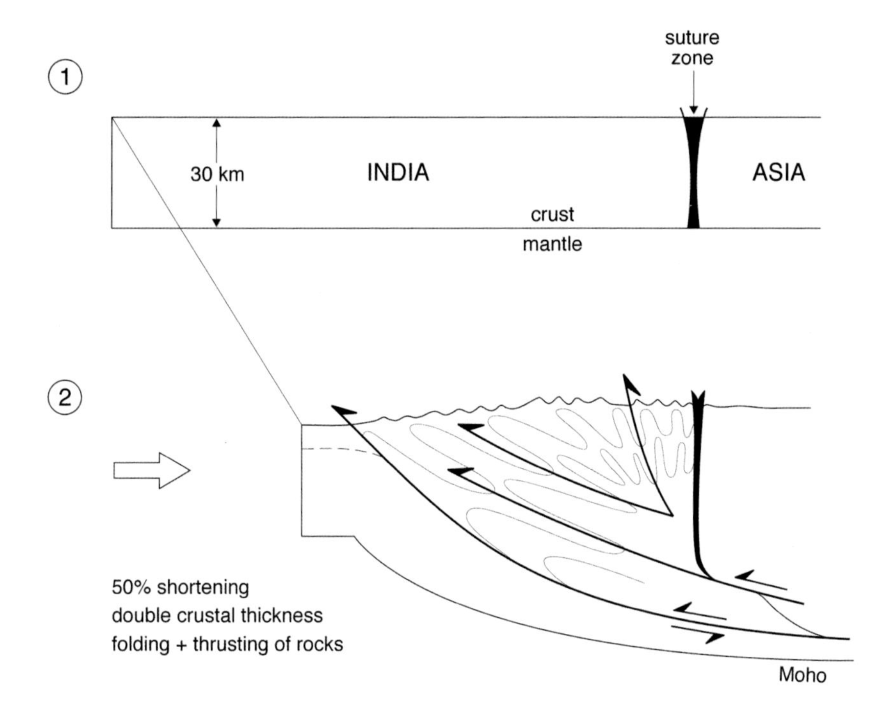
1
suture zone
30 km
INDIA
ASIA
crust
mantle
2
50% shortening
double crustal thickness
folding + thrusting of rocks
Moho

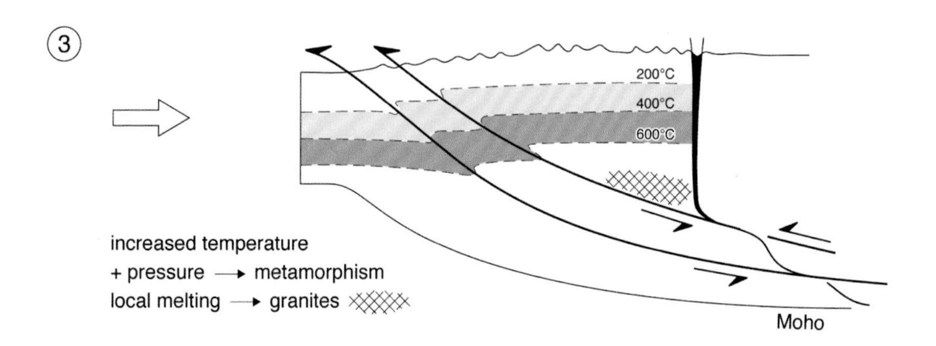
3
200°C
400°C
600°C
increased temperature
+ pressure → metamorphism
local melting → granites
Moho

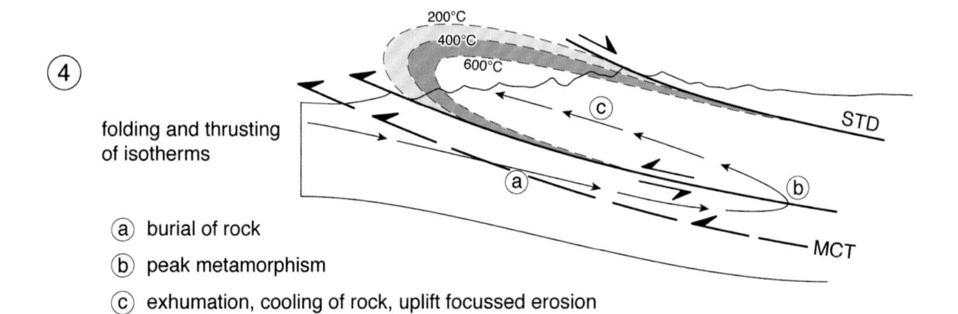
4
200°C
400°C
600°C
folding and thrusting
of isotherms
c
STD
a
b
MCT
a burial of rock
b peak metamorphism
c exhumation, cooling of rock, uplift focussed erosion

Figure 2.18 (*left*) Layered leucogranites intruding metamorphic rocks of the Greater Himalaya, Chenab River gorge, Kishtwar Himalaya. Note the main trail along the top.

Figure 2.19 (*above*) The peak of Ghumbur Ranjung (5,320 m) in eastern Zanskar composed of leucogranite formed by partial melting at deeper levels of the Greater Himalaya. Note the nearly horizontal layers of dark-coloured metamorphic rock xenoliths (enclaves) still preserved within the granite.

I had been trekking for three long months and had now a good reconnaissance idea of the geological structures of the entire northern part of the Himalaya between the suture zone along the Indus Valley to the north and the High Himalayan Range to the south-west. The central ranges of the Zanskar–Ladakh Himalaya appeared to be a giant thrust sheet more than 20 kilometres thick, with a central core of partially molten migmatite rocks thrusting out to the south-west, bounded by a low-angle normal fault above and a major thrust fault below. The metamorphic grade decreased both upwards towards the Zanskar shear zone, and downwards towards the Main Central Thrust. This was a very unusual geological structure and I could not find anything similar in the literature from other mountain belts across the world. The next step was to refine the Pressure–Temperature distribution across the Greater Himalayan slab and try to date the age of crustal thickening and regional metamorphism, melting, and granite formation.

Royal Society Conference

On my way back from India I stopped off in England to spend a few weeks at home. In London I attended an international conference on the geological evolution of the Himalaya and Tibetan Plateau held at the Royal Society. Tibet had recently opened up to Western scientists and travellers after the horrors of the Red Army invasion, the wanton destruction of the Cultural Revolution, and the occupation by the Chinese, who had annexed the entire plateau region. Several large groups of geologists had already carried out reconnaissance expeditions on the vast Tibetan Plateau, and many of these famous geologists converged on the Royal Society for the conference. The Tibetan Plateau is a key geological region, being the largest area of high elevation (on average just over 5 kilometres above sea level) and thick crust (between 70 and 85 kilometres thick) anywhere in the world. It was generally believed that the plateau was formed as a result of the collision of India with Asia and that its extreme height and crustal thickness, double that of normal continental crust, was affecting not only the mantle convection systems of continental Asia but also possibly inducing major climate change. It had been proposed that the recent rise of the Himalaya and Tibet resulted in a huge high-pressure system developing over the plateau during the summer months that sucked in warm, moist air from the Indian Ocean and caused the Indian summer monsoon.

During the conference there were several fascinating presentations about the geology, geomorphology, and geophysics of Tibet, but there were very few about the Himalaya. The great pioneering Himalayan geologist, Professor Augusto Gansser from Switzerland, gave a fascinating and historic perspective of his epic expedition with Arnold Heim in 1935, during which he actually crossed the Indo-Tibet border in disguise in order to see the famous ophiolitic rocks of the Jungbwa–Amlang-la ophiolite in southern Tibet.[6] John Dewey gave a succinct summary of the findings of the Royal Society expedition across the plateau.[7] Another lecture was given by a scholarly, urbane Frenchman, Patrick LeFort from the University of Nancy in France. Patrick had studied the Manaslu granite in western Nepal for several years and presented some field and geochemical data from these rocks.[8] I had seen similar tourmaline-bearing granites high on Langtang Lirung in 1980 and on my treks across Zanskar and Lahoul in the following three years. At the end of Patrick's talk I stood up from the back of the lecture theatre and asked a question concerning the amounts of crustal melting at Mansalu compared to the Zanskar Himalaya.

I had been a post-doctoral research fellow at Memorial University of Newfoundland in Canada but my three-year grant was nearly ending and I was looking around for other opportunities. After the Royal Society meeting presentations, during a drinks reception, I was approached by Brian Windley, a professor from the University of Leicester who had recently started a major geological field-based project in northern Pakistan. Brian quizzed me for several minutes on my recent geological work in Ladakh and Zanskar, and then surprised me by offering me a post-doctoral fellowship in Leicester to work in Pakistan. It seemed serendipitous; if I had not asked that simple question at the Royal Society conference, I would not have met Brian Windley and would not have landed a research fellowship that really enabled me to have a serious go on a longer-term basis at understanding how the Himalaya developed. I now had the relative security of a three-year research grant so I could devise a more ordered plan of campaign. In 1984 I moved back to England and spent the following five summers working on the geology of both the Ladakh Himalaya in India and the central Karakoram Range in Pakistan.

Mapping the Zanskar Himalaya

During the early 1980s there was a surge of international interest in the geology of the Himalaya. Major advances in interpreting the structure of fold- and thrust-related mountain belts had been made, most notably in the Rocky Mountains and the Moine Thrust belt of north-west Scotland. One such advance was the construction and restoration of cross-sections across mountain belts using line or volume balancing methods. These enabled geologists to determine minimum amounts of crustal shortening (the horizontal compression and thickening of the crust during mountain-building) and to figure out the evolution of the mountain-building process. After my first two field seasons in Ladakh and Zanskar I realized that no single person could work on the multitude of geological processes that were needed to understand orogenesis. We needed more manpower and expertise in other fields. Brian Windley proved extremely adept at obtaining research grants from the government to carry out exciting research projects, so in 1985 two more post-doctoral fellows joined our Himalayan research group at Leicester. Tony Rex had just finished his PhD working on the mineralized granites in the Chilean Andes and was an expert on the geochemistry and isotope chemistry of granites. David Cooper was an old friend and colleague of mine from Oman days. He was a sedimentologist and had recently finished his PhD from the University of Edinburgh, working on the deep-sea sediments immediately beneath the ophiolite in the Oman Mountains. David, Tony, and I now had three years' funding to carry out similar studies

Figure 2.20 The author mapping large-scale folds in high-grade kyanite-bearing gneiss (dark rocks) and marbles (pale colour) between the Kashmir and Zanskar Valleys.

in the Indian Himalaya so we spent a considerable period of each summer mapping and sampling across the Indus suture zone and the sedimentary rocks of the Zanskar Himalaya, the upper crustal levels of the Indian plate (Figure 2.20).

During the early 1980s Tony Rex and I trekked back into the heart of Zanskar, mapping and sampling the high-grade metamorphic rocks and crustal-melt granites of the deep crust along the high peaks of the Himalaya south of Zanskar. David Cooper and I trekked all over the Zanskar Range logging stratigraphic sections and mapping the complicated folds and thrust structures of the Tethyan Himalaya. Making geological sections along every one of the trekking routes between the Indus Valley and Zanskar we tried to piece together the three-dimensional structure of the mountain range. Because of the lack of vegetation cover and the excellent exposures along the canyon walls we came across so many textbook perfect examples of geological forms, such as recumbent folds, fold-nappe structures, sheath folds, and thrusts that we became quite blasé about them.

In 1987 David Cooper and I made another long trek across the Ladakh and Zanskar Himalaya, traversing from the Indus Valley across the Tethyan sedimentary rocks of the northern plate margin of India. These rocks were almost identical to the Mesozoic limestones and deeper water sedimentary rocks we had both mapped in Oman, but here in Ladakh they were completely deformed by dramatic folds and thrust faults. After four weeks of mapping and trekking through the gorges of Zanskar, and across the Dat Desert of western Rupschu, we finally reached the monastic settlement of Phugtal Gompa (Figure 2.21).

Figure 2.21 Phugtal gompa, a major gelupa (yellow hat) monastery in eastern Zanskar above the Tsarap Lingti *chu*, originally built in the 12th century. Alexander Csoma de Körös, the Hungarian pilgrim who wrote the first Tibetan–English dictionary stayed here in 1826–7.

Phugtal Monastery

Most of the Lamaistic monasteries and traditions of Tibet were destroyed by the Red Army during the Chinese invasion and the disastrous 'Cultural Revolution' that followed. Ladakh and Zanskar, part of the state of Jammu and Kashmir, has a Tibetan culture and heritage, with over sixty living and working monasteries and some nunneries, many dating back to the 10th and 11th centuries. They are usually built on tops of prominent hills or buried into caves halfway up a cliff face. Because Ladakh, Zanskar, and Kashmir were part of the British empire, they survived the ravages of the Han Chinese armies and so today they are some of the few places left where Tibetan Buddhism is still practised as a living religion. The monasteries today are oases of culture, full of wonderful *thankas* (paintings), books, and artefacts, and always with the all-pervasive smell of burning yak butter candles.

Perched high on an almost inaccessible cliff face, Phugtal Gompa is one of the most amazing monasteries in all Ladakh. The monastery is a collection of white-painted mud-brick cells built in and around a large cave plastered onto the side of a vertical cliff face above the Tsarap Lingti Chu River. It is about eight days' trekking from the civilization of the Indus Valley villages in Ladakh to the north, and about six days' walk across the Himalaya to the nearest hill station town of Manali in Himachal Pradesh to the south. Phugtal is inhabited by about thirty monks belonging to the Gelukpa 'Yellow Hat' reformist sect of Buddhism, although many monks were away wandering around the Himalaya and India seeking enlightenment. The monks meet regularly on the rooftop of the monastery and are summoned by banging drums and blowing conch shells to debate, chant mantras, eat, and drink potent *chang*, barley beer.

* * * * *

After five years of trekking and geologizing in the mountains of Ladakh, Zanskar, Kulu, and Kashmir I thought we had a pretty good idea of what the general structure of the Indian plate side of the collision looked like. We now had good maps along the Indus suture zone and around the Spontang ophiolite, and had constructed structural sections across the upper crustal fold and thrust structures of the Zanskar Ranges. Restoration of these fold and thrust structures in the upper crust enabled us to place minimum constraints on the amounts of crustal shortening during the collision process. We had mapped a huge ductile shear zone bounding the top of the middle crust, the Greater Himalayan metamorphic slab, and discovered a thick area of partially molten rocks, migmatites, and leucogranites, along the highest peaks in the structurally deepest part of the Himalaya. We had mapped the inverted metamorphic field gradient above the Main Central Thrust along the base of the extruding slab, and documented an upward decreasing metamorphic field gradient beneath the Zanskar shear zone. The next major part of the puzzle was to find out what the Asian plate margin looked like. Much of the southern margin of Asia lies in the high and politically difficult part of Tibet, but in the west, north of Ladakh, lies the Karakoram Range, the largest concentration of high peaks and glaciers that straddle the disputed borders of India, Pakistan, and the Chinese province of Xinjiang.

CHAPTER 3

The Dreaming Spires of the Karakoram

All experience is an arch wherethro'
gleams that untravell'd world, whose margins fade
for ever and for ever when I move.

Eric Shipton

Travelling by bus across the northern areas of Pakistan on my way back to England after our first climbing expedition to Kulu in 1978, I remember it being hot, dry, and dusty down in the plains of the Peshawar basin, but the distant sight of glinting snowfields way to the north of Swat and Gilgit heralded the mightiest mountain range of them all. The Karakoram Range has the highest concentration of mountains over 7,000 metres anywhere in the world including K2, at 8,614 metres high the second highest peak, and three other mountains which are over 8 kilometres above sea level (Broad Peak 8,047 m, Gasherbrum II 8,034 m, Gasherbrum I also called Hidden Peak, 8,068 m) (Figure 3.1). Literally hundreds of peaks over 6 kilometres high are clustered along the length and breadth of the range, which spans the borders of Afghanistan, Pakistan, and the Chinese province of Xinjiang, also just clipping the far northern Indian state of Ladakh. The Karakoram Range contains the longest continental glaciers outside the polar regions, the four longest being the Siachen (73 km long), and the Hispar, Biafo, and Baltoro Glaciers, all about 60 km long. In the middle of the Karakoram is a huge continental icecap, Snow Lake or the Lukpe-lawa, surrounded by glistening, improbably steep and high granite spires. The mountains here leap out of the glacier like the wildly imaginative lines of a child's drawing.

Figure 3.1 Landsat satellite photo of the Karakoram Range. The two major glacier systems are the Baltoro Glacier in the west and the Siachen Glacier draining into the Nubra and Shyok Valleys in the east. The dark band of rocks at the bottom is the Indus suture zone in Ladakh separating the granites of the Ladakh Range to the north from the sedimentary rocks of the Zanskar Range to the south.

First Explorations of the Karakoram

During the later stages of the Great Trigonometrical Survey of India, the chaotic array of contours and mountain ranges around the north-western Himalaya was surveyed. The main Himalayan Ranges extend west into the Zanskar and Kashmir regions, but to the north of the Indus River lie another whole series of ranges, the Ladakh Range (the 'Transhimalaya' of Sven Hedin, the greatest of all Tibetan explorers), the Karakoram, and the Pamir Ranges. In 1856, Colonel T. G. Montgomery first spied the great peaks bordering the Baltoro Glacier from the distant Kashmir foothills over 150 km away. Two giants stood above the rest, K1 (Masherbrum) and K2. Everest had just been computed as the highest mountain at 29,002 feet (8,829 m), later increased to its now widely accepted height of 29,064 feet (8,848 m). In 1858 the height of K2 was computed as 28,287 feet (8,621 m), two rope lengths lower than Everest and just a little higher than Kangchenjunga (8,598 m), the third highest peak in the world.

K2 was not visible from any permanent habitation from either side of the range, so the peak never had a local name. The name 'Chogori', which is sometimes used, simply means 'big mountain' in the local Balti dialect, and is generally applied to a hundred other Karakoram peaks. The first Westerner to see and survey K2 was Colonel Henry Godwin-Austen of the Survey of India who heroically managed to ascend the Baltoro Glacier as far as Masherbrum and possibly even Concordia. Francis Younghusband crossed the Old Muztagh Pass to the west of K2 in 1887 on his incredible journey overland from Peking to Kashmir, and in 1892 Martin Conway was the first to reach the base of K2 during his explorations around the central Karakoram.

The first real attempt to climb K2 was a heroic effort made on the north-east ridge in 1902 by Oscar Eckenstein and Alastair Crowley.[1] They spent sixty-eight days on the mountain in appalling weather conditions and managed to reach 6,600 metres. In the early 1900s several large Italian expeditions led by Luigi Amedeo, Duke of Abruzzi (1909) and Prince Aimone, the Duke of Spoleto (1929) made deep incursions into the Karakoram.[1] The former first identified the south-east ridge as the most favourable ascent route, a route that became known as the Abruzzi Ridge. On the Duke of Abruzzi's expedition was a young photographer, Vittorio Sella, who published some of the most beautiful and evocative black and white photographs of the Karakoram, and did much to inspire future explorers.

During 1937–9 a small British expedition made a major contribution to the exploration and mapping of the whole central Karakoram. Eric Shipton, Bill Tilman, J. B. Auden, and the surveyor Michael Spender, together with their team of seven trusty Sherpas from Nepal covered a huge amount of territory attempting to fill in those 'blanks on the map'.[2] The topographic maps produced by Michael Spender were not only amazingly accurate but also works of art. Their expedition was finally cut short when they turned on their radio and heard of the outbreak of World War Two. Two American expeditions led by Charles Houston and Fritz Weissner in 1937 and 1938 both failed but got high. Charles Houston returned to lead a major expedition to K2 in 1953. Art Gilkey fell ill high on the mountain and in a desperate retreat four climbers fell and were all miraculously held by Pete Schoening.[3]

The year after the first ascent of Everest in 1953, the geologist Ardito Desio led a large Italian expedition that finally managed to climb K2 for the first time. On 31 July 1954, Lino Lacedelli and Achille Compagnoni reached the summit by climbing the Abruzzi Spur, supported by Walter Bonatti and the Hunza porter Mahdi. Desio was a geologist and during the expedition he also made the first important observations and maps along the Baltoro Glacier and around K2 itself.[4]

* * * * *

Brian Windley's Leicester University project was mainly working in the Kohistan region of northern Pakistan. In the barren, wild, and windswept mountains there, a sequence of rocks revealed that, in between the Indian and Asian plates, caught up in this great continental collision, was an ancient island arc tilted on end. The Karakoram Highway, the major road that traverses the Himalaya, Kohistan, and Karakoram Ranges to cross the watershed of Asia over to the Pamir and Takla Makan deserts, actually cuts right across this sequence of rocks.

Some of the better-known modern-day island arcs include the Bonin and Philippine arcs of the western Pacific Ocean, the Lesser Antilles in the Caribbean, and the Aleutian arc linking Alaska and the Kamchatka peninsula of far-eastern Russia. Only the tops of these, giant volcanic edifices are exposed, and the base of the structures can only be known by remote means, like seismology and geophysics. Here in northern Pakistan, however, it was possible to see exactly what rocks made up the base of such oceanic island arcs. Together with Mike Coward, a structural geologist from the University of Leeds, and their students, Brian had started to study the rocks that made up this island arc. By the time I arrived at Leicester University in 1984, several PhD students and post-doctoral fellows were busy working on the geochemistry and structures of the Kohistan region. I was drawn more to the high peaks of the Karakoram, north of Kohistan, and started working there during the summer of 1984. The geology of the Karakoram was very poorly known, with the only previous study being the reconnaissance work of Ardito Desio, the geologist who led the first successful expedition to climb K2.[4]

The Pakistani Karakoram

The geology of the Karakoram Range is similar to that beneath the Tibetan Plateau to the east, but whereas Tibet is high and flat with an extremely arid climate and very little erosion, the Karakoram has a similar average elevation of about 5 kilometres, but has enormously high relief, deep glacial erosion and exposes mainly metamorphic and granitic rocks that were formed at great depth and subsequently uplifted during the India–Asia collision. Whereas the geology of Kashmir, Ladakh, and Zanskar revealed the fate of the Indian side of the great collision, the Karakoram would show the geological results of the collision along the Asian margin.

Flying into Skardu on my early trips to Pakistan, as the plane banked sharply to descend into the Indus Valley, a massive white wall of huge snowy mountains suddenly appeared out of the aeroplane windows to the north (Figure 3.2). This seemed to mark out the northernmost outposts of a chaos of contours that marks the Karakoram. If the Himalaya were the highest and longest mountain range in Asia, then surely the Karakoram was the most spectacular, the wildest, and the remotest of all. It was the final mountain range of the North-West Frontier of the British empire, a formidable natural barrier which divided the Russian tsarist empire to the north from British India to the south. It was thought that no army could cross the Karakoram and survive, let alone fight.

Figure 3.2 Aerial view of the Karakoram Range looking north from above Skardu, with K2, Broad Peak, and the Gasherbrum Range prominent on the horizon.

Figure 3.3 The Baltoro Glacier from the summit ridge of Lobsang II spire. Gasherbrum IV, Hidden Peak and Baltoro Kangri form the headwall.

The Baltoro Glacier rises high on the southern slopes of K2 and the even more remote mountains to the north, and carves a course right through the middle of the Karakoram (Figure 3.3). It is, for me, the greatest natural wonder of all. The Baltoro rises in a riot of icefalls and glaciers that tumble down from the peaks around the upper Godwin-Austen Glacier, Skyang Kangri, Broad Peak, and the formidable south face of K2. Cascading over icefalls, the glacier flows down to Concordia, the point at which five major glaciers all converge, piling down from the southern ramparts of Broad Peak and the Gasherbrum Range. From Concordia, the Baltoro cuts across the entire Karakoram geology from the sedimentary rocks of the Gasherbrum Range through the giant granite batholith to the metamorphic rocks of the Southern Karakoram. At Paiyu, some 55 kilometres from K2, the glacier disgorges into the Braldu River from massive cliffs of carving ice.

Trekking along the Baltoro Glacier is a mountain feast. Every hour the views change dramatically. At the snout of the glacier at Paiyu the massive granite ramparts of the Trango Towers leap out of the ice. Tall, elegant spires of gleaming granite rise everywhere, the castle-like Paiyu Peak, Shipton Spire, Nameless Spire, and Uli Biaho (Figure 3.4), to name a few of the more impressive ones. South of

Figure 3.4 Uli Biaho spire (6,417 m) viewed from the Baltoro Glacier. Granite crags of the Cathedral spire in the foreground.

Figure 3.5 The north face of Masherbrum (7,821 m) from the Baltoro Glacier.

Figure 3.6 Chogolisa (7,654 m) rising high above Concordia, the meeting place of the Baltoro, Vigne, Gasherbrum, and Godwin Austen Glaciers, view east from K2 Base Camp.

Figure 3.7 The classic view of the south face of K2 (8,611 m) from Concordia on the Baltoro Glacier.. The Abbruzzi Ridge is the skyline ridge on the right.

the Baltoro Glacier, Masherbrum, slightly under 8,000 metres high, rises head and shoulders above everything else (Figure 3.5). Lining the side of the glacier is a row of granite domes like a line of Yosemite's El Capitan's. Two days' trekking up the Baltoro and the improbably steep Muztagh Tower comes into view to the north, then the massive icy ramparts of Biachedi and the bewildering array of mountains around Concordia. The beautiful Matterhorn-like Mitre Peak and the huge icy lump of Chogolisa rise directly out of the southern end of Concordia (Figure 3.6). Gasherbrum IV Peak stands as a sentinel guarding the inner sanctum, or the 'Throne Room of the Mountain Gods' as Galen Rowell eloquently put it. Here, at Concordia, one can look north at K2's south face (Figure 3.7), swing east to the twin summits of Broad Peak and Gasherbrum IV, further along the range to the two giants of Gasherbrum II and III. At the head of the glacier Baltoro Kangri stands above the high passes of Indira Col and the access passes over to the Siachen Glacier.

Biafo Glacier and the Latok-Ogre Ranges

In the summer of 1984 I made a reconnaissance trek along the Braldu Valley up to the last Balti village of Askole, where I hired a couple of ibex hunters to guide me up the Biafo Glacier. The trek was wonderful, passing from the luscious green barley fields and apricot orchards of the lower valleys up to the high and barren glacial country of the high Karakoram. Half a day's trek out of Askole and the Biafo Glacier cascades down to the Braldu Valley. Massive seracs of ice push out into the valley diverting the Braldu River. We negotiated the icefall and managed to get up onto the ice of the Biafo Glacier. Unlike most Karakoram glaciers this was as smooth as a road and the crevasses were easily visible. It was a real pleasure trekking up into the heart of the Karakoram, just the three of us with two hunting dogs. Soon the foothill spires revealed the massif granite ramparts of the Latok Range, culminating in an incredible mountain named the Ogre (Figure 3.8). Spectacular vertical walled spires of pristine white granite rose straight out of the ice. During the first ascent of the Ogre, Chris Bonnington and Doug Scott had one of the most remarkable escapes in mountaineering legend.[5] Abseiling from the summit after days of hard climbing, Doug Scott pendulumed across the summit granite cliffs and slammed into the wall breaking both his legs. Chris Bonnington broke several ribs in another abseiling accident. During the following days of painful and difficult retreat, Doug managed to crawl back along the glacier on shredded knees. The pair finally made it back to base camp, where after several days an army helicopter picked them up, only to crash-land back at Skardu military base attempting a landing.

We camped in the beautiful meadows beneath the Ogre and Latok Peaks for a week whilst I scrambled around studying the rocks around the ablation valleys.

Figure 3.8 The northern bank of the Biafo Glacier with Conway's Ogre above. The dark rocks are blocks or xenoliths of metamorphic rocks enclosed in Baltoro granites.

The main Latok granite was pristinely fresh, with glinting red garnets, white flaky muscovite, and the occasional black tourmaline crystal. The southern margin of the granite was a steep, intrusive contact. Along the margin hundreds of fragments of gneisses, the metamorphic country rock formed under high pressure and temperature conditions, were entrained in the granite. These field relationships clearly showed that the granite had been intruded after the metamorphism. I collected large samples of both rock types from the base of the Ogre so that we could obtain precise ages of both the gneisses and the granite (Figure 3.9a, b, c).

My two Balti friends meanwhile were busy hunting ibex. They had the sharpest eyes I had ever known and were able to spot ibex hundreds of metres above us on the high slopes. Once the herd was selected they released the two dogs which raced up the hillsides with incredible speed, making a pincer movement behind the ibex. The ibex were unwittingly being herded down to the valley where one of the hunters lay hidden behind a boulder. As soon as an ibex appeared within range he let off his enormous rifle and felled it. The ibex was gutted and strung up to dry, to be taken back to Askole where it would provide meat for the entire village. We spent nearly a month exploring the Biafo Glacier and the side glaciers along the southern edge of the great granite mountains, and then retraced our steps back to Askole. On the way back I had a tantalizing glimpse up the Baltoro Glacier to the stupendous granite peaks of the Trango Towers. Masherbrum reared up nearly 8 kilometres into the sky, pure pinkish white granite cliffs gleaming beneath the hanging glaciers. I decided then and there that this was the mountain range I wanted to work in for the next few years.

Geologizing along the Baltoro

Back in Leicester I excitedly regaled Brian Windley with what I had found up in the Biafo Glacier region and then suggested that we organize a major expedition the following summer all the way up the Baltoro Glacier to K2 itself. This would provide us with a geological transect that sliced right across the mountain range so we could map the structures and collect all the samples we needed for dating the timing of metamorphism, granite emplacement, and structural evolution. I set about organizing the trip, and as time went by the whole expedition mushroomed in size and scope. I soon decided that we needed a climbing component so that not only could we climb high but also collect samples from a wide range of altitudes, something needed in order to constrain uplift rates. I decided that Masherbrum was going to be our mountain target and applied to the Pakistani authorities for permission to climb it. In retrospect it was a massively ambitious mountain and we really stood little chance of getting to the summit. However, I would be happy

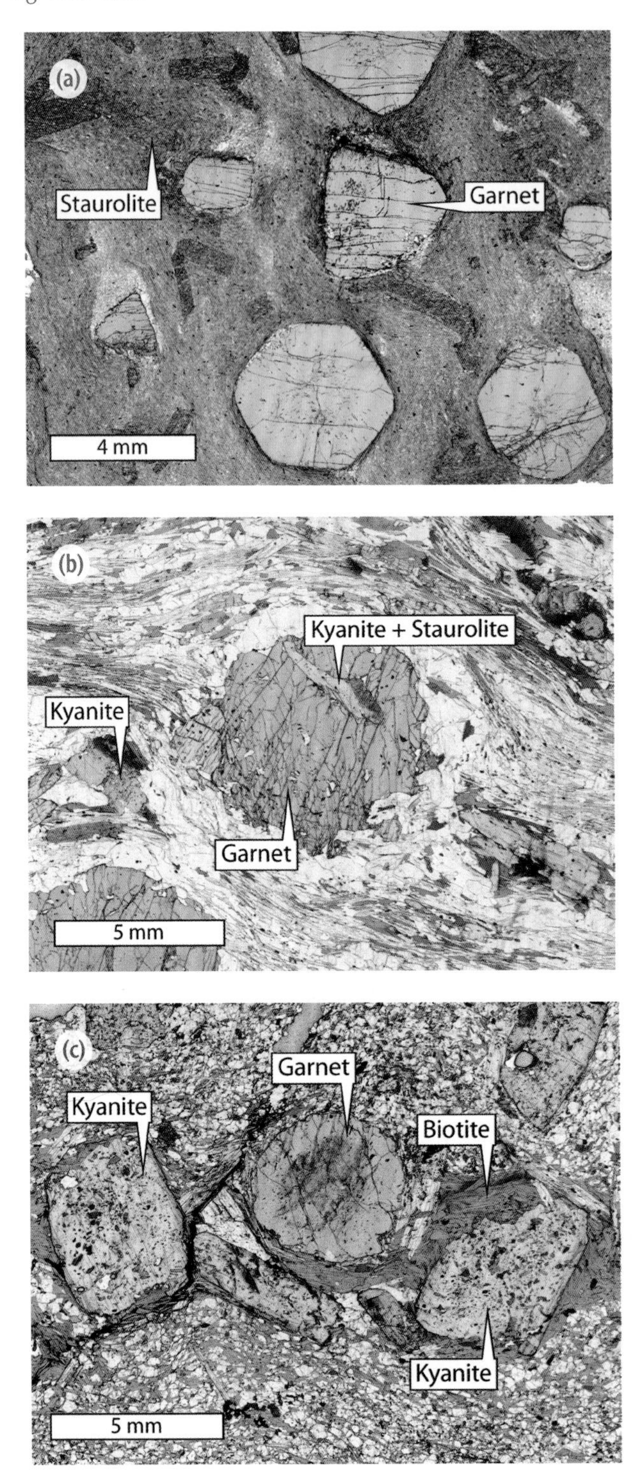

Figure 3.9 Microscope thin-sections of (a) a gneiss from the Hunza Karakoram showing garnet and staurolite crystals in a graphite-rich matrix; (b) a deeply buried metamorphic rock from the Baltoro Karakoram showing garnet, staurolite, and kyanite crystals; (c) a kyanite-garnet gneiss that was buried to over 35 km depth. Note the scale in millimetres.

simply to get high so that we could map the geology and collect those all-important samples for dating the granite.

Tony Rex had just completed his PhD at Leicester working on Andean granites from Chile, and had just accepted a two-year post-doctoral fellowship to work on the Karakoram with me. Tony was also a very proficient rock climber and I was keen that he and I not only carry out the geological mapping work along the Baltoro Glacier but also try and climb high on Masherbrum. I invited an old climbing friend from America, Dana Coffield, who had several Yosemite big-wall climbs to his credit to join us. We had done several new rock climbs on the limestone walls in the mountains of Oman. Tony asked one of his regular partners, Dave Mould, along. The plan was that we would spend one month trekking along the Baltoro Glacier mapping the geology en route all the way up to K2, and then we would meet the climbers at Urdukas, the last green grassy meadow on the Baltoro and a day's trek below the Masherbrum Base Camp. Looking at the map one day I realized that I had taken on a truly daunting prospect—mapping an entire mountain range in central Asia. We needed more like-minded field geologists.

The previous summer I had met a Canadian geologist, Paul Hoffman, at a conference in Moscow and we had been on the same field trip across the Soviet Tien Shan Mountains along the southern border of what is now Khirghizia. Paul was running a huge and very efficient field campaign from the Geological Survey of Canada, mapping the remote Wopmay and Cape Smith fold belt in the far north of Canada. On impulse I phoned him up from my office in Leicester and asked whether he would like to join us in the Karakoram for three months the following summer. He first regaled me with all his plans and commitments for the summer in the Northern Territories and then said he'd think about it and ring me back. Half an hour later he rang back and said yes he would come, and could he bring two other field geologists along with him, Marc St-Onge and Rein Tirrul. I was elated, as these three were the most impressive of all field geologists in a country overflowing with them. Finally we had a team to carry out this great Karakoram idea.

Permission eventually arrived from the Pakistani government and the geology team flew out to Islamabad in June 1986. We spent a frustrating week in Islamabad and Rawalpindi doing all the beaurocracy and buying up supplies. Finally we were off in a hired bus along the Karakoram Highway across the desolate barren hills of Kohistan to Gilgit, thence cutting east through the deep gorge of the Indus River plunging between the 8,000-metre-high peak of Nanga Parbat and the 7,500-metre-high peaks of the Haramosh Range. At Skardu we hired almost a hundred porters and a *sirdar* to manage them, gave them an advance and a pair of shoes, and sent them ahead to Dassu, the end of the road along the Shigar Valley. We calculated that we needed this number of porters to carry all the food and fuel to keep six geologists and another four climbers up in the K2–Masherbrum region for three months.

The K2 hotel is the mountaineering base of the Karakoram climbing world. It is located in a magnificent setting on top of a river terrace with spectacular views north over the Indus River to the foothills of the high Karakoram. From here it was possible to drive along a dirt road, cross the Indus on a rickety swaying suspension bridge, and weave through the sand dunes at the confluence of the Shigar and Indus Valleys into the Shigar Valley. At Dassu we camped in the Rest House gardens where our motley crew of Balti porters was waiting.

The first day's trek started with about three hours of shouting and arguments sorting out the loads. Finally all the porters were loaded up ready to go and, with a huge roar of 'Allah u Akhbar' ('God is great'), they strode off along the Braldu Valley towards the high Karakoram. It was a fantastic moment. After months of planning we were finally off, complete with a rag-tag army of friendly and hard-working Baltis and two live goats. It was a military operation, given the number of people. At the end of day one we paid off two porters as the other porters and the expedition members had eaten the food supplies they were carrying. After five days' walking we had a rest day at the campsite of Paiyu just before the snout of the Baltoro Glacier. The two goats met their grizzly end here and provided a meat feast for the porters before the rigours of the glacier trek ahead. From Paiyu we had the first views of the granite spires of the Trango Towers and the excitement started to mount.

In 1985 it took fourteen days to walk from the end of the road at Dassu in the Shigar Valley up to K2 Base Camp, the last eight all on glaciers. After Paiyu the trail ended and we climbed onto the ice of the Baltoro Glacier then across to the ablation valley along the south side. This led to a magnificent campsite at Urdukas, a collection of huge granite boulders set in grassy meadows with awe-inspiring views across to the Trango Towers. This was to be the last sight of green grass for ten weeks. After Urdukas the trail again went onto the ice and followed east towards Concordia, the meeting point of four glaciers cascading down from K2, Broad Peak, and the Gasherbrum Range. Every day brought stunning new mountain views. Trango, with its outlying slender Nameless Spire was the epitome of Karakoram elegance, surrounded by the impossibly steep-looking granite spires of Uli Biaho, Lobsang, and the castle-like ramparts of Biale. Further east the Muztagh Tower dominated the north flank of the Baltoro and the huge icy ramparts of Masherbrum rose up to the south.

During this most spectacular of all treks it was impossible to keep one's eyes locked on the trail with all those incredible mountain views in every direction. During the last day's trek to Concordia where we planned to base for the first two weeks, there was one spot where you had to jump across a glacial stream. During the summer months these glacial streams on top of the ice were common, and it was frequently a problem crossing the larger ones. Brian Windley was jumping across one of these when he slipped and became completely immersed in the freezing water. With increasing horror I saw that he was slowly losing his grip and being swept downstream. Fifty metres further and the glacial stream disappeared into a big hole plunging down through the ice. I had visions of my professorial boss being swept to his certain death in my first year at Leicester University. Dropping my rucksack I raced across the ice and hauled him out, but by then he was completely soaked and rapidly becoming hypothermic.

That night it was an exhausted crew that camped at Concordia. We paid off the porters and established a camp. At last we had relative peace after our motley army of Baltis left down the valley. We had eight weeks' worth of food and we were established right in the middle of the Baltoro. We could finally start unlocking the secrets of the Karakoram. Next morning, however, Paul and Marc decided that the Baltoro was just too difficult to access, too time-consuming, and too formidable a place to work; they wanted to leave and return to the Northern Territories, where Canadian money and efficiency made for an altogether more comfortable working environment. Brian Windley was also showing signs of worsening health after his impromptu dunking in the glacial stream, and we were forced to evacuate him by helicopter back to Skardu. After months of planning, three weeks of hassling in Islamabad, and trekking across the Karakoram, that evening we were down to three geologists, Rein Tirrul, Tony Rex, and myself.

The three of us split up in order to study as much of the rocks of the upper Baltoro as we could before the climbers arrived, and we moved camp down the glacier to Masherbrum. The geology proved spectacular although the access was indeed a problem. Just to get to the side of the Baltoro we had to climb and descend as many as three or four moraines, each with steep icy cliffs that necessitated crampons and ice axes, and most had fast-flowing rivers on top of the ice. Tony and I trekked up to K2 Base Camp and beyond to the icefall beneath Skyang Kangri. At base camp we were invited to dinner by a friendly Swiss expedition including some of the best-known professional Himalayan climbers around, including Erhard Loretan and Jean Troillet. We asked their Pakistani liaison officer if we could ascend a short way up the Abruzzi Ridge of K2 to collect rock samples. We were given permission to go as far as we wanted in two days and be back at base camp, so we lost no time leaving early next morning for the climb through the K2 icefall.

Geo-Mountaineering on K2

Climbing the 'Savage Mountain' is a much harder prospect than all other 8,000 metre peaks. As of 2010 only 302 people have summited K2, as opposed to more than 2,700 who have climbed Everest. At least eighty climbers have lost their lives attempting to climb K2. The normal route up K2 is the Abruzzi Ridge, which rises steeply from the upper Godwin-Austen Glacier between K2 and its mighty neighbour, Broad Peak. Avalanches poured off the western slopes of Broad Peak, some actually yo-yo-ing all the way across the glacier and up the foot of the east face of K2. The icefall was a dangerous place, creaking and groaning with its continuous movement. We arrived at the foot of the Abruzzi Ridge, not actually much of a ridge at all but more like a huge face of rock and ice sweeping up to the south-eastern shoulder of K2. After a couple of hours' rest, we roped up and climbed fast. Conditions were excellent, with ice axes and crampon front points biting nicely into the ice. We climbed higher and higher, with views in every direction getting better and better. At the top of the sweeping rock face, cliffs of gneiss and granite rose into the upper reaches, leading to the Shoulder. At about 4 p.m. we finally stopped, belayed to a suitable cliff and looked up in awe at the impressive upper ramparts of K2. The Swiss were in base camp and we had the entire mountain to ourselves; it felt like our lifeline was extended about as far as it could possibly go.

This was geology at its most extreme. Tony and I were at an altitude of about 7,000 metres, hanging on the ends of a rope attached precariously to some rock towers. We got out the hammer and collected some samples for dating, both of the granites and the gneissic rocks they intruded. Before retreating we photographed the south face of K2 sweeping up in all its incredible majesty to the icy summit pyramid, the western flanks of Broad Peak with its enormous collapsing hanging glaciers, and up to the farthest reaches of the Godwin-Austen Glacier, where Skyang Kangri and its outlying peaks marked the final frontier with Chinese Xinjiang. It took several abseils down the face, and inward-facing down-climbing to regain the Godwin-Austen Glacier, then another four hours delicately winding through the icefalls and crevasses to get back to base camp. When we finally staggered back, the Swiss took pity on our meagre one-tent base camp and rice and *dhal* food supplies, and asked us in for a wonderful dinner of cheese fondue and wine all the way from Zurich. Six weeks later, Erhart Loretan, Marcel Ruedi, Norbert Joos, Pierre Morand, and Jean Troillet went on to make the eighth ascent of K2 via the Abruzzi Ridge in an astonishing feat of Alpine climbing, without the back-up of big expedition tactics that the first ascent by the Italians Lino Laccadelli and Achille Compagnoni had back in 1956.

Tony and I trekked back down to Concordia with our precious load of rock samples from K2. Rein had left in order to map out the lower part of the Baltoro Glacier region and we had arranged to meet at Urdukas. Our climbing buddies, Dana Coffield and David Mould, duly arrived, and after a few days of rest and relaxation we headed east again, this time to Masherbrum Base Camp and the second part of our expedition.

Masherbrum: Out on a Limb

Our chosen route, the east ridge of Masherbrum, was approximately 5 kilometres in length, mostly above 6,000 metres, serrated, and extremely dangerous. Seracs broke away from the corniced ridge and the resulting avalanches poured all the way down to the Biarchedi Glacier. The snow was also deep powder and we soon found ourselves floundering in waist- or sometimes even shoulder-deep soft powder. It was far longer, harder, and more dangerous climbing than any of us had ever done before and we soon realized that we were woefully inadequately prepared for Masherbrum. The geology in the cliffs was, however, revealing, with beautifully fresh orange-pink granite, sometimes enclosing enormous rafters of older metamorphic rocks and older granites. We traversed to the sides of the glaciers whenever possible to study and collect samples of these granites.

We did manage to climb to a col between the Baltoro and Hushe Valley watersheds, overlooking the eastern approaches of Masherbrum, and thence a short way along the ridge itself. However, it very soon became apparent that there was no way we were even remotely going to gain access to the summit pyramid of Masherbrum. Once again we had bitten off far more than we could chew. It was a revealing and fascinating climb, however, and we managed to collect some more samples, photograph the amazing field relationships of the great granite intrusions, and generally map the region. It was during the final stages of our climb on the east ridge of Masherbrum that I looked north and saw an imposing castle-like mountain sandwiched between the improbably steep spires of the Trango Towers and the magnificent stately mountain of the Muztagh Tower. Later I found out that the castle-like peak was just below 7,000 metres high and had a name—Biale.

We had about a week's food left so when the climbing team packed up to leave, Tony Rex and I decided to do one last trek up to Hidden Peak and Gasherbrum Glacier to fill in the blank on the geological map of the Baltoro. We left Urdukas and, skirting through the maze of glacial valleys and streams, made our way to Concordia and then east along the Baltoro, Abruzzi, and Gasherbrum glaciers. After two further days we arrived at Hidden Peak (Gasherbrum I) Base Camp below the Conway Saddle at the very top of the Siachen Glacier. We thought that we would continue as far as we could towards the wall of rock and ice that forms the watershed between the Baltoro Glacier on the Pakistan-controlled side of the ceasefire line and the Siachen Glacier on the Indian side. Suddenly we stumbled into a small encampment of white tents. A small team of Pakistan army soldiers all kitted out in full Arctic gear and a big Bofors gun manned the camp. They were very wary of us, but did make us a cup of tea before turning us away. It turned out that these men's job was to let off the big cannon every morning at 6 a.m. They were trying to shell Indian Army positions on the Siachen Glacier by firing live shells across a 6,400-metre-high ridge. The only effect was to set off waves of avalanches. The soldiers suffered from altitude sickness and frostbite but never actual combat. This proved to me the absolute futility and complete stupidity of war up here in the remote Karakoram. The army spies had reported our movements to their superiors down in Skardu so two weeks later as we finally walked out at Dassu in the Shigar Valley, we were approached by the army and asked questions as to our movements. They were not convinced by my insistence that we were only mapping the rocks along the Baltoro Glacier and trying to convince them that we were not Indian spies required all my very limited diplomatic skills.

Hushe Valley

The Pakistan–India border dispute in the Siachen region was the most absurd conflict ever fought and had more to do with bravado and face saving than anything else. There was, however, a large contingent of Pakistan army troops camped at the ends of several of the glaciers. In 1987 Rein Tirrul and I returned to Pakistan with a larger team of climbers, including my friends Steve Razzetti, Mark Miller, Simon Yates, Jon Tinker, and Craig Kentwell. My plan this year was to base in the Hushe Valley south-east of the Baltoro and close to the front line of the Indo-Pakistan Siachen Glacier war. We established a base camp on the Masherbrum Glacier south of the same spectacular mountain we had tried to climb from the northern Baltoro Glacier approach the previous year.

The Hushe Valley cuts right across the Karakoram Range and, like the Baltoro provides an ideal cross-section to map the geological structures. Leaving Javed, our local cook cum *chawkidar,* to look after the camp, we set off north to get up to the Gondoro-la, the only reasonably accessible pass connecting over to the upper Baltoro Glacier. There is a trekking peak called Gondogoro Peak just above the pass that provides a spectacular vantage point. From here you can peer upwards to the icy ramparts of Chogolisa and the stunning array of high mountains along the Siachen watershed. The almost perfectly shaped mountain of Layla Peak, surely one of the most striking mountains anywhere, dominated the view south (Figure 3.10). The trek up to the Gondogoro-la was spectacular for the geology as well. Whereas the rocks along the Baltoro Glacier I had mapped the previous year traversed the deeper structural levels of the batholith, the rocks around the summit ridge of the Gondogoro-la showed the very top of the Baltoro granite batholith intrusion. Here the granite intruded into the black shales, in which had formed large crystals of the characteristic low-pressure high-temperature mineral, andalusite. The heat from the granite had baked and metamorphosed the shale.

Figure 3.10 Layla Peak (6,096 m), the most beautiful mountain in the Hushe Karakoram.

After our group ascent of Gondogoro Peak we split into ropes of two and spent the next four weeks exploring and climbing a whole series of mountains, almost every one a first ascent. It was a great privilege to be the first to explore these remote mountains in such a wonderful setting. The local Baltis of course thought we were mad; their only reasons for climbing were to hunt ibex or gather juniper scrub. The idea that we were looking at rocks or climbing for the sheer fun of it was patently ridiculous in their view, but in their good nature they simply laughed and shrugged. Craig and I climbed two new mountains further along the ridge overlooking the Vigne Glacier, one a dramatic thin spire of a summit with a vertical drop of over 2,000 metres. Mark and Simon explored new routes on Trinity Peak and some granite spires along the K7 Glacier. I was slowly but surely filling in my new geological map of the Karakoram and it was exciting to watch as each small piece of the puzzle fell into place.[6]

We had spent the last two months along the upper reaches of the upper Hushe Valley and around the south side of Masherbrum and Chogolisa. We had climbed altogether eight mountains, seven of them first ascents, and I had managed to collect another hundred or so samples for petrology and dating work. It had been another very good field season. I thought the following year it would be good to get right into the heart of the Baltoro granite around the Trango Glacier area, and the peak Biale that I spotted from high on the north face of Masherbrum in 1986 looked just about right for our climbing capabilities. Whereas the Trango Towers and Muztagh Tower were some of the hardest rock climbs on the planet with

Figure 3.11 The south face of Biale Peak (6,730 m); the route taken was through the icefall below then up the granite cliffs to the left.

no easy ways up, Biale was a great fortress-like mountain, actually the northern extension of the Cathedral Ridge, with steep ice fields winding up in between giant vertical granite faces. From Masherbrum, 10 kilometres away, I could even see potential climbing lines winding up these granite faces and ice gullies (Figure 3.11). The thought occurred to me that I would ask some of my Himalayan climbing friends along and we would try to combine the geological mapping of the northern Baltoro with a more efficient and better-organized climbing expedition to Biale.

Metamorphism and Melting

Returning to Peshawar in our dilapidated Daihatsu jeep after the Baltoro expedition I shipped about 200 kg of rock samples back to the UK, and spent the winter months drafting up maps, analysing the samples and attempting to figure out the three-dimensional geological structure of the Karakoram. Gradually the picture began to emerge and with mounting excitement we discovered that the entire southern part of the Karakoram was composed of high-grade metamorphic rocks, marbles, amphibolites, and metamorphosed shales or pelites that contained an abundance of minerals that we could use for thermobarometry (Figure 3.12). In order of increasing temperature and pressure, the characteristic index minerals biotite, garnet, staurolite, kyanite, and sillimanite appeared. By mapping the first incoming minerals on the ground it was possible to map out a series of metamorphic isograds, or chemical reactions, in three dimensions. At the highest metamorphic grade, the bladed blue mineral kyanite indicated particularly high pressures (around 10 kilobars or more) and sillimanite indicated high temperatures. At the highest temperatures, around 750°C, the rocks actually start to melt as muscovite mica breaks down to form potassium feldspar and small melt fractions. This process resulted in widespread partial melting, a process called migmatization—and eventually to the formation of granites. Unlike the Andean-type granites, these crustal-melt granites contain the characteristic minerals, tourmaline, garnet, muscovite, and biotite, with the occasional gem quality aquamarine (beryl) and topaz crystals.

Most of the highest peaks along the Karakoram Range from the Hunza Valley across the Baltoro and Muztagh Ranges to the Hushe Valley and beyond into the Siachen–Nubra region of Ladakh are composed of granite. The spires of the Latok Ranges, Trango Towers, Biale, and Muztagh Towers are all solid granite, occasionally loaded with tiny red garnets or black tourmalines. These granites resulted from the highest temperatures and deepest levels of the thickened Asian plate crust, and intruded upwards, ballooning into giant igneous masses or plutons that separate the metamorphic terrane to the south from the predominantly sedimentary terrane to the north. Unlike the Himalayan granites we had seen and mapped throughout Zanskar and the Indian Himalaya, these granites were of huge

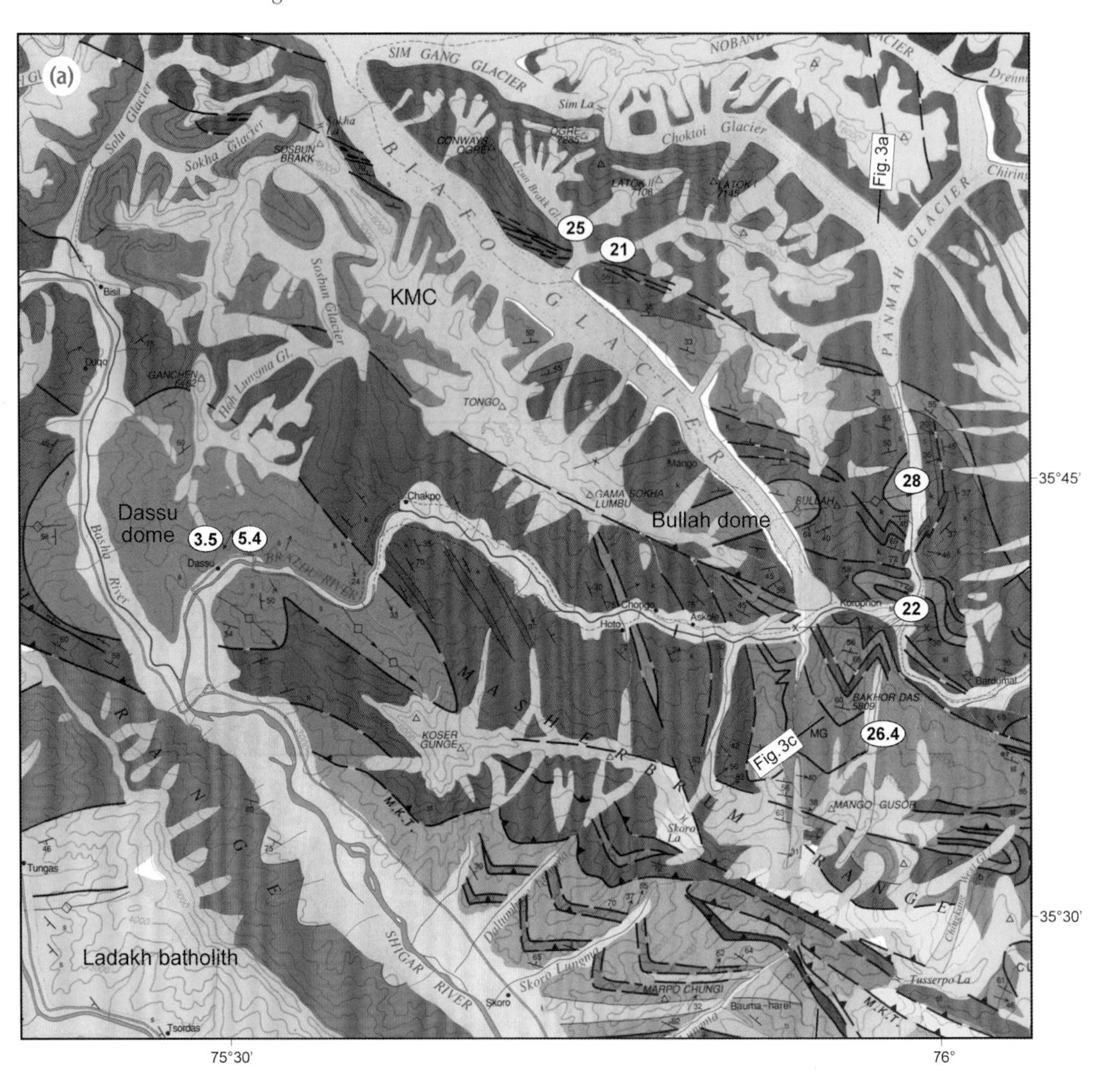
(a)
SIM GANG GLACIER
Sim La
Choktoi Glacier
Fig. 3a
PANMAH GLACIER
Solu Glacier
Sokha Glacier
Sokha La
SOSBUN BRAKK
CONWAYS OGRE
OGRE 7285
LATOK II 7108
LATOK I 7145
Uzun Brakk Gl.
BIAFO GLACIER
25
21
KMC
Bisil
Duqo
Sosbun Glacier
Hoh Lungma Gl.
GANCHEN 6462
TONGO
Mango
GAMA SOKHA LUMBU
Chakpo
Dassu dome
3.5
5.4
Dassu
BRALDU RIVER
Bullah dome
BULLAH
28
Korophon
22
Chongo
Hoto
Askole
Bardumal
Bashu River
MASHERBRUM RANGE
BAKHOR DAS 5809
MG
26.4
Fig. 3c
KOSER GUNGE
MANGO GUSOR
M.K.T.
Skoro La
Tungas
Ladakh batholith
SHIGAR RIVER
Skoro
Skoro Lungma
MARPO CHUNGI
Bauma-harel
Tusserpo La
Chingkang West Gl.
Tsordas
35°45'
35°30'
75°30'
76°

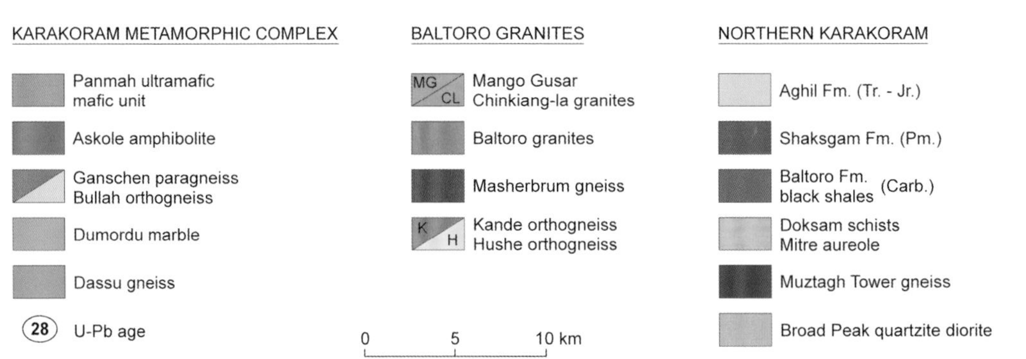
KARAKORAM METAMORPHIC COMPLEX
Panmah ultramafic mafic unit
Askole amphibolite
Ganschen paragneiss
Bullah orthogneiss
Dumordu marble
Dassu gneiss
28 U-Pb age
BALTORO GRANITES
MG Mango Gusar
CL Chinkiang-la granites
Baltoro granites
Masherbrum gneiss
K Kande orthogneiss
H Hushe orthogneiss
0 5 10 km
NORTHERN KARAKORAM
Aghil Fm. (Tr. - Jr.)
Shaksgam Fm. (Pm.)
Baltoro Fm. black shales (Carb.)
Doksam schists
Mitre aureole
Muztagh Tower gneiss
Broad Peak quartzite diorite

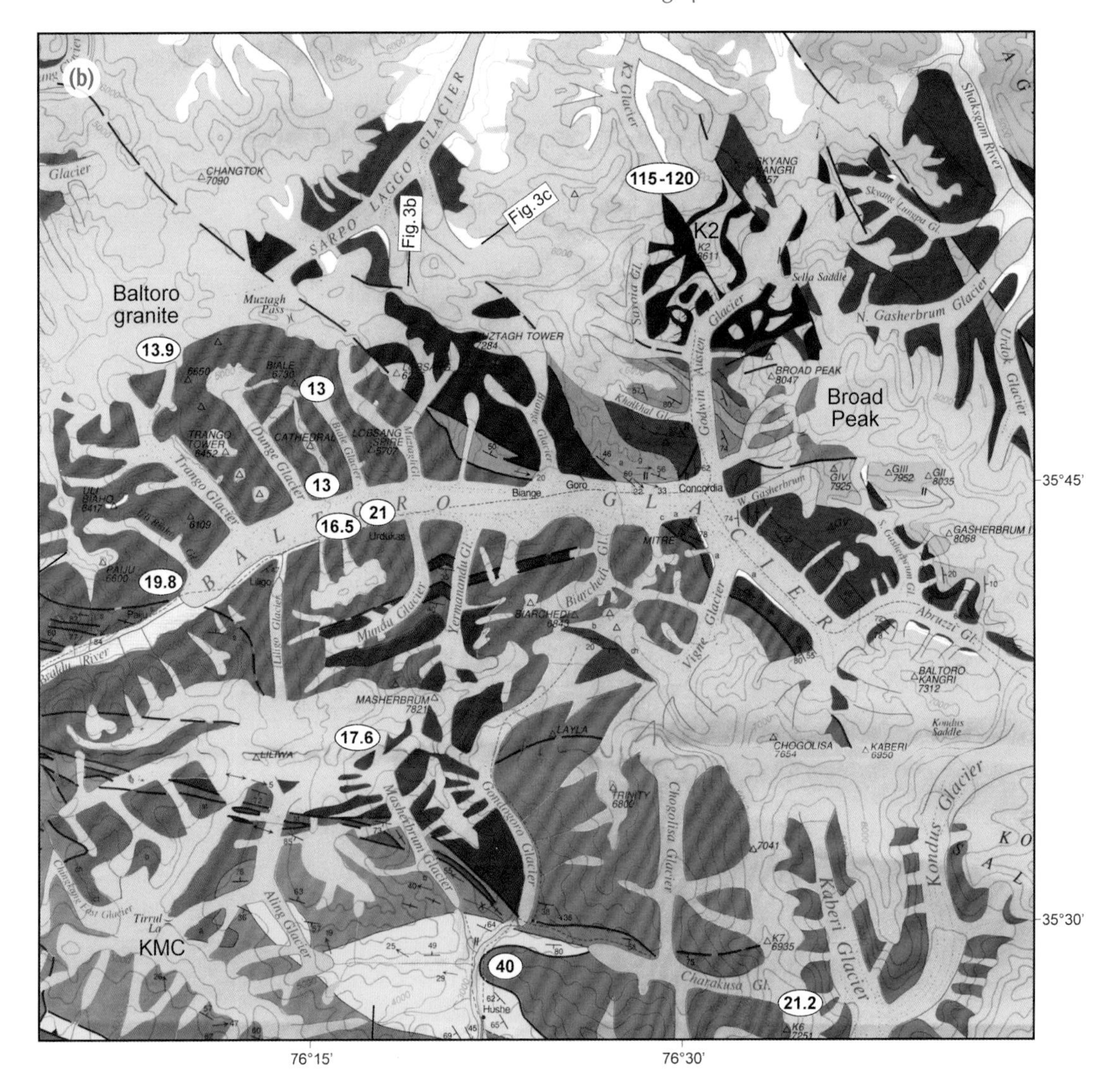

Figure 3.12 (a, b) (*opposite and above*) Geological map and cross-sections of the Baltoro Karakoram showing the ages of the different granitic and metamorphic rocks (in millions of years).

batholithic proportions and had clear intrusive contacts, showing how the granite had pushed up through the country rocks along both margins.

I soon realized that the southern Karakoram exposed the deepest crustal levels of the Asian plate rocks, equivalent to those buried at depths of 30–70 kilometres beneath the Tibetan Plateau, but because of the low erosion they were never exposed in Tibet. Clearly the key to the processes of crustal thickening and timing of thickening and uplift lay in these deep crustal rocks of the Karakoram, not in

Tibet at all. It was crucial to find out the age of formation of these rocks, both the timing of the metamorphism, when the rocks were buried at the deepest levels, and the age of high-temperature melting, when the Baltoro granites were formed.

Biale and the Trango Towers

After the Masherbrum expedition of 1986 and the Hushe expedition of 1987 we had a large collection of rock samples back in England, but there remained many unanswered questions. In the summer of 1988 I returned with a climbing team so that we could gain access to the high ground in the middle of the great granite batholith. I particularly liked the look of Biale, the great fortress-like mountain that I had spied out across the Baltoro Glacier from Masherbrum the year before, so I applied to the Pakistani mountaineering authorities for permission to climb the mountain and carry out geological research. I asked my old climbing partner Nick Groves and his New Zealand girlfriend Maryrose Fowlie, also a highly accomplished climbing instructor from Mount Cook, and my climbing companions from the previous expedition, Mark Miller, Simon Yates, and Sean Smith, three of Britain's leading Alpine climbers, to join me. Their laid-back approach and impressive mountaineering skills in extreme climbing were exactly what we needed (Figure 3.13a, b, c). The six of us flew out to Rawalpindi in June 1988 and started the lengthy bureaucratic process required by the Pakistan Mountaineering Association. I caught a bus to Peshawar and picked up our trusty little Daihatsu jeep from the university. My good friends at Peshawar University, Asif Khan and Qasim Jan, were, as always, very hospitable and keen to help in any way. I tried to persuade some of the younger faculty members to at least come as far as Skardu with us but they were not swayed.

Back in Rawalpindi, Mark Miller borrowed the jeep to drive into Islamabad to pick up new visas from the Chinese embassy and on the way back in a torrential rainstorm managed to roll the jeep off the side of the Grand Trunk Road. He emerged with a few cuts and bruises but the entire upper half of the Daihatsu leaned to one side at fifteen degrees. It was impossible to beat back into shape so we drove around north Pakistan for the next three months in this lop-sided jeep attracting wide stares from everyone we passed. We loaded all our supplies onto a local bus and Simon, Mark, Nick, and Maryrose climbed aboard for the two-day journey to Skardu. Sean and I drove the jeep along the Karakoram Highway, and then turned east on a spectacular road that follows the Indus Gorge where it cuts a chasm over 6 kilometres deep. The river tumbled down the gorge in a froth of white-water rapids and thundering waterfalls. In Skardu we set about hiring a *sirdar* and a team of tough Balti porters. Our route to base camp left as the previous year from Dassu at the junction of the Shigar and Braldu rivers and retraced the route along the Braldu

Figure 3.13 (*opposite*) (a) Simon Yates and Sean Smith in a flying-fox bridge crossing the Panmah River; (b) Tony Rex crossing a glacial stream on the Baltoro Glacier; (c) Simon Yates crossing a swollen river by Tyrolean traverse.

(a)

(b)

(c)

Gorge past Askole and the Biafo Glacier snout. After Paiyu, however, instead of following the normal Baltoro Glacier route to Urdukas, Concordia, and K2, we needed to break out a new trail along the northern ablation valley of the Baltoro Glacier. After crossing several large side glaciers, the Trango and Dunge Glaciers flowing down from a range of spectacular granite peaks along the Chinese Xinjiang border, we found a small patch of flat ablation valley at the confluence of the Biale and Baltoro Glaciers suitable for a base campsite. The sheer southern ramparts of Cathedral rose up in waves of pale granite directly above, the Lobsang Spire protruded vertically out of the ice field across the valley, and to the north the Biale Glacier tumbled in a maze of tottering ice seracs in a scary looking icefall (Figure 3.14). We paid off the porters, our cook Hakim from Hushe built a sturdy kitchen from boulders and a tarpaulin and we set up tents. Base camp was established and we were right in the middle of the Baltoro granite batholith with enough food to last six weeks and the weather looking stable and clear. We had a whole summer of exploration, climbing, and geology ahead. It was an exciting prospect.

One of the main geological objectives of this expedition was to map out the Baltoro granite batholith and to collect samples for geochemistry and dating. Our first foray was an attempt to climb the southern Lobsang Spire, a sharp needle-like granite spire that rose directly above our base camp. Leaving before dawn one day, carrying a heavy rucksack full of ropes and climbing gear, we clambered up the increasingly steepening gully that split our spire from the main Lobsang Spire. At about 10 a.m., after five gruelling hours, we arrived at a narrow col with stunning views east to the ramparts of the Muztagh Tower. This was the site of British climber Martin Boysen's accident during the epic first ascent when his knee became jammed in a narrow crack high up on the Tower and he was left hanging for several hours before he finally cut his own knee free to escape.

The temperature at midday on our precarious col on Lobsang was almost intolerably hot. We were in the middle of the Karakoram at about 5,000 metres altitude, surrounded by glaciers and hanging ice fields draped over the great array of granite spires, and the sun was fierce and unrelenting. We huddled into a tiny bivouac tent attached by a few pitons to the vertical granite wall and panted the afternoon away. Next morning at 3 a.m. we roped up and started climbing by the light of the moon and stars reflected off the glinting snowfields (Figure 3.15). The climbing was steep and hard but eventually we reached the knife-edge summit ridge, with striking views down to the motorway-like moraines of the Baltoro Glacier directly below. Across to the west were vertical tiers of gleaming orangey pink granite cliffs soaring up to the heavens. The cliffs of the Cathedral Spires extend right up to massive ramparts of Biale on the Chinese border, with the vertical granite towers of Uli Biaho, Paiyu, and Shipton Spire rising even more improbably

Figure 3.14 The Lobsang Spire (5,707 m) from the Dunge Glacier.

Figure 3.15 Nick Groves climbing on the Lobsang Spire (5,707 m) with the Muztagh Tower (7,276 m) behind.

beyond. To the east, the huge fortress-like Muztagh Tower dominated the view, but beyond was the most spectacular high range anywhere in the world. K2, Broad Peak, Gasherbrum IV, Hidden Peak, and Baltoro Kangri rose in a display of savage grandeur along the watershed of the Pakistani Baltoro Glacier, with the Shaksgam Valley to the north and the Ladakhi Siachen Glacier and Nubra Valley to the east.

Back in base camp, Hakim cooked up mounds of *dhal bhat* and curry and we had a few rest days to recover and prepare for the main climb ahead, what we hoped would be the first ascent of the south face of Biale. Mark, Simon, and Sean were climbing as one rope, and Nick and I teamed up as another. We left base camp on a perfect cloudless day for the five-hour climb through the Biale icefall to our proposed site of advanced base camp on the ice plateau beneath the granite cliffs that rose to the north. This was dangerous ground, with a constantly moving river of ice breaking up into giant seracs with huge gaping crevasses opening in front of us. We moved together, roped up all the time. Every fifteen or twenty minutes the leader would inadvertently step through a snow-bridge and suddenly plummet into a hidden crevasse, a frightening experience. As the lead climber fell into the crevasse the rope would suddenly reel out and go taut as the second held the fall. The trick was to be alert and twist the rope around an ice-axe to hold the fall. The shock of suddenly falling through the snow, to be left dangling in a yawning chasm of ice was jangling on the nerves but just another occupational hazard for the Karakoram geologist. Luckily in those days we did not have any health and safety regulations to worry about.

Our first foray into the bowels of Biale resulted in us establishing an advanced base camp, a single bivouac tent full of supplies from where we could launch our climb up the steep granite walls. We climbed through a steadily increasing snowstorm to the ice shelf, dumped our loads, and returned to base camp. After two weeks of acclimatizing on the Lobsang Spire and the Cathedral granite cliffs above camp and establishing our attack camp full of supplies we were ready to go for the summit. I was keen to climb a line up through the granite cliffs in order to map out and sample a vertical profile through the granite batholith. These Karakoram granite spires provide a unique opportunity to map and sample over 3 kilometres deep into such a batholith. We left once again at 3 a.m. for the dangerous plod through the icefall, and arrived at the ice-shelf camp at about 10 a.m. As soon as the sun came on to the glacier, freezing night-time temperatures soared up to incredibly hot temperatures above 35°C. Frozen icicles dripping off the rock face turned into trickles of water and then into torrents. Huge avalanches of powder snow exploded down the steep granite faces all around us. This was nature in the raw: powerful, frightening, but at the same time immensely beautiful.

Next morning we left at the usual 3 a.m., roped up, and started climbing the steep ice face on the south face of Biale. Very soon the ice petered out and we were on vertical solid granite. Climbing vertical granite walls with a 20 kg rucksack, a rack full of slings and nuts and big plastic double boots was not easy. I was trying to record geological observations in the granites and put these onto a map at the same time; the sample collecting was to be done on the way down. After two days of this steep and scary climbing we finally broke out of the cliffs onto a large snowfield that led up to a knife-edge ridge. As we approached the ridge the most spectacular mountain panorama I have ever seen unfolded in front of us (Figure 3.16). We were right in the middle of the Karakoram, with huge glaciers flowing all around us, separating ridiculously steep cliffs of pure granite. To the north the whole mighty ridge comprising K2, Broad Peak, Hidden Peak, and the Gasherbrum Range rose above the nearer Muztagh Tower (Figure 3.17). To the west, the vertical cliffs of the Trango and Nameless Spires rose sheer from the Trango Glacier, and beyond were more spires rising skywards in every direction (Figure 3.18). Nick and I stood in dumbstruck silence for several moments. To the south the summit of Cathedral appeared right below us along our ridge and the spectacular shapely mountain of Masherbrum rose in the distance (Figure 3.19). Mark, Simon, and Sean bivouaced just above us. The summit cliffs of Biale were right above and we reckoned that with another day, given good weather, we could pick out a possible line to the summit.

That night it snowed. Around midnight we heard claps of thunder and streaks of lightning lit up the mountains around us. By morning we were snowed in, with about a metre of fresh snow buckling the tent. The summit cliffs were completely covered, and with that we knew that our chances of summiting were slim. We were consoled by the fact that we had actually climbed Cathedral without realizing it. Our ridge dropped gently down to the south to the summit of Cathedral, which was actually a point on the main ridge leading up to Biale. We braved it out for one final day but conditions deteriorated more, so we were left with no option but to retreat. Ten abseils later we finally made it back to the ice shelf and next day it was a weary party that retraced our route through the icefall back to base camp.

After the snow storms of the last two days the melt-water had flooded down the valley, so by the time we arrived back at base camp a scene of devastation greeted us. As the deluge cascaded down the ablation valley, Hakim and Maryrose

Figure 3.16 *(opposite, top)* Panorama of the Baltoro Karakoram: K2 is largely hidden behind the Muztagh Tower (7,276 m), Broad Peak (8,047 m), and the Gasherbrum Range from high on Biale.

Figure 3.17 *(opposite, bottom)* Panorama of peaks along the Upper Baltoro Glacier: Gasherbrum IV (7,925 m), Hidden Peak (8,068 m), and the Baltoro Kangri Range from high on Biale (6,730 m). The Biale icefall leads down to the Baltoro Glacier, with the Lobsang Spires rising directly above.

Figure 3.18 *(overleaf)* (a) The Great Trango Tower (6,452 m) above the Trango Glacier from high on Biale; (b) Nameless Spire, 2,000 metres of flawless granite, from high on Biale.

(a)

(b)

Figure 3.19 Masherbrum (7,821 m) rising above the cloud-filled Baltoro Glacier with the summit of Cathedral Peak in the foreground, taken from high on Biale.

had madly raced around tearing down tents as our little stream of glacial melt water swelled to a flash flood. After the flood, our camp was a sorry sight, but next morning the sun broke through and we dried everything out and relocated our tents. Despite not bagging the summit of Biale, I had managed to collect a suite of granite samples from the summit ridge down to the Baltoro Glacier so I was happy. It had been an incredible week's climbing, and certainly we were completely spoilt for the wonderful mountain panoramas from high on Cathedral and Biale.

Shimshal and the Lukpe-la

After three months around the Baltoro Glacier our supplies were finally exhausted and we trekked out back to Askole. It had been a wonderful experience, we had done some amazing climbing, and I had managed to map a section right across and down into the Baltoro granite batholith. Back in Skardu, Mark, Simon, and Sean left to return to Rawalpindi and then fly to Kathmandu. They had been invited to join Doug Scott's expedition to Makalu and were about to embark on another three-month expedition in Nepal. Nick and I were in no great hurry to return to England. In Skardu we met an old trekking friend, Steve Razzetti, who had just led a group of trekkers over Snow Lake. Steve had left a barrel of food hidden up at the top of the Biafo Glacier and was keen to try an audacious new trekking route across Snow Lake from the Shimshal Valley in the north-west to the Biafo Glacier over a high pass, the Lukpe-la. Steve asked us to join him and we jumped at the chance. This traverse would give me an ideal opportunity to look at the rocks that formed the northern Karakoram Ranges, beyond the great granite batholith.

I drove our small jeep back to Gilgit and then up the beautiful Hunza Valley to Karimabad and Baltit. It was apricot season so the roofs of all the houses were laden with yellow apricots drying in the strong sun. It was magical camping in the apricot and cherry orchards, with views from the tent door straight across the valley to Rakaposhi. Golden orioles flashed through the trees, and hoopoes picked at the earthen fields. We spent a few days around Karimabad doing some short trekking routes along the southern flank of the Ultar Range and up the base of the granite spire of Bub-lu-mo-Ting. The views across the Hunza Valley to Rakaposhi are some of the most glorious in Pakistan. Leaving the jeep at the Passu Inn, a small lodge at the snout of the Batura Glacier, Steve, Nick, and I hiked in to the remote and hidden valley of Shimshal, a veritable Shangri-la. This trek just to get to Shimshal village involved crossing two swaying rope bridges over the turbulent Shimshal River and crossing the bare white ice of the Malanguti Glacier. The Shimshalis are totally different, ethnically and linguistically, from the Baltis. Whereas the Baltis speak a curious dialect of Tibetan, the Shimshalis speak Wakhi, a dialect that sounds like a fusion between Pushtu, Farsi, and Turkish.

In Shimshal village we were invited into the *Lambadar's* (headman) house for tea and we asked if anyone in the village would be willing to guide us over the Lukpe-la to the Biafo Glacier and Askole. The son of the *Lambadar* was an extremely intelligent and pleasant character called Shambi Khan and he immediately volunteered. Two others, Tafas Shah and Dolat Kazi, said they had crossed the Braldu River during a hunting expedition and gone into the Shaksgam Valley north of K2, a valley that was actually in China. A fourth man by the name of Faisal Baig was another very interesting character who had roamed all around the upper reaches of the Shaksgam and northern parts of Snow Lake. Clearly in this remote corner of the Karakoram international borders were simply lines drawn on somebody else's maps. These four Shimshalis turned out to be the strongest and most efficient porters we had ever hired and they all did a magnificent job on our twenty-four-day trek.

Leaving Shimshal we crossed to the north bank of the river and followed the valley up to a beautiful area of high steppe and pastures called Zardgar-bin. These are the summer pastures for the flocks of yaks, pashmina goats, and sheep and almost the whole village migrates up here for the summer. The entire population came out to greet us, and Shambi Khan's relatives dragged us into their mud house for tea. The women were dressed in multi-coloured dresses and pillbox hats and lined up to have their photos taken with a couple of handy yaks. The following day we continued towards the enormous North Braldu Glacier that drains northwards from the ice cap of Snow Lake. Our path was blocked by a fast-flowing muddy river that was at least 80 metres wide: a river that we had to cross in order to gain access to the glacier.

The Shimshalis immediately strapped all the gear onto the yaks then told us to climb aboard. Steve, Nick, and Shambi Khan jumped up on their yaks and Faisal Baig herded them into the river. After a few strides the yaks were out of their depth and lost their footing. The yaks started swimming as the current swept them downstream towards China. After ten minutes of frantic paddling the three yaks reached the opposite bank. Shambi Khan unloaded them, turned them around, and herded them back into the river for round two. Faisal Baig, Dolat Kazi, and I then mounted our yaks, which again leapt into the river (Figure 3.20). Yaks are giant balls of hot air so are unsinkable, but when they lose their footing the whole ship starts rolling and it takes all one's concentration to stay aboard and not slip into the swollen river. We ended up on the other bank, but about a hundred metres downstream. Once again we unloaded the yaks, turned them around and herded them back across the river throwing rocks to get them to re-cross. Tafas Shah then took the yaks back to Zardgar-bin and we were left alone on the far bank with no yaks and no way back. Now we had to cross the Lukpe-la or, if the icefalls proved impenetrable, we would be forced to descend down to the Shaksgam Valley and China.

For the next six days we trekked up the horrendous crevassed glaciers of the north Braldu. The glaciers here are over a hundred metres wide and completely chopped up with crevasses (Figure 3.21). It was dangerous ground and we had to move roped up at all times. We had one 50-metre rope between us and barely enough gear to get anyone out if they did happen to fall into a crevasse. After getting lost several times we finally crossed the Lukpe-la, and ahead of us lay the enormous Antarctic-like snow field of Snow Lake. This is the largest snow-covered area outside the polar regions and when first discovered by Martin Conway was thought to be a continental ice cap, rather than the huge amalgamation of about ten different glacier systems.

Figure 3.20 (*top*) The author crossing the North Braldu River on a yak, on the way to the Lukpe-la during the crossing of Snow Lake.

Figure 3.21 (*bottom*) The icefalls on the North Braldu Glacier which drains north of Snow Lake into the Shaksgam Valley, in the Chinese province of Xinjiang. Shambi Khan and Steve Razetti are in the foreground.

Traversing Snow Lake

Snow Lake is an astonishing place for many reasons. It looked as though someone had collected the highest, steepest, and most spectacular rock spires on the planet and dumped them all around an Antarctic snow and ice cap. To the south the great granite ramparts of the Latok Peaks and the infamous Ogre Peak rose straight from the ice fields (Figure 3.22). It took us two days to cross Snow Lake with only a single pair of snowshoes for the leader. By mid-morning when the sun started to melt the ice, there was a layer of water on top of a thigh-deep layer of soft powder snow. All the crevasses were completely hidden and it was impossible to know when you were directly over one. After Shambi Khan had taken several metre-deep falls into a crevasse and cut his head, Nick and I took turns leading. Every twenty minutes or so the leader would suddenly disappear as he stepped through the snow bridge above a crevasse. These crevasses on Snow Lake are hundreds of metres deep and some as wide as a bus. It was frightening to suddenly find oneself dangling on the end of taut rope in an icy prison with no sign of the bottom. When the leader fell through the others would rush forwards and pull him out, nerves jangling. It was a relief when we made the crossing and finally landed on terra firma on the rocky moraines of the upper Biafo Glacier. In between the crevasse falls and ploughing through the sometimes waist-deep snow I tried to map out the northern contacts of the granite, but at the time there were a lot of other things like simple survival, on my mind, so I cannot vouch for the accuracy of this part of my map.

By this time our food had run out and we were all anticipating Steve's sealed barrel he had left several weeks before on his Hispar–Biafo trek. After an hour's hunting around the rocks where he had left it we found the torn and shredded remnants of a blue barrel, but no sign of any of its edible contents. The bears had found it and torn it apart. The last two days' trekking down the Biafo Glacier were hungry days. When we finally arrived back at Askole, Haji Mehdi the headman recognized us from our Masherbrum and Biale expeditions and ordered a huge meal of roasted ibex, freshly baked *rotis* and eggs. Replete at last we trekked on the final stage down to Dassu in the Shigar Valley and back to Askole. We had been on the go trekking, climbing, and geologizing for four months, and had seen some of the most wonderful mountain scenery in the world and certainly some of the most impressive geology anywhere too. It was time to go home.

Hunza and the Hispar Glacier

The final year of my Leicester University grant in the summer of 1989 I returned to the Hunza Valley with a new PhD student, Mark Crawford. Mark was going to work on the geochemistry of all the granites, and our job was to map out all the different

Figure 3.22 Crossing the Sim Gang Glacier and Snow Lake with the Latok (7,145 m)–Ogre (7,285 m) Range rising to the south.

phases of granite intrusion to attempt to work out the petrological and chemical variations within the batholith. This year my friend from Peshwar University, Asif Khan, came with us and we drove around the entire North-West Frontier from Gilgit over the Shandur Pass to Chitral and back to Peshawar. I had arranged with my climbing friends Simon Yates, Sean Smith, and Simon Richardson to meet up in Gilgit. The plan was to spend a month along the Hispar Glacier and climb several of the spectacular mountains that make up the Pumari Chhish and Khunyang Chhish Range, the western extension of the Baltoro granite peaks. Trekking up through the swaying barley and hashish fields of Nagar, we followed the horribly loose moraines of the Hispar Glacier up to beautiful meadows, the final grazing grounds of Bitanmal.

The critical area south of the Hispar Glacier and north of the Nanga Parbat–Haramosh massif is where Indian plate metamorphic rocks practically touch the metamorphic rocks of the Karakoram. Here the original suture zone had closed and the upper 20 kilometres or so had been eroded off, so there remained an almost unique area where the deepest structural levels of the zone of collision could be studied. This line goes right across a range of 6,400–6,800-metre-high mountains

Figure 3.23 The vast expanse of glaciers draining Snow Lake; looking down the Hispar Glacier, west of Snow Lake.

called Makron Chhish. Simon Yates and I decided to attempt the previously unclimbed neighbour, a shapely peak we called Mikeron Chhish. The ascent took two days and the summit offered a superb panoramic view eastwards up the Hispar Glacier to Snow Lake and south to the giant icy ramparts of Haramosh (Figure 3.23). The most extraordinary views north were the surging ice of the Khunyang Glacier which drained south from the 7,800-metre-high peaks of Distaghil Sar and Khunyang Chhish. Unlike almost all the Himalayan glaciers that were slowly receding, the Pumari Chhish Glacier (Figure 3.24) had surged about 5 kilometres southwards, and waves of ice were overthrusting the main westward-flowing Hispar Glacier. The Pumari Chhish Glacier had grown at least 5 metres higher than the lateral moraines, and bits of ice were falling off the rapidly expanding seracs and hitting grazing yaks on the adjacent meadows.[7] It was an extraordinary sight.

The Dreaming Spires of Oxford

When I started my geological work in the Indian and Pakistani Himalaya I thought it would be a good idea to find out if there was anyone else also working in the Himalaya or surrounding mountain ranges. I convened an informal workshop–

Figure 3.24 The surging Pumari Chhish Glacier flowing out onto the Hispar Glacier; photo taken in August 1989.

conference at the geology department in the University of Leicester in 1985 and gave it the grand title of 'International Himalaya–Karakoram–Tibet Workshop' or HKT for short. I sent an open invitation around and about fifty people arrived from all across Europe, a few from the US, and we flew in some colleagues from Pakistan and India. It was a great success with lots of friendly and constructive discussions, and after the meeting we all convened for a conference dinner in one of Leicester's famous curry houses. Here we decided that there was sufficient new data and field studies that it would be worthwhile holding an annual HKT in different universities. The following year Patrick LeFort organized it at the University of Nancy in France and subsequent conferences were organized by Aymon Baud in Lausanne, Switzerland, and Maurizio Gaetani in Rome. After these first meetings, the HKT was held annually in many other countries, including Nepal, Pakistan, India, and China.

It was at the Lausanne conference that I met up again with John Dewey and, after the meeting, walking around the lakeside, he quizzed me about all my studies in the Karakoram as well as Ladakh-Zanskar. John had recently led a major Royal Society-funded expedition across the Tibetan Plateau and was becoming interested

Figure 3.25 The immaculate view of Great Trango (6,452 m) and Nameless (6,650 m) spires from the summit ridge of Biale with Paiyu Peak (6,600 m) behind.

in the Himalayan side of the collision. I told him that I was applying for a five-year grant from the Natural Environment Research Council (NERC) to continue my studies in the western Himalaya and the Karakoram of North Pakistan. John immediately invited me to transfer to the University of Oxford where he had just the year before been elected as the new Professor of Geology. This was a dream come true and I accepted immediately. After five years at the University of Leicester it was time to move on, and in Oxford I would have the added advantage of running my own research project and being able to get first-class PhD students to work with. Oxford, of course, still has one foot firmly in the Middle Ages so PhD students here are actually called D.Phil students (Doctors of Philosophy).

CHAPTER 4

Pressure, Temperature, Time, and Space

The truth resides in the rocks.

John Dewey

There is nothing as sobering as an outcrop.

Francis Pettijohn

After the summer field season of 1989 in the Pakistani Karakoram, I drove to Oxford, the 'city of dreaming spires' and arrived in the Department of Earth Sciences. In those days Oxford was probably the best field-geology 'hard-rock' department in the country and one of the best in the world. It was a wonderful place for me, buzzing with excitement and full of talented geologists working on projects all over the world. John Platt had post-graduate students working on several projects in the European Alps and the Spanish Betics, Simon Lamb was starting a major new field project in the Andes of Bolivia, and the department had some of the world's leading igneous petrologists working on volcanic and granitic rocks all over the world. The department was overflowing and I was given an office on the top floor of the 'annexe' a wonderful old Victorian building at 62, Banbury Road. My office was up in the attic and I called this grandly the 'Oxford Centre of Himalayan Research'. Right across the Banbury Road was an excellent public house, the Rose and Crown on North Parade, and we used to congregate there regularly for discussions on geology, and the world in general over a pint or two of traditional real ale. It was an excellent life.

In the 1830s the first Professor of Geology in Oxford was the Reverend William Buckland who naturally came with a lot of religious baggage. Buckland was a bit of an eccentric in many ways including living with and eating a whole variety of wild animals and doing his geological fieldwork dressed in full academic gown.[1] Following Buckland the department settled down to a more conventional geological approach, studying the stratigraphy and palaeontology of Oxfordshire. By the 1950s Oxford had become one of the leading departments of geology and mineralogy in the world. The head of department was Lawrence Wager, who had made his name studying the classic Skaergaard igneous intrusion of Greenland. Wager had earlier joined the 1933 Everest expedition climbing to 27,500 feet on the north ridge and collecting an extremely useful set of samples from the north slopes of Everest.[2] Three previous British expeditions to Everest had included the reconnaissance expeditions in 1921 and 1922 and a full-scale attempt at the summit in 1924 culminating in the tragic loss of George Mallory and Andrew Irvine. The 1933 Everest expedition was led by Hugh Ruttledge and included the best Himalayan climbers of the time, Jack Longland, Frank Smythe, Wyn Harris, and a young Eric Shipton. On 30 May 1933 Wager and Harris made a gallant attempt on the summit climbing above the Yellow Band but failed to climb the 'Second Step', an infamous band of cliffs on the north-east ridge. On the way back they discovered Mallory's lost ice-axe and peered down the unseen east face, the Kangshung Face, of Everest.

When I arrived in Oxford in 1989 I found a department full of characters almost as strange as in Buckland's day. There was the slightly mad Professor of Mineral Physics whistling away to the walls, there was the Buckland-esque Professor of Palaeontology buried in his enormous office in the Natural History Museum surrounded by fossils of every description, there were two geophysicists who never looked at rocks and always argued with the geologists, and there was my old friend John Dewey, the newly appointed Professor of Geology, always in animated discussions with anyone who would listen about cricket, politics, and his latest theory of tectonics of somewhere on the globe. It was a wonderful place to be. I was elected a Fellow of Worcester College, one of the older, more traditional colleges in Oxford, and this opened up an entire new dimension to my experience at Oxford University.

Worcester College is a magnificent medieval series of Cotswold stone buildings with huge Gothic arches around grassy quads and sunken gardens. The college was established by Royal Charter in 1714 but its origins went back to 1283 when it was a Benedictine monastery. The old cottages in the front quad were built during this time and still stand to this day. There is a large lake amid the gardens complete with mallard ducks and Canada geese, and further down the grounds a county-size cricket pitch with tennis courts at the end. The gardens are a picture of quintessential English beauty, changing with the seasons but always a place of complete tranquillity.

Traditions have been maintained since the beginnings of the college and academic gowns are required for formal dinner. Fellows are allowed to bring guests so there is never any shortage of fascinating people to meet and talk to.

On Sundays and Wednesdays, guest nights, following grace, the fellows retreat to the wood-panelled, candle-lit Senior Common Room for port, desert wines, fruit, and more stimulating conversation. It is a magnificent remnant of great British tradition and ceremony and an experience I never tire of. Personally I have found the college system probably the most stimulating aspect of life in Oxford.

Pressure, Temperature, and Time

In order to interpret the structures across the deeper parts of the Himalayan orogen we needed to gain information on the temperature, pressure, and depth of formation of the rocks. Once we had these data it was also crucial to find out the timing of peak metamorphism and melting. My Himalayan work now expanded and with a new grant I was particularly keen to recruit expertise in fields that were beyond mine. Foremost in these were the fields of metamorphic petrology, geothermobarometry, and geochronology. The year after I arrived in Oxford, Dave Waters was offered, and accepted, the job of Lecturer in Metamorphic Geology in the department and I immediately persuaded him to join me in my Himalayan project. Dave had previously worked on the Precambrian rocks of Southern Africa in the University of Cape Town and was a superb mineralogist and metamorphic petrologist. He has a brilliant encyclopaedic mind and is one of the leading experts in the new field of thermobarometry, obtaining precise pressure and temperature estimates using geochemical data from analysing very fine slices, known as thin sections, of metamorphic rocks with a scanning electron microscope (SEM). This large microscope scans the sample with a beam of electrons that interact with the atoms in the rock sample to give chemical composition information. This was exactly the sort of data we needed to obtain across the High Himalayan Ranges and I was elated that Dave immediately took a deep interest in my Himalayan project. We started writing proposals to get some D.Phil students to work on various aspects of the Himalayan deep crust.

Randall Parrish was a good friend from the Geological Survey of Canada in Ottawa. Randy is one of the world's leading geochronologists and had developed the technique of dating rocks using uranium and lead (U-Pb) isotopes. He had dated my Baltoro granite samples from the 1985 expedition to K2 and Masherbrum, and it was then that I realized the enormous potential of this new field of radioactive isotope dating of rocks. In the early 1990s he moved from Canada to head the Isotope dating laboratories at the British Geological Survey in Nottingham, and I immediately got him interested in collaborating on the Himalayan and Karakoram samples I had collected. Steve Noble, another Canadian friend and dating expert, was also enlisted, and between us we covered all aspects of fieldwork, structure, metamorphism, thermobarometry, and dating. An exciting prospect lay ahead!

In Oxford I drew up plans to systematically study all the geology of the western Himalaya in Kashmir, Ladakh, Zanskar, and Himachal Pradesh, and also to continue my work in the Karakoram of Pakistan, geologically part of the Asian side of the great India–Asia collision zone. The first expedition I organized from Oxford in 1990 was to the Zanskar Himalaya in northernmost India, and I took two new D.Phil students out with me, Jon Burton and Malcolm Dransfield. The following year we recruited Ben Stephenson, a Cambridge graduate, who was given the task of deciphering the inverted metamorphic sequence around the Main Central Thrust—the huge shear zone showing an entire sequence of metamorphic rocks upside down. Ben's area was in the eastern Kashmir and Kishtwar Himalaya of India, and we flew out to Delhi and Srinagar during the summer of 1991. Unfortunately the Kashmir conflict had started to erupt again, with Kashmiri separatists agitating throughout the valley. Kishtwar seemed to be a stronghold of the separatists, and the main road access from Srinagar was full of Indian army paratroopers. We decided to trek in from Ladakh in the north, along the Suru Valley in Zanskar, and across the Greater Himalayan Range to access the region exposing the inverted metamorphic isograds above the Main Central Thrust.

The following year we recruited James Walker, a Leicester University geology undergraduate, to work on the nearby High Himalayan rocks of Zanskar and Lahoul. In 1993 we managed to get two further students, both Cambridge graduates, Christian Walker and Richard Corfield. Chris was given the task of trying to obtain precise pressure–temperature estimates from the superb metamorphic lithologies I had earlier found around the Suru Valley in western Zanskar. Richard was given the Spontang Ophiolite area in Ladakh to map and unravel the complex structural evolution. Together we tried to reconstruct the evolution of the whole collision process from the earliest ophiolite obduction phase, when the ophiolite was thrust onto the continental margin (as seen for example in Oman, where continent–continent collision has not yet occurred) through to the post-collision folding and thrusting that affected the entire northern margin of the Indian plate. We did puzzle over the fact that all the fold and thrust structures we mapped occurred in the sedimentary rocks of the upper crust—the so-called Tethyan rocks of the northern Himalayan zone. There appeared to be no involvement of the deeper Greater Himalayan metamorphic 'basement' rocks at all. Over a period of seven years we had trekked and mapped along almost every trail across the Ladakh, Zanskar, Lahoul, and Kulu Himalaya of northern India, and had collected hundreds of rock samples for analysis back in Oxford.

Stratigraphy, Structure, and Time

In interpreting geological structures and finding out how mountains were made, time is the most important dimension, and finding out the precise age of rocks is absolutely fundamental. In 1785 James Hutton, the 'father of geology', realized that sedimentary rocks were laid down in layers and that one could make a stratigraphic column starting with the oldest rocks at the bottom and ending with the youngest at the top. He was the first geologist to get to grips with the immensity of geological time. On the north-east coast of Scotland he made the critical field observation of unconformities—sharp breaks in the stratigraphic record where younger sediments are deposited across earlier structures, truncating the older sediments beneath along an angular discordance. 'Hutton's unconformity' on the east coast of Scotland became an iconic geological locality. It was here that he figured out that the folded rocks in the lower part of the cliffs had been deformed first and later planed off by erosional forces before the sedimentary rocks were laid down horizontally above the unconformity plane. This observation laid the foundations for stratigraphy and the science of geology. Geologists such as John Playfair, Roderick Murchison, and Charles Lyell all produced great tomes detailing the stratigraphic rock record of the British Isles.

Hutton's views on the immensity of geological time clearly did not sit well with the Reverend Buckland's biblical views that the world was created in seven days. Buckland promoted 'catastrophism', the concept that the Earth had been shaped by catastrophic events such as Noah's Flood. One of Buckland's first students was Charles Lyell who had a brilliant and independent mind. Lyell believed that the changes deduced from the geological record were uniform and that 'the present is the key to the past'. This was the principle of uniformitarianism that was used to unravel the stratigraphic record of the rocks. Lyell went on to write an epic work, *The Principles of Geology*, which ran into twelve editions. Charles Darwin took a copy of Lyell's book on the voyage of the Beagle and it was this geological concept of vast time that shaped his revolutionary ideas on biological evolution.

The birth of geological mapping can be firmly laid at the feet of William Smith, who almost single-handedly made the first geological map of the British Isles.[3] Smith mapped the sedimentary layers of rocks throughout England, and his remarkable map now hangs in Burlington House, the Geological Society of London's headquarters in Piccadilly. The British Geological Survey set about mapping the whole country using the basic concepts developed by Smith, Hutton, and Lyell. In many parts of England and Wales where sedimentary rocks are widespread, the mapping was relatively straightforward. Geologists traced the beds across the country and collected fossils from each horizon. Soon it was realized that each sedimentary layer had a specific fossil fauna that was distinctive, and each horizon could be assigned to a relative time. Thus the Cambrian, Ordovician, and Silurian epochs were defined initially from the geology of the Welsh hills, and the younger rocks of the Devonian period were mapped out in the hills of Devon in the south-west of England.

In Scotland, however, vast tracts of the Highlands were made of strange crystalline metamorphic rocks that were folded and in places intruded by granites. Many of the great ideas on stratigraphy, structure, and time originated from geological fieldwork carried out in the Highlands of Scotland. Along parts of north-west Scotland it even appeared that these metamorphic rocks overlay the Cambrian and Ordovician sedimentary rocks. Roderick Impey Murchison, the director of the British Geological Survey, had simply assumed that the higher metamorphic rocks, the Moine schists, must be younger than the lower Cambrian and Ordovician sedimentary rocks, so he assigned the schists a Silurian age.[4] It took a giant leap forward by geologists like Charles Lapworth in 1882 and Henry Cadell in 1896 to realize that Murchison's stratigraphic view of the Scottish Highlands was not correct, and that giant thrust faults could place exotic older rocks on top of younger rocks. Lapworth was a close observer of the rocks and made detailed observations around Loch Eriboll in the far north-west of Scotland. Here he discovered a new rock type—mylonites, rocks that were formed by the grinding of rock masses during shearing. Mylonites were always located along the actual zone of contact with the overlying metamorphic rocks, the Moine Thrust. Lapworth proposed that this giant fault plane had pushed the metamorphic rocks of the central and eastern Highlands westwards up and over the unmetamorphosed Cambrian and Ordovician rocks of the foreland.

It took the very detailed and intricate mapping of a team of geologists led by the talented field geologists Ben Peach and John Horne over fourteen years' (1883–1897) mapping across the northern Highlands of Scotland to show for certain that thrust faults did exist and were largely responsible for creating mountain belts through crustal shortening and thickening processes. The *Northern Highland Memoir* published in 1907 remains one of the most intricate and detailed descriptions of the geology of the Moine Thrust zone and the Scottish Highlands and the beautiful coloured geological maps they produced are still regarded as being mostly correct.[5]

The idea that huge masses of rocks could be thrust over other undisturbed regions soon gained credence in other mountain belts around the world, particularly in the Rocky Mountains of Canada and along the Appalachians. In the Alps, large-scale thrust faults were mapped in three-dimensional splendour around the Helvetic nappes and Glarus thrust sheet in particular. Arnold Heim visited the Scottish Highlands on a field trip organized by Peach and Horne in 1900 where he saw the Moine Thrust at first hand. Immediately he realized that similar thrust faults existed in the Alps and he published the first detailed accounts of these. Later, Heim exported these ideas to the Himalaya in the 1930s when he made a traverse of the Indian Himalaya with the then young Augusto Gansser and proposed that the entire metamorphic hinterland of the Greater Himalayan Ranges were thrust southwards over unmetamorphosed rocks of the Lesser Himalaya and Indian shield.[6] It was Heim and Gansser who first proposed the name 'Main

Central Thrust' to describe this giant fault zone that runs the entire length of the Himalaya from Kashmir through the mountains of northern India, Nepal, and Sikkim, to Arunachal Pradesh in the far north-east of India. Having established the structural geometry of these mountain belts the next big question was: how did these crystalline metamorphic rocks form?

Deep heat: Metamorphism and Thermobarometry

Metamorphic rocks form as a result of increasing temperature and pressure during crustal thickening and mountain building. A limestone heated up and buried will become a marble, and sedimentary sandstones or shales become metamorphic schists or gneisses. George Barrow (1853–1932) also working in the Highlands of Scotland, first made the crucial discovery that certain index minerals showed specific temperature and pressure conditions of formation. He found that in pelitic (shaley) rocks with increasing metamorphism the index minerals occurred in order: chlorite, biotite, garnet, staurolite, kyanite, and sillimanite. Barrow mapped out all the different mineral zones across the Highlands of Scotland and this typical sequence of mineral assemblages became known as the 'Barrovian' series. For good measure, he also discovered a second metamorphic series along the Aberdeenshire coast of north-east Scotland, one that was the result of metamorphism at lower pressures and higher temperatures, called the 'Buchan' series. These rocks were characterized by minerals such as andalusite (rather than the high-pressure Al_2SiO_5 polymorph, kyanite), sillimanite, and cordierite. At higher temperatures, above about 700°C, it was possible even to melt the rocks, producing migmatites, partially molten rocks showing both a 'melt phase' represented by the first granitic liquids (mainly quartz and feldspars) and a 'restite phase' (mainly the darker *mafic* minerals, rich in magnesium and iron) representing the unmelted part.

Later work defined several different tectonic settings where distinct metamorphic rocks could be found.[7] In Japan, blueschists, characterized by the blue amphibole glaucophane, and eclogites consisting mainly of garnet and pyroxene, were found to be characteristic of subduction zone settings where rocks were buried to high pressures. Blueschists and eclogites are used to map out ancient subduction zones or plate boundaries such as the Indus suture zone along which the Indian and Asia plates collided. Around intrusive hot granite plutons, different types of metamorphic rocks were commonly formed where static metamorphism results from high temperatures but low pressures. These rocks are called *hornfels* and characterized by minerals such as andalusite or sillimanite (Figure 4.1).

Regional burial metamorphism is usually accompanied by deformation, so linking the metamorphic minerals to the structural fabrics is also very important. With later advances in thermobarometry it became possible to quantify the pressures (P)

Figure 4.1 A thin section of a typical pelitic rock showing andalusite crystal with distinctive 'hour-glass' texture replaced by kyanite and sillimanite. By analysing the chemical compositions of minerals in rocks like these it is possible to obtain information of the pressure and temperature of formation.

and temperatures (T) at which a specific mineral assemblage in the rock was formed. Later, development of scanning electron microscopes made it possible to work out how P–T conditions evolved through time. For example, as a garnet forms in a metamorphic rock it grows in some sort of chemical equilibrium with its surrounding minerals. The garnet entrains tiny mineral inclusions as it grows from core to rim over time, and the chemical compositions of coexisting phases can give us information on the P–T evolution of the rock. It was now possible to determine the path along which a specific rock had evolved through time (a pressure–temperature–time path, or PTt path for short), and to work out its burial and exhumation history.

The mapping of the Scottish Highlands clearly showed that the deeper structural levels were composed of high-grade metamorphic rocks. The first incoming index minerals or isograds could be mapped out with great difficulty and a sharp eye. Crustal thickening, achieved by folding and thrusting, resulted in higher pressures and temperatures with increasing depth. In the Himalaya, however, Arnold Heim and Augusto Gansser discovered that the metamorphism appeared to be upside down (Figure 4.2). As they trekked further north towards the highest peaks of the Himalaya, going up structural section, they encountered metamorphic rocks of increasingly higher grade and eventually, near the crest of the highest peaks along the Himalayan watershed, even granites—rocks that had been heated up so much that they actually melted. How could this inverted

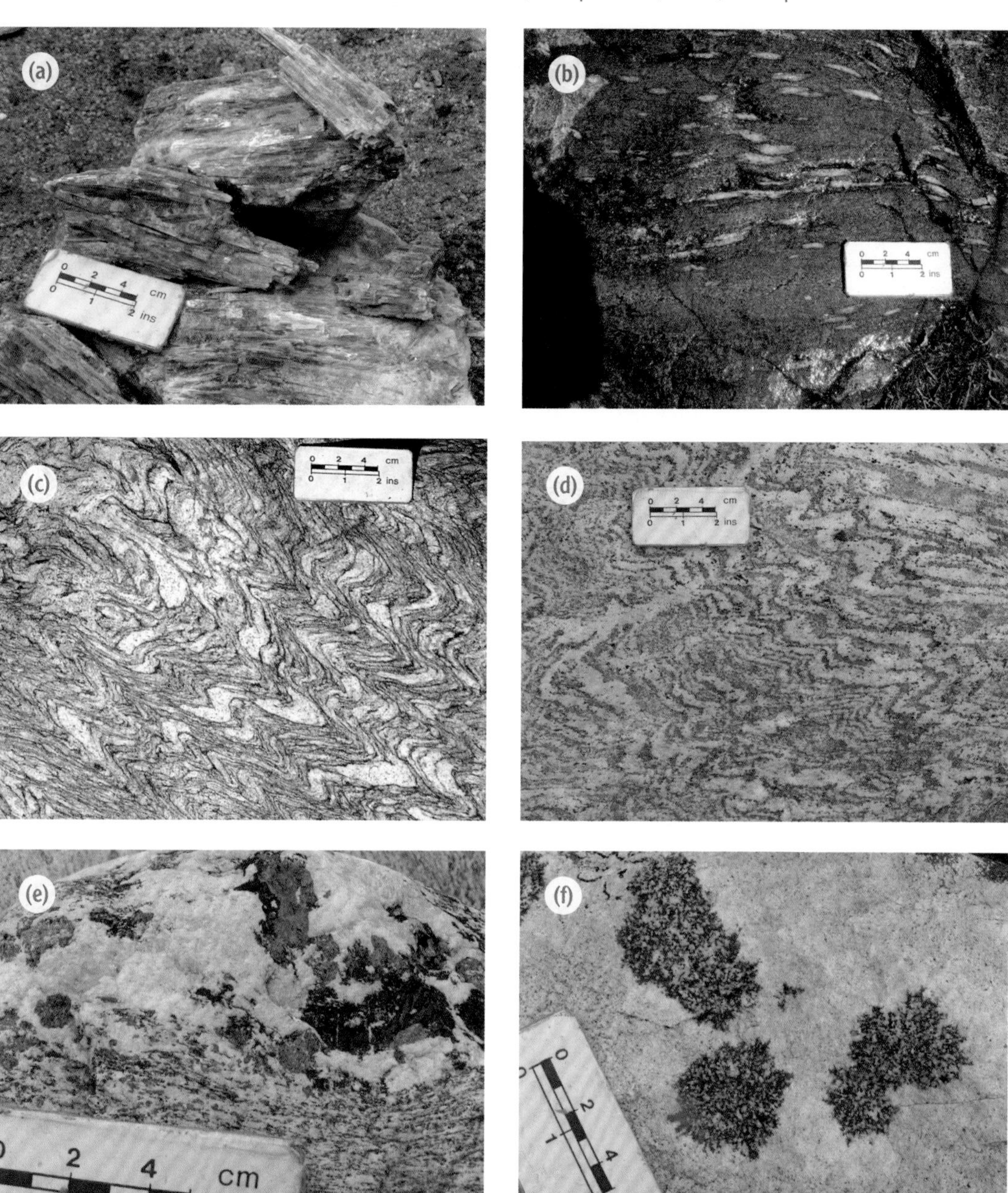

Figure 4.2 Rock types across the Greater Himalayan Ranges showing the inverted metamorphism with higher grade rocks above lower grade rocks. (a) Kyanite crystals above the Main Central Thrust pass up to (b) sillimanite gneisses, (c) migmatites, showing beginnings of partial melting, (d) segregation of pale-coloured melts from dark-coloured restite, (e) garnet, cordierite bearing granites, and (f) tourmaline leucogranites the end-product of crustal melting.

sequence of metamorphic rocks have formed? Either the hotter metamorphosed rocks of the Greater Himalaya were thrust over the colder lower rocks of the unmetamorphosed Lesser Himalaya, or the metamorphic isograds were folded by later deformation events after metamorphism.[8]

Clearly it was now of vital importance to quantify the pressures and temperatures of formation across the Himalaya. A major part of our Oxford Himalayan project involved our D.Phil students doing painstakingly detailed thermobarometric work on samples collected along well-chosen profiles across the Indian and Nepalese Himalaya. This work involved hours of peering down microscopes looking at thin sections, recording the petrology, and then probing points across individual garnets and adjacent minerals to find out their chemical compositions, sitting in front of a scanning electron microscope. Chris Walker and Ben Stephenson worked on these P–T profiles across the Zanskar Himalaya, and James Walker did similar work in the Kulu-Lahoul region of India.[9] In Nepal, Micah Jessup worked on the profile across the Everest region[10] and Mike Streule did a parallel profile across the Arun Valley south of Makalu.[11]

These P–T profiles were a crucial data set that allowed us to interpret the evolution of crustal thickening through time along the Himalaya. We now had data on temperature, pressure, and depth of formation of the rocks during Himalayan mountain-building. The next step was to find a method to actually date the rocks. We needed to quantify precisely the timing of peak metamorphism and the timing of the melting events that formed granites.

Radioactivity and the Dating Game

The other branch of the geological sciences that was developing rapidly during the 1980s and 1990s was that of geochronology, the dating of rocks. The original discovery of radioactivity is attributed to Henri Becquerel, a French physicist, and his students Marie and Pierre Curie working in Paris at the end of the 19th century. They discovered that the elements uranium and thorium gave out some mysterious invisible rays that were capable of passing through photographic paper similar to recently discovered X-rays. Later they discovered two new radioactive elements, polonium and radium, that were to become crucial for the determination of the age of the Earth.[12] It was a young New Zealander, Ernest Rutherford, working in Canada, who found that radioactive elements actually decayed through time to form another element. The unstable 'parent' uranium 238 (^{238}U) decays at a constant rate through time to radium, radon, polonium, and other intermediate elements, eventually forming the stable 'daughter' product, lead 206 (^{206}Pb) isotope. Thorium 232 (^{232}Th) decays to the isotope lead 208 (^{208}Pb). As radium decayed, two gases were produced—radon and helium, and then radon itself decayed, so the 'decay chain' that started with a 'parent' atom of uranium evolved into a 'daughter' product of radon, and so on. As electrons were emitted from the atom, energy was constantly being given out in the form of heat. The recognition of radioactivity from these heat-producing elements was the cornerstone for all nuclear physics.

Uranium 238 then was the vital clock that geologists needed to tell the time. Uranium is locked up in minerals such as zircon, monazite, allanite, and rutile that are found in minute quantities, but ubiquitously in many granitic and metamorphic rocks. The rate of decay of uranium 238 is so slow—its half-life being four and a half billion years, almost the age of the Earth itself—that its decay rate can be used for measuring time. The number of atoms that decay is dependent on the number originally present. Rutherford was able to measure the amount of parent isotope and the amount of the daughter isotope in a rock sample and was then able to make the first age determination from a rock. Previously geophysicists thought that the Earth was only about 20 million years old, following Lord Kelvin. Despite his obvious brilliance in every other field he studied—theoretical physics, thermodynamics, electromagnetism—Kelvin was unable to calculate the correct age of the Earth. Kelvin's estimates for the age of the Earth came down from an initial educated guess of 100 million years, to 50 million years, and finally to 24 million years. There was nothing in physics known at the time that could explain how a body like the Sun could burn continuously for more than a few tens of million years without running out of energy. Kelvin did hedge his bets by saying that the discovery of some other source of heat would change his calculations. Rutherford had now discovered that extra source of heat. Radioactive isotopes clearly showed that the age of the Earth was several billions of years, not a few million.

Arthur Holmes, then at Durham University, developed the use of uranium-lead isotopes to date rocks, and for the first time it was possible to put actual numbers on a geological timescale.[13] In 1946 Holmes calculated that the Earth was 3,350 million years old. Others started using meteorites as proxies for dating the age of the Earth. Meteorites preserve a pristine chemistry from their original source whether they originate from the outer part of the Solar System, the Moon, or from Mars. U-Pb dating of meteorites, turned out to be close to 4.45 billion years old, similar to the U-Pb zircon ages from some of the oldest rocks known on Earth (in Northern Canada, Greenland, southern Africa, and western Australia).

The uranium–lead dating system is based on two separate decay chains: the radioactive decay of ^{238}U to ^{206}Pb (with a half-life of 4,510 million years) and the decay of ^{235}U to ^{207}Pb (with a half-life of 713 million years).[14] The amount of radiogenic lead, resulting from the decay of uranium, must be distinguished from naturally occurring lead, and this is calculated by using the ratio of ^{206}Pb with the stable isotope ^{204}Pb. If the mineral has remained as a closed system, after correcting for original lead, the ratios of ^{235}U to ^{207}Pb and ^{238}U to ^{206}Pb should agree.

If so, the age is said to be 'concordant' and can be plotted on a 'Concordia' curve with the age determined being the true age. If the two ages do not agree they are discordant and some lead loss must have occurred, possibly during a subsequent heating event.

U-Pb dating of zircons and monazites, therefore, could solve the problem of dating the timing of peak metamorphism and melting to form granites (Figure 4.3). Other elements were also useful for dating the subsequent cooling history of the rock during its exhumation to the Earth's surface. The rubidium–strontium (Rb–Sr) dating system is based on the radioactive decay of ^{87}Rb to ^{87}Sr: ^{87}Rb decays in a single step to ^{87}Sr by the emission of a low-energy beta particle (electron); ^{87}Sr/^{86}Sr is plotted against ^{87}Rb/^{86}Sr to produce a straight line or an 'isochron' from which the age can be determined. This method cannot be wholly relied upon in a situation in which high-temperature fluids are circulating, as both elements can be mobile in some situations.

The element potassium (K) decays to the gas argon (Ar); ^{40}Ar is an inert gas that does not combine chemically with any other element and so escapes from rocks when they are heated. When the rock cools, the radiogenic ^{40}Ar is trapped within

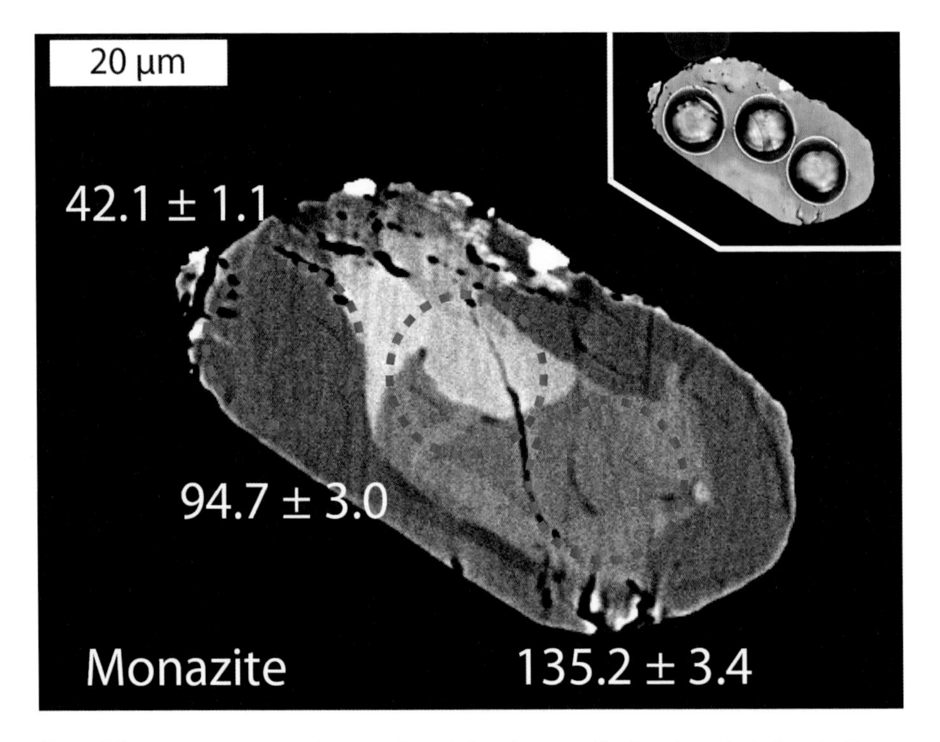

Figure 4.3 A single monazite crystal from a pelitic rock from the eastern Himalaya viewed in the Scanning Electron Microscope showing three pits where laser analyses for uranium isotopes have been carried out. The ratios of isotopes give us ages that the monazite crystal grew, which can be interpreted in terms of the metamorphic history of burial and exhumation. Courtesy of Richard Palin.

the crystal structure and accumulates with time. The time for ^{40}K to decay to ^{40}Ar in a mineral like hornblende or mica could be used to date the time of mineral growth. However, the temperature at which the mineral cooled as determined by dating, known as the 'closure temperature', of minerals like hornblende (500°C), and the micas muscovite (350°C) and biotite (300°C), is lower than the known peak temperatures of high-grade metamorphism and melting to form granite. So these K-Ar or $^{40}Ar/^{39}Ar$ 'ages' only date a time some way into the process of cooling of the rock. Even lower-temperature chronometers such as fission-track dating of apatites and zircons or U-Th-He dating likewise give only final cooling times during exhumation of the rock (Figure 4.4).

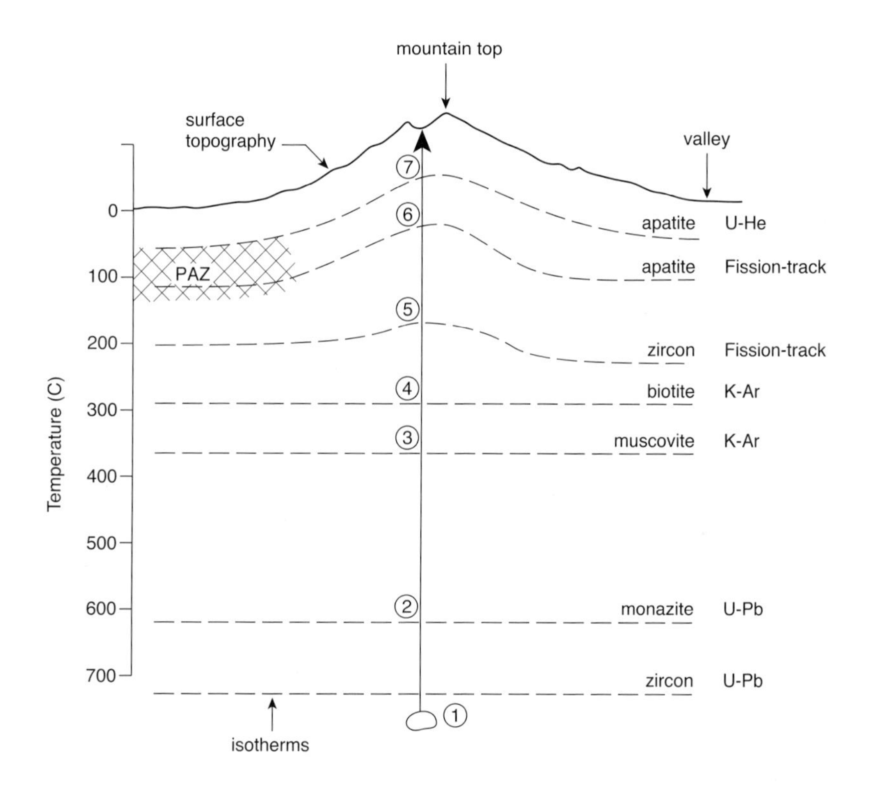

① ② timing of crystallisation of granite > 700C
③ ④ timing of cooling through 350 - 300C isotherms
⑤ ⑥ timing of cooling through 200 - 120C isotherms
PAZ fission tracks annealing within Partial Annealing Zone
⑦ sample cools through 75C isotherm – U-He apatite closure temperature

Figure 4.4 Temperature versus Time diagram illustrating the isotherms corresponding to closure temperatures of various minerals in different isotopic dating systems (on right) and the stages recorded by individual minerals during its exhumation towards the Earth's surface.

Clocks in the Rock: Fission-Track Dating

One useful method of determining low-temperature ages is with the use of fission tracks, or radiation damage tracks, that form in uranium-bearing minerals from the spontaneous fission of atoms of ^{238}U. The numbers of fission tracks increase with time, allowing the calculation of fission-track ages.[15] Heating anneals the tracks, causing reductions in track length and in apparent fission-track age by amounts controlled mainly by the maximum temperature experienced. The concentration of uranium can be measured by counting tracks produced by fission of ^{235}U as a result of the irradiation of the sample with thermal neutrons in a nuclear reactor. The main minerals used in fission-track dating are apatite and zircon. The closure temperature for fission tracks in zircon is 200±50°C, and for apatite even lower at 120°C. During the movement of the rock sample towards the Earth's surface, either by tectonic forces pushing it up, or erosional forces removing the overburden, or both, the rock will pass through the partial annealing zone where fission tracks are formed. As erosion removes material from above, the rock cools and retains its 'frozen-in' fission tracks. In high mountains the fission-track ages of apatite, and zircon will normally increase with increased elevation. The youngest ages will be at the base of the mountain and the summit area should have the older ages. Where samples collected over a wide altitude range have similar ages, it must mean that both uplift of the rocks and erosion have been very rapid.

One final dating method is an even lower-temperature geochronometer, that of Uranium-Thorium-Helium (U-Th-He) dating. The accumulation of radiogenic He from the decay of U and Th in apatite crystals has a closure temperature of only 75±5°C. In a normal geothermal gradient the zone of partial helium retention occurs only about 2 kilometres below the Earth's surface, shallower even than the fission-track partial annealing zone. In great mountain ranges like the Himalaya these low-temperature ages can be used in conjunction with the high-temperature geochronological methods to deduce exhumation rates of the rocks, and both fission-track and U-Th-He dating can be used to infer the evolution of landscapes. One other tool in the box is dating by cosmogenic radionuclides.

Cosmogenic Radionuclide Dating

Cosmic ray particles continually bombard the Earth's surface from the atmosphere. Cosmic radiation comes from solar flares and solar winds, as well as from galactic sources outside the solar system. Nuclides are produced through nuclear reactions, inducing high-energy cosmic radiation. Radioactive species produced by cosmic rays, known as cosmogenic radionuclides, that are commonly used for dating purposes are beryllium-10 (^{10}Be), aluminium-26 (^{26}Al), and chlorine-36 (^{36}Cl); ^{10}Be is produced from ^{18}O, and ^{26}Al is produced from ^{32}Si. Silica and oxygen make up the common mineral quartz (SiO_2), so the usual method of radionuclide dating is to analyse quartz boulders collected from glacial moraines or river terraces.[16] The production rates of cosmogenic radionuclides at 3 kilometres elevation are about seven times higher than at sea level, so corrections have to be applied for altitude. Cosmic rays also hit the Earth's surface from all angles, steered by the magnetic field of the Earth, and are charged. Stronger concentrations of cosmic rays occur at high geomagnetic latitude than at low latitude.

The cosmogenic radionuclide dating technique has been applied mainly to geomorphology, in particular the dating of alluvial fans, moraines, terraces, and debris flows.[17] The half-lives of these nuclides are long relative to the ages of most landforms. The dating of geomorphological features offset along an active fault can provide the age constraint necessary to derive a short-term slip rate for the fault. However, there are pitfalls in the interpretation of this dating method. Moraines and debris flows are dynamic features, with boulders continually being buried by landslides, then exposed, eroded, and re-buried. In complex geomorphological systems, one is never precisely sure exactly what the derived age is actually dating.

If a single rock sample contained several of the datable minerals described above it would then be possible to plot the ages on a temperature versus time diagram. Minerals with successively lower closure temperatures should plot at successively younger ages as the rock cools during its exhumation from deep levels in the crust towards the Earth's surface. The line joining all the ages is referred to as the cooling curve.

Uplift, Erosion, and Exhumation

All these geochronological methods applied to dating specific closure temperatures of minerals in specific isotopic systems can be used to infer the uplift of the rocks. This information, however, tells us nothing about the uplift of the Earth's surface and the formation of mountains and plateaux as we see them today. A steep cooling curve implies a period of rapid exhumation of the rock, erosion or removal of overburden, but not the timing of surface uplift of a Himalayan mountain. Dating surface uplift is very tricky. Exactly when the great plateau of Tibet uplifted is a major source of research, not only for tectonics but also for the effects of plateau uplift on the climate of the northern hemisphere. We can use the geological record to find out when Tibet was last at or below sea level. Limestones containing fossils of the tiny marine organisms called foraminifera dated as Mid-Cretaceous (about 110 million years ago) are the final marine sediments known across the plateau area, after which the area was above sea level. But when did Tibet and the Himalaya attain its extreme elevation seen today, at an average elevation of just over 5 kilometres above sea level?

River profiles can be used to infer bedrock uplift, although precise dating remains problematic. River incision determines the gradients of hill slopes. Abandoned straths or terraces are left when the river cuts deeper into the valley. Rates of river incision can be calculated using the age of the terrace (dated by cosmogenic radionuclides) and its height above the river. Oxygen isotopes can also be used to infer 'palaeo-altimetry', heights from the past. The isotopic composition of groundwater from rain and snowfall (known as 'meteoric water') changes with increasing elevation so that ratios of $^{18}O/^{16}O$ decrease with altitude.

Finally, palaeobotany can also be a useful tool. Fossil leaves have been found in continental sedimentary deposits known as red-beds because of their oxidized red colouration, in small, restricted basins such as the Lunpola basin in Tibet. Interbedded volcanic ash layers provide an ideal dating tool: $^{40}Ar/^{39}Ar$ ages from the ash beds show that the plant horizons were formed 13 million years ago. The leaves suggest a far moister climate probably at lower elevation than the present high-altitude desert. The shape and physiognomy of leaves is dependent on altitude and climate. Small, simple leaves occur in dry, cold conditions above 3,000 metres, moderately sized leaves with toothed margins occur in cool, wet climates at 2–3,000 metres, and large leaves with smooth margins and pointed tips are characteristic of hot, wet conditions below 1,000 metres. The Lunpola leaves were generally typical of dry, cold conditions, suggesting that 13 million years ago Tibet was near the height that it is today.

Determining exactly when the Himalaya rose to their present-day altitude has been a major source of geological research. We had a reasonably good idea when the major periods of folding, thrusting, and crustal thickening occurred after the collision of India with Asia, 50 million years ago. Increased crustal thickening led to increased pressure and temperature and the rocks were metamorphosed as a result of this burial. We had a fairly accurate idea of the timing of peak high-pressure (about 35–45 kilometre depth) kyanite grade metamorphism (35–30 million years ago) and peak high-temperature sillimanite grade metamorphism associated with partial melting producing the Himalayan leucogranites (24–15 million years). All this recorded the uplift of the rocks from deep in the crust. But when did the mountain surfaces rise to their present elevations of 7–8 kilometres above sea level? What we needed now was a suite of granite samples collected from precise elevations vertically up as large a Himalayan cliff face as possible. This was the perfect excuse to climb a major Himalayan peak, preferably one with 2- or 3-kilometre-high cliff faces of granite.

Garhwal Himalaya

There are many iconic mountains around the world—the Matterhorn, Machhapuchare, Mount Kailas, K2, and Everest to name just a few. For rock climbers the list would have to include the Aiguille du Dru and the Grand Jorasses in the Alps, Cerro Torre and the Torres del Paine in Patagonia, and the Trango Towers in the Karakoram. The peaks around the Gangotri Glacier in the Garhwal Himalaya of Northern India have a dazzling array of beautiful mountains, and the Gangotri Glacier carves a course right through the middle of them. The source of the Ganges River is Gaumukh, the 'Cow's mouth' at the snout of the Gangotri Glacier, and immediately above is the stunning peak of Shivling standing at 6,543 metres' altitude. Shivling has the quintessentially perfect mountain shape, an upward-tapering cone laced in hanging glaciers and icefields in between narrow serrated ridges of pristine white granite (Figure 4.5).

To the north of Shivling the entire Bhagirathi Range has four peaks lined up along the north shore of the Gangotri Glacier (Figure 4.6). Each one shows an incredible 2,000-metre profile through a single Himalayan leucogranite sill from the summit region, where the granite has intruded black shales, to the base, where high-grade metamorphic rocks crop out. Around the margins of the granites, blocks of country rock have been plucked up by the force of the intruding granitic liquid and frozen in as the granite cooled. To the east of Shivling lie the massive

Figure 4.5 Shivling (6,543 m) rising above the Gangotri Glacier in the Garhwal Himalaya of northern India.

Figure 4.6 Bhagirathi III (6,856 m) south face showing the entire 2,000-metre granite sill intruding metamorphic high-grade gneisses along the base and low-grade shales along the top.

snow peak of Kedar Dome and a wild array of icy mountains over towards Tibet. To the south lies Meru, an outlandish curving fin of pure white granite that sails up for over one thousand metres above the glacier, and further away the great granite fortress of Thalay Sagar (Figure 4.7). It is no wonder that these mountains play such a prominent role in Hindu mythology and that the Ganges River that springs from them is a holy river for Hindus.

Shivling has several faces separated by serrated ridges but the most prominent one is the unclimbed north face that rises directly above the Gangotri Glacier. I decided that the north-west face of Shivling would provide not only a magnificent first ascent climb but also a suitable route for collecting granite samples up a 2,000-metre vertical profile through a Himalayan leucogranite. In 1986 I received permission to climb Shivling from the Indian Mountaineering Federation and set about recruiting a team of geologists and climbing friends. I invited Patrick LeFort and Arnaud Pêcher from France as they had all the experience of working on the Manaslu leucogranite in Nepal. They recruited a new PhD student from Belgium, Bruno Scaillet, and I recruited a new PhD student, Paul Metcalfe, who would carry out the geochemical analyses of the granite samples. Tony Rex and I would carry out the geological mapping and sampling and also attempt to climb to the summit. My old climbing friends Nick Groves and Maryrose Fowlie, and two young, hotshot Alpinist friends, Mark Miller and Jon Tinker, made up the team.

Figure 4.7 Meru (6,672 m), the Sharks Fin, a 1,500-metre cliff of perfect granite.

In early April we hired a small bus and drove from Delhi to Rishikesh then along the Bhagirathi Valley to Uttarkashi. It was a relatively short trek to base camp accompanied by several half-naked *saddhus* (Hindu ascetics) with their long matted hair, Shiva's tridents, and begging bowls. The *saddhus* were headed for Gaumukh, where the glacier calved away to feed the great Ganges River, a very holy site in Hinduism, where they would sit cross-legged meditating on a boulder surrounded by icebergs and tumbling waters (Figure 4.8). We crossed the glacier and headed up the ablation valley to the green and flowering meadows of Tapovan. Our tents looked straight up at the daunting north face of Shivling and across the glacier to the Bhagirathi peaks. It was a spectacular setting. That night a storm rolled in and it snowed continuously all night, great big heavy snowflakes falling silently out of a black sky. Early in the morning the snow had completely buried our tents up to the roof and we had to dig ourselves out. The snow was so soft and so deep that without snow-shoes or skis it was almost impossible to move.

Our first forays from Tapovan were reconnaissance treks further up the Gangotri Glacier, across to the base of the Bhagirathi peaks and up the Meru Glacier towards the incredible sweeping granite face of the 'Sharks Fin'. Mapping out the granite bodies we discovered that they formed giant lenses or flattened boudins within the upper part of the Greater Himalayan metamorphic series. The granite appeared to have intruded laterally along the foliation planes and then subsequently had been flattened and stretched, both along the strike in an east–west direction

Figure 4.8 A Hindu ascetic *saddhu* meditating on the meaning of life beneath Shivling, the source of the Ganges River.

and also down the dip in a northerly direction. The rocks above the granites from the summits of the Bhagirathi peaks dipping down to the lower glacier levels were unmetamorphosed Palaeozoic sedimentary rocks so the structural position of the granites appeared to be similar to ones in the Zanskar region to the west (Figure 4.9). Right next to the granite contact we even found delicate crystals of andalusite showing the distinctive hour-glass texture, indicating a low-pressure metamorphic overprint.

Figure 4.9 The south face of Bhagirathi III (6,856 m) north flank of the Gangotri glacier.

Climbing Shivling

After a few weeks of geologizing and acclimatizing we were ready for the climb. One of my main objectives was to do continuous rock sampling as high as possible so that we could date zircons, apatites, and micas from the granite in order to constrain the rates of exhumation and uplift of the rocks. Tony and I planned a route up the north face but keeping close to the ridge so that we could access the rocks; we would obviously collect rocks first from our highest point, then on the descent. Jon and Mark were attempting a bold new line direct up the north face and Nick and Maryrose were trying another line close to ours up the north-east ridge. First we climbed a little way up the Meru Glacier and onto the face itself. Near a prominent buttress we dug an enormous snow cave, complete with bedroom, kitchen and storage compartments that we would use as an advanced base camp. Sleeping in the snow cave was surreal, cocooned completely from the gale force winds and driving snow outside on the face. On our first night in the snow cave several small avalanches thundered down right overhead. The drifts usually blocked the narrow entrance, so frequently we woke to a lack of air and had to furiously dig out the overlying snowdrifts to breathe again.

Early one morning in mid-May we finally set off, racked up with slings, nuts, pitons, and ice axes, and roped up. Straight above our snow cave we found ourselves on 50° snow slopes overlying icy granite slabs. Tony and I front-pointed up on our crampon tips, ice axes swinging from both hands, alternating leads. Every time the 50-metre rope length ran out the leader would place an ice-screw and belay, then bring up the other who would climb past for the next pitch. We climbed all morning from 5 a.m. to about 11 a.m. By 10 a.m. the ice on the face was starting to melt and small powder avalanches were set off. Small rocks became loosened by the melting ice and whizzed down past us. By midday the face was far too dangerous, with avalanches pouring down all around us. We climbed on faster, encased in spindrift, running out 50-metre pitches with no protection between belay points to try to access the relative safety of the rock ridge. It was extremely dangerous ground but some of the most exhilarating climbing I had ever done (Figure 4.10). Finally, at about 2 p.m., we reached the ridge and found a relatively safe location to place some rock protection gear and get out of the way of the thundering avalanches. It was easy to see why no one had climbed our route before! That night we had a very exposed and extremely uncomfortable bivouac. We had, however, climbed about 1,000 metres and were halfway up the face.

It was impossible to get any sleep on our hanging bivouac despite the clear, cold starry skies, so around 2 a.m. we made a brew and then set off again in the dark, using our head torches and the light from the stars. It was a magical experience

Figure 4.10 Climbing on the north-west face of Shivling. The ice cave was at the bottom of the ridge in the lower right of the photo.

with just us, way out on a limb, climbing steep ice and rock, high on the north face of Shivling, with the summit getting closer and closer (Figure 4.11). As dawn broke, we could clearly see the summit right above us, but the immensity of space below our crampon tips told us what a precarious position we were in. Sunrise over the vast ocean of snowy mountains to the east was awe-inspiring but soon afterwards, clouds welled up the Gangotri and Meru Glaciers and engulfed us. By mid-morning the winds started battering us and the melting ice loosened canon-balls of rocks bouncing down the face. Climbing past mid-morning was far too dangerous, so

Figure 4.11 Tony Rex climbing on the north-west face of Shivling.

we found an even more exposed place to bivouac by a solitary rock buttress poking out of the steep ice face. We banged two pitons into the rock and fixed a couple of ice screws, but there was nowhere to even sit, let alone lie down. We spent the next fourteen hours hanging from these slender threads, hoods pulled down to try to avoid the spindrift and powder snow. In the middle of the night an explosion scared us silly as a huge rock fall and avalanche roared down the face barely 30 metres away from where we were standing. By dawn we were completely spent, snow had penetrated right through our climbing suits and we were frozen to the core. It was only three or four more pitches to gain the summit, but our nerves had been shattered by the night's avalanche, so we decided to retreat.

The summit of Shivling is a rounded snow cone right at the top of the granite intrusion, so the rock that we were belayed to was close enough to the uppermost part of the granite sill. We had about 1,500 metres of dangerous steep icefields and rocky buttresses to abseil down and we had to collect rock samples all the way down. Our upper bivouac was so precarious we could not even get our small stove out to make a brew. I swung the hammer and collected one sample from the buttress and then leaving one of our precious pitons to abseil from, we started our long descent. Launching out in space on the first abseil I thought how thin that rope was. Luckily the winds eased off so we carried on throughout the day, abseil after abseil, stopping whenever we could to collect a small granite sample. After what seemed an inordinate amount of time with energy levels depleted we arrived at some flat-looking ledges. At last we were able to get out our stove and rehydrate with a brew. We had a rucksack full of precious rock samples and abseiling with this load on our backs was unbalancing. There were still another 500 metres or so to descend to our snow cave. Tony suggested we just pack one rucksack tightly with all the rocks and glissade it down the slope. We could then down-climb more easily and pick it up on the Meru Glacier at the bottom of the face. Simple!

We filled the small rucksack with our rock samples, tied it tight and threw it off. The rucksack slowly gathered speed, then started bouncing down the slope and after a short time hit the only rock on that part of the face and exploded in tatters. The rock samples, all in small plastic bags, skidded off in all directions. We had to spend the next three hours climbing down and traversing the face to retrieve as many as we could. Finally we regained our ice cave and after four days and three nights of shredded nerves had a long sleep in its safety. Once again we woke to slow suffocation of our air supply as powder snow avalanches blocked out the entrance. Next day we staggered back down to Tapovan and the luxury of base camp.

Dating Shivling

The samples we brought back from the Shivling expedition were subjected to the whole array of geochemical and dating methods that were available. Steve Noble at the British Geological Survey geochronology laboratories extracted tiny flawless monazite crystals from the granite samples and ran them through a mass spectrometer. The U-Pb age he obtained was precisely 23 million years old, which we interpreted as the age the magma solidified and the granite crystallized.[18] Tony Hurford and Andy Carter from the fission-track dating laboratory at University College London extracted zircon and apatite crystals from the same samples. The ages of zircons from samples collected above 5,000 metres on Shivling ranged between 14.2 and 8.8 million years, with apatite ages even younger, between 3.5

and 2.6 million years. These new data suggested that these Himalayan rocks had experienced continual burial and metamorphism throughout the time period from continental collision, 50 million years ago, to 23 million years ago, when the rocks became so hot that they melted to form granite. Maybe these data were telling us that the climax of the Himalayan mountain-building phase was around 23-20 million years ago when the peak burial and highest temperatures were recorded by the ages of the granites along the main Himalayan Range. Following the crystallization of the granite the rocks were then exhumed, initially very rapidly from 23 to 14 million years ago, but then more slowly from 14 million years ago until the onset of the Quaternary glaciations about two and a half million years ago, when another phase of intense erosion resulted in rapid final exhumation to give the landscape that we can see today. The results from Shivling proved to be extremely important, but posed another question: was this thermal history unique to Shivling and this part of the Himalaya or was it common along the whole length of the range? The next part of the jigsaw puzzle required finding another mountain to climb in another part of the Himalayan chain in order to test the model.

A Mountain in Tibet: Shisha Pangma

Shisha Pangma (8,013 m) is the fourteenth highest peak in the world, and is the only 8,000-metre peak wholly in Tibet. It is a huge massif located north of the Langtang Valley along the southern margin of the plateau. Despite being one of the easier 8,000-metre peaks, it was the last one to have been climbed. The easier northern route was first climbed by a Chinese team in 1962 and the much steeper south face first climbed by the Brits Doug Scott, Roger Baxter-Jones, and Alex McIntyre in 1982. In 1994 our annual Himalayan Workshop meeting was held in Kathmandu. Later that season, my old climbing friend Jon Tinker was leading an expedition to climb both Shisha Pangma and Cho Oyu (8,188 m), the sixth highest peak in the world, from the northern Tibet side, and he invited me along. We drove from Kathmandu, crossed into Tibet over the Friendship Bridge at Zhangmu, and only two days after leaving Kathmandu arrived at base camp at 5,000 metres on the Tibetan Plateau. Already most of us were getting splitting headaches from the altitude. After two days sorting out gear and acclimatizing, a team of about fifty yaks appeared to transport all our gear the one day's march up to our main base camp site alongside a spectacular row of ice penitents, beneath the north face. Several of the yaks were jittery and a few waited until they were fully loaded before bolting off, scattering our gear across the plateau.

The *drokpa* nomads would then have to run off to herd them back and load them up all over again.

The climb up Shisha Pangma is a long gut-wrenching snow plod, crossing enormous icefields and skirting large areas of crevasses. The route almost followed the line of the South Tibetan Detachment, the low-angle normal fault that I had seen and mapped in Zanskar, and along the Bhagirathi Range in Garhwal (Figures 4.12 and 4.13). We set up three camps up the mountain, the final one being at the base of the summit ridge. I was climbing with Babu Chhiri, a quiet and friendly Nepali high-altitude Sherpa who had already summited Everest three times, and once spent a record 21 hours bivouacking on its summit. He went on to climb Everest ten times before he was tragically killed in the Western Cwm in 2001. Babu Chhiri held the record for the fastest ascent without bottled oxygen (16 hours, 55 minutes), a record that has since been trumped by young Pemba Dorjey (an incredible 8 hours, 10 minutes). Whilst I was puffing and panting my way up the icefield, for Babu it was a stroll in the park.

At the top camp beneath the final ridge we were gearing up at 4 a.m. preparing for another Alpine start to climb to the summit when I noticed that my vision was extremely blurred. I asked Babu to look into my eye and he reported that it was very red. Immediately I thought of the symptoms of high-altitude oedema. The first symptoms are blinding headaches, then loss of eyesight, then water on the lungs (pulmonary oedema) and water on the brain (cerebral oedema), culminating in death. All this can happen extremely quickly, as it did to an old climbing friend and doctor, Peter Thexton on Broad Peak in the Karakoram a few years previously. The only solution is to descend as fast as possible. I went straight into survival mode, threw my sleeping bag into my rucksack and descended as fast as I could. Babu wanted to go with me but instead I asked him to collect my precious altitude samples for me up the summit ridge, noting exactly each site from which he collected them. Back in base camp the following day, the expedition doctor confirmed I had numerous haemorrhages in both my eyes, and he advised me to return to the UK as soon as possible.

The samples that Babu and I collected from Shisha Pangma were once more subjected to thorough dating. Randy Parrish extracted tiny uranium-bearing crystals of xenotime and monazite from the older biotite granite phase and obtained a U-Pb age of 20.2 million years old. Most of the Shisha Pangma granite was composed of a later tourmaline, muscovite, and garnet-bearing granite that was dated at 17.3 million years old, an age we could interpret as the timing of granite crystallization.[19] Kip Hodges dated the muscovites using potassium-argon techniques at 16.74 million years, recording the timing the granites cooled

Figure 4.12 View across the north face of Shisha Pangma (8,013 m). The South Tibetan Detachment separates pale-coloured granitic rocks below from black shales above.

through 350°C, and Tony Hurford and Andy Carter obtained fission-track ages from apatites extracted from our altitude samples up to the summit. These ages, from samples collected between 5,800- and 8,000-metres' altitude, were 14.8 to 12.3 million years old, only slightly younger than the main granite crystallization age. From all these data we could conclude that crustal melting occurred between 20 and 17 million years ago, remarkably similar to the ages we had obtained from Zanskar[20] and Shivling.[18] Now we knew that the timing of metamorphism and melting was very similar along most of the Himalayan chain. We also used our ages to calculate that steep cooling rates (100–180°C/million years) followed crustal melting and rapid exhumation (4 millimetres/year) during the period 17–14 million years ago, that resulted in removal of at least 12–15 kilometres of rock during that time. If high erosion and high exhumation rates correlated with high topography as is likely, then our data suggested that the Himalaya reached their maximum elevations around 20–17 million years ago, very similar to the timing we deduced from Zanskar and Shivling.

Figure 4.13 North face of Shisha Pangma (8,013 m) from the Tibetan Plateau.

Both the Shivling and Shisha Pangma expeditions had been extremely revealing. We had combined exploration mountaineering with geological mapping and vertical sampling and had used a variety of modern analytical techniques using scanning electron microscopes and mass spectrometers to constrain the age and depth of burial of the rocks and their subsequent exhumation from deep in the crust to their present icy heights along the crest of the Himalaya. We had used almost every branch of geology, sedimentary, igneous and metamorphic, and mapped at every scale from the kilometre-sized granite cliffs of Shivling to the micron-scale chemical compositions of minerals to determine their pressure, temperature, depth, and time of formation.

CHAPTER 5

Frozen Rivers and Fault Lines

Beacon Highway sign on Khardung-la: 'Here you can have dialogue with God.'
Sign on a Jammu and Kashmir bus to Leh: 'Oh God save us and the bus.'

Winter along the Zanskar Gorge

After seven summer field seasons working in the north-western Himalaya in India, I had heard of a winter trade route that must rank as one of the most outlandish journeys in the Himalaya. The largely Buddhist Kingdoms of Ladakh and Zanskar are high, arid, mountainous lands to the north of the Greater Himalayan Range and in the rain shadow of the summer monsoon. Whereas the southern slopes of the Himalaya range from dense sub-tropical jungles and bamboo forests to rhododendron woods and magnificent alpine pastures carpeted in spring flowers, the barren icy lands to the north are the realm of the snow leopard, the yak, and the golden eagles and lammergeier vultures that soar overhead. The Zanskar Valley lies immediately north-east of the 6–7,000-metre-high peaks of the Himalayan crest and has about thirty permanent settlements, including about ten Buddhist monasteries. I had seen the Zanskar Ranges from the summit of White Sail in Kulu and later spent four summer seasons mapping the geology along the main trekking routes. In summer, trekking routes cross the Himalaya westwards to Kashmir, southwards to Himachal Pradesh, and northwards to Leh, the ancient capital of Ladakh. Winter snows close the Zanskar Valley from the outside world for up to six months a year when temperatures plummet to minus 38°C.

Central Zanskar is a large blank on the map, virtually inaccessible, with steep-sided jagged limestone mountains and deep canyons (Figure 5.1). The Zanskar River carves a fantastic gorge through this mountain range and for only a few weeks in the middle of winter the river freezes. The *Chaddur*, the walk along the frozen Zanskar River, takes about ten to twelve days from Zanskar to the Indus Valley and, in winter time, was the only way in or out before the road to Kargil was constructed. I mentioned this winter trek to Ben Stephenson during our summer fieldwork in Kishtwar and he stopped suddenly, turned around, and said 'Mike we just have to do this trek!' So the idea of a winter journey into Zanskar was born, and four of us set off from Oxford in January 1995. Ben Stephenson, Joe McCarron, Ollie Hassall and I flew into Leh and met up with my old friend Fida Hussein (Figures 5.2 and 5.3). All the hotels and tea-shops closed during the winter as no tourists came, so Fida invited us to stay in his house in the old part of town beneath

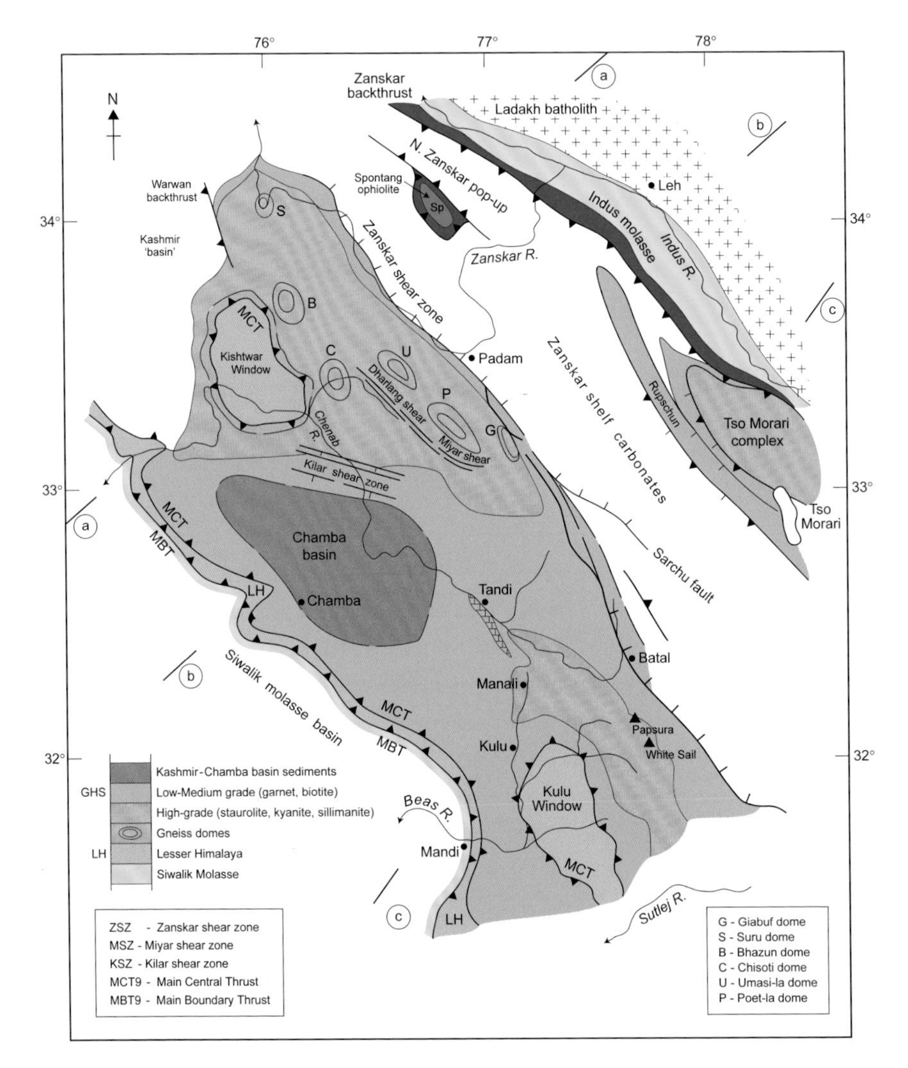

Figure 5.1 Geological map of the central Zanskar-Ladakh Himalaya in India. The Spontang ophiolite (green) lies above the Zanskar shelf carbonates, the northern margin of the Indian plate with the metamorphic rocks of the Greater Himalaya to the south-west. The Zanskar River cuts through a series of gorges from the Zanskar Valley north to the Indus River.

Figure 5.2 The Zanskar-Kashmir Himalaya in winter.

Figure 5.3 Lamayuru gompa in winter.

the crumbling castle. Fida also introduced us to six wonderful Zanskaris from the village of Pishu at the beginning of the gorge—Lobsang, Tashi, Tondup, Motup, Dorjey, and Tanzin—who would act as our porters.

Our plan was to walk the frozen river into Zanskar and then Nordic ski out along the Zanskar Valley, cross the Pensi-la (4,400 m), then descend the Suru Valley to Kargil, on the border with Kashmir. We would have skinny skis and carry pulks, or Norwegian sledges, loaded with all the food and gear. The Pishu wallahs knew of a few caves along the *Chaddur* where we could sleep, but we also brought two tents for camping directly on the ice. The idea was that we would also try to map the geology along the way, construct a balanced cross-section through the dramatic folds and thrusts of the northern Himalaya, and work out the amount of crustal shortening. Fida told us that the fast-flowing Zanskar River usually only remains frozen for a few weeks so it was vital to leave Leh when temperatures could be relied on to stay below freezing. If the ice broke up when we were in the middle of the gorge we would be in big trouble. Although we could get out of the river valley in several places, there was no way out over the cliffs to Leh or to Padam (Figure 5.4).

Figure 5.4 The confluence of the Zanskar and Indus Rivers in winter. The Zanskar River is blue with icebergs and the Indus is a muddy brown.

We set off in mid-January carrying 25–30 kg rucksacks when temperatures were at their lowest and the river as frozen as it was likely to get. At Chilling, the start of the gorge, the ice was between 10 and 30 cm thick and plastered along the sides of the raging torrent of the river. In places it was frozen all the way across and like a skating rink; in others, where the river was fast-flowing, only ragged bits of ice were hanging precariously to the vertical sides of the rocky canyon. We climbed gingerly around the edges hoping that the ice would hold. If the ice collapsed, it would have pitched us helplessly into the river. Our waterlogged clothes and heavy rucksacks meant that the river current would have swept us under the ice and survival would have been impossible (Figures 5.5–5.9).

Figure 5.5 Walking on thin ice along the margin of the Chaddur, the frozen Zanskar River. Photo courtesy of Ben Stephenson.

Figure 5.6 Folded limestones of the north Indian continental margin, along the frozen Zanskar River in winter.

Figure 5.7 Sheet ice in the middle part of the Zanskar gorge, winter Chaddur.

Figure 5.8 Winter conditions along the frozen Zanskar gorge. Photo courtesy of Ben Stephenson.

Figure 5.9 The Zanskar River flows through narrow deep canyons dissecting the limestone mountains of Ladakh.

Lobsang and our Pishu wallahs knew the best route although the conditions changed every day and every winter with the snowfall and state of the river. It was unnerving walking on the ice. You could look straight through to the flowing waters beneath, and occasionally the whole ice slab would crack or sink a little and water would flood across the slippery surface. Lobsang and his men would sometimes take their boots off to wade these shallow water sections, but their feet were made of solid leather compared to ours. Sometimes we had to climb short crags to negotiate particularly bad stretches of the river where the ice was not strong enough to hold our weight. Temperatures during the day in the bright sun hovered close to zero but during the nights would plummet to minus 20°C; the lowest we recorded was minus 30°C. Life after dark revolved around the juniper and yak dung fire, shivering, sipping Tibetan salt tea, and longing for the mid-morning sun to finally hit us. Winter drives the wildlife down to the valleys and we frequently saw magnificent ibex silhouetted against the high precipices, herds of *bharal* (blue sheep), and always the recent tracks of the elusive snow leopard (Figure 5.10).

Figure 5.10 The iconic symbol of the high Asian mountains, the snow leopard in the trans-Himalayan mountains. Though rarely seen, the snow leopard's existence is severely threatened throughout its range. Photographed with 'self-triggered' camera by David Willis.

After eight days' shuffling along the icy river, clambering over frozen waterfalls and skirting around cliffs walking on thin ice, we finally emerged at Hanumil, a tiny settlement of three houses. From here we could put on our Nordic skis and plough through the increasingly thick snow as we approached the high peaks of the main Himalayan Range to the south. At our porters' home village of Pishu, we must have looked like the pied piper's raggle-taggle army of scruffy urchins arriving in a triumphal procession. Much to the delight of the local children we skied into the Kingdom of Zangla, where the bemused inhabitants gave us a fabulous reception, plying us with the infamous yak butter tea and *tsampa* (roasted barley flour). As we approached Zangla we had a particularly raucous welcome from some *annis* (nuns) lined up on the roof of the *gompa* in their crimson cloaks. The *Giapo*, the King of Zangla, again invited us into his ricketty castle home for more *chang* and stories; we were becoming regular visitors to his wonderful remote Himalayan kingdom (Figure 5.11). The final day's skiing across the flood plains from Karsha *gompa* to Padam, the main village in Zanskar, was particularly memorable—clear blue skies, deep soft powder snow, and the whole Himalaya spread out before us (Figure 5.12). We had managed to map the section along the Zanskar River Gorge and fill in that bit of the blank on the map.[1]

Figure 5.11 Ben Stephenson, Oliver Hassel, and Mike Searle with the *Giapo*, the King of Zangla. Photo courtesy of Joe McCarron.

Arriving in Padam it was wonderful to be greeted by old friends from earlier summer journeys to Zanskar and one of them, Ghulam Murtaza, invited us to stay in his house with characteristic Zanskari generosity. Despite living in some of the harshest conditions on Earth, the Zanskaris and Ladakhis are the happiest, most contented, and friendly people I have ever known. During winter months the snows are frequently up to at least one storey high so the houses have to be continually dug out and the locals must dig narrow paths through the snowdrifts just to get next door. The children had a unique form of play: they would don local 'skis'—bits of sawn off rubber tyres strapped beneath their boots, grab the tail of the nearest yak, whack the yak on the backside, and off they would go. This new sport of 'yak skiing' was reasonably effective, but only as long as the yak was cooperative. Ben Stephenson decided that what the Zanskaris really needed was simply to learn to ski to get around in winter. The following summer he raised money and collected some hundred pairs of skis from all around Britain, persuaded Air India to transport them to Ladakh, and went back to Zanskar. Ben even raised enough money to buy a small Ladakhi house above Padam and, together with our good friends Amin Zanskari and Urgen Dorjey, he started the 'Zanskar Ski School' teaching the locals to ski.[2] The local children soon became adept at skiing and the winters around Padam became alive with skiers, laughter, and fun.

Figure 5.12 Arctic conditions skiing across to Padam in the Zanskar Valley in winter.

Extruding Tibet

In the mid 1970s the earliest black-and-white photos of the Earth's surface taken from the LANDSAT satellites became freely available to the scientific community. This new tool revolutionized the interpretation of crustal structure all over the world. Two prominent geophysicists interested in Tibet, Peter Molnar and Paul Tapponnier from the Massachusetts Institute of Technology in the US, started stitching together a mosaic of satellite photos of the plateau. The most obvious features to stand out were a series of giant fault lines that stretch across the entire plateau for hundreds of kilometres. Along the northern margin of the plateau, the Altyn Tagh fault runs for over 1,800 kilometres from the eastern Pamirs to the Gobi Desert. Other large-scale faults stretch along the Kun Lun, the Xianshui-he and Jiale mountain ranges of the eastern plateau (Figure 5.13). Along the south-western margin of the plateau the most prominent fault of all, the Karakoram fault runs for over 800 kilometres from the eastern Pamir south-east to the Mount Kailas region of south-west Tibet (Figure 5.14).

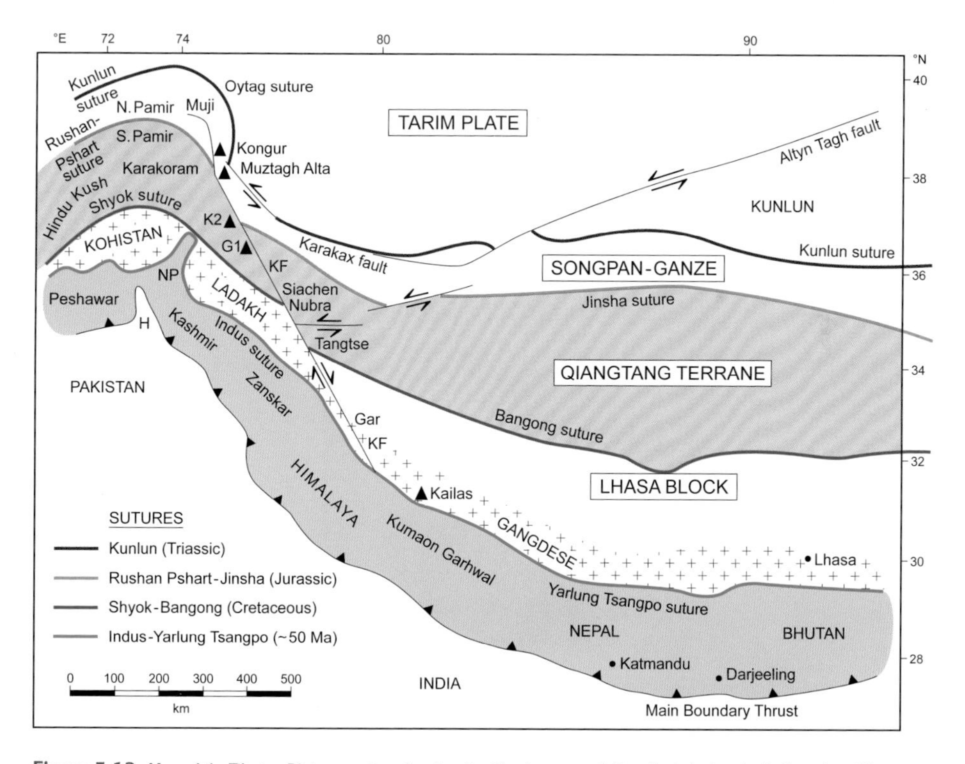

Figure 5.13 Map of the Tibetan Plateau region showing the Karakoram and Altyn Tagh faults, the Indian plate Himalaya (green) and the Asia plate in south Tibet (Lhasa block and Ladakh terrane) in white and central-north Tibet and the Karakoram terrane in blue.

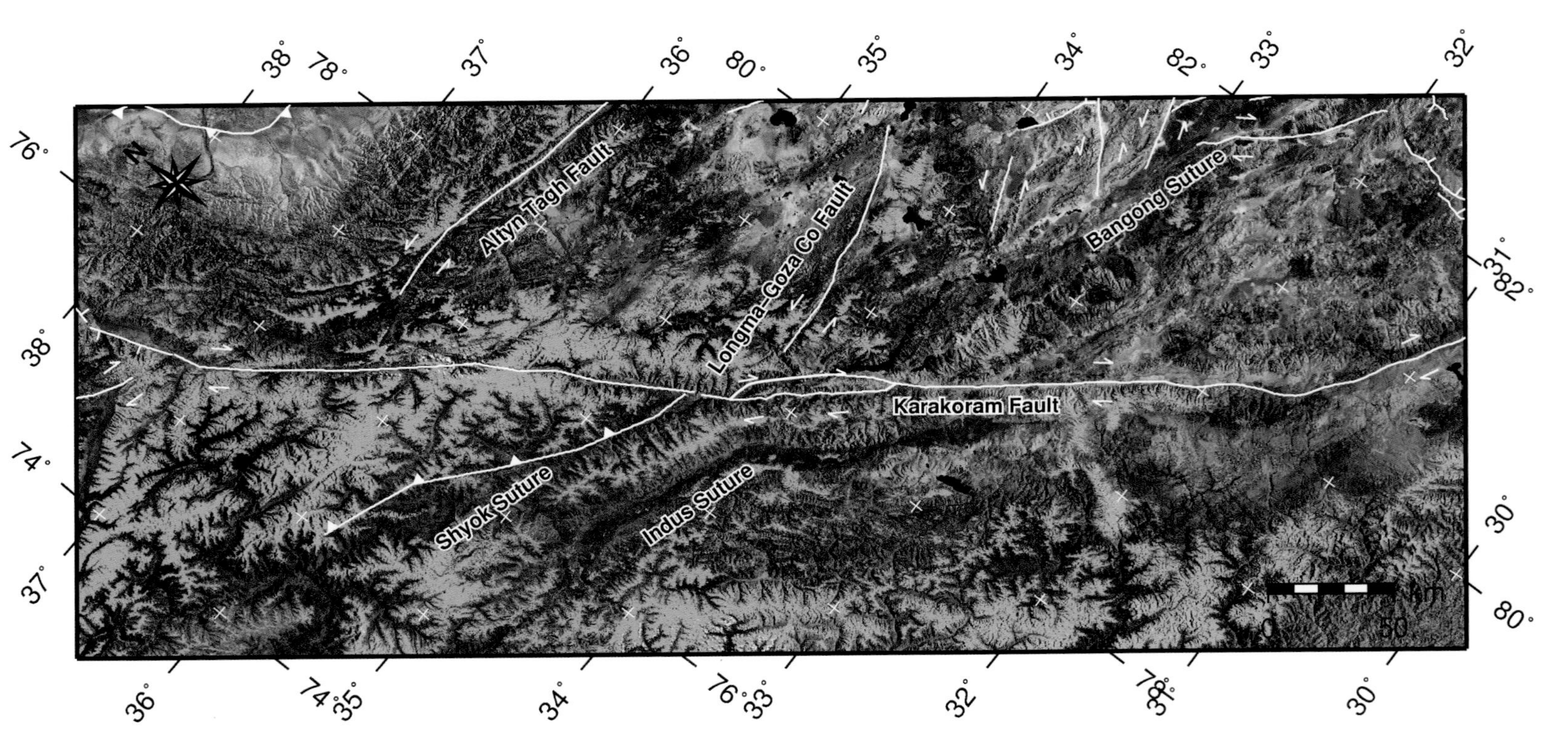

Figure 5.14 Landsat satellite photo showing the central part of the Karakoram fault and geological offsets along it. Note the drowned river valley of the Pangong Lake dammed by uplift during uplift of the Pangong Range.

All these fault lines are strike-slip faults, in which the Earth's crust has moved laterally, and the fault slip has resulted in major earthquakes. Peter Molnar and Paul Tapponnier studied these LANDSAT photos in great detail and in the mid-1970s came up with a major new tectonic theory, 'continental extrusion'.[3] They suggested that, as India continued to penetrate northwards into Asia, the thickened crust of the Tibetan Plateau region was pushed eastwards out of the way of the Indian indentor. The faults along the northern part of Tibet, the Altyn Tagh and Kun Lun, showed left-lateral or sinistral motion (Tibet moving east relative to the Tarim–Gobi crust to the north). The faults along the southern margins of Tibet were all dextral (Tibet moving east relative to the Karakoram and Himalaya to the south). This showed that the plateau region was extruding eastwards, squeezed between the pincer-like faults of the Karakoram and Altyn Tagh.

Having mapped out these spectacular fault lines using satellite photographs, the next step was to determine how much offset had occurred and when the faults were initiated. These were easy questions to pose but more difficult to solve. Most of these areas are in the remotest parts of Tibet, far from roads or civilization. Molnar and Tapponnier proposed that the faults had very large offsets, with up to 1,000 kilometres along the Karakoram fault. Such substantial offsets supported the theory of large-scale continental extrusion. The offset markers they used to come to this conclusion were somewhat dubious, because it was not possible to actually go and map these remote areas of Tibet. The answers lie in the field, in this case the frozen highlands of Tibet. Clearly it was vital to map the fault in detail and determine what features could be used as offset markers. Looking at the maps of Tibet, I noticed that the Karakoram fault lay mainly in the far western province of Xinjiang and the south-western parts of Tibet, now under Chinese control. The middle section of the fault, however, along the Siachen Glacier and Nubra Valley, south-east to the Pangong Lake region, just clips the far northern part of Ladakh in India, and is well exposed in these valleys. In 1995, after our winter trip along the Zanskar River, I decided to take a look at the Nubra Valley.

The Nubra Valley in Winter

Back in Leh after our winter trip along the frozen Zanskar River, Fida told me that the Indian government had just opened the Nubra Valley and Pangong Lake areas in the Ladakhi Karakoram to the north for tourists. To my surprise Fida said that we could rent a jeep and drive to the area there and then, in February, in the middle of winter. Although the temperature was well below zero the precipitation was low, the region being in the rain shadow of the Himalaya. The road across the Khardung-la Pass at 5,600-metres' altitude was frozen and icy but there was only about half a metre of snow on the high peaks of the Ladakh Range (Figure 5.15). Fida and I drove over to Nubra and stayed with one of his relatives in the monastic village of Diskit. East of here the army were busy stockpiling supplies for the summer skirmishes with Pakistan along the Siachen Glacier and the de facto border, the 'Line of Control'. We managed to cross the Shyok River and drive north along the Nubra Valley as far as Panamik, a small village with a series of hot springs coming up right along the fault itself. A few bedraggled dromedary camels grazed in the scrubby juniper bushes. The line of the Karakoram fault was exposed along the side of the Nubra Range as traces of fault gouge and a line of hot springs, but where the fault crossed higher mountains I found some spectacular outcrops of mylonites, the highly sheared rocks that are typical of major faults.

The following year, 1996, I returned to Ladakh and spent four weeks walking all over the Karakoram fault line in the remote eastern Karakoram Ranges (Figure 5.16). First I went over to the Pangong Lake area along the border with Tibet and Xinjiang. A newly constructed dirt road crossed the Ladakh Range over the Chang-la to the village of Darbuk. Along the north-eastern margin of the Ladakh Range, the trace of the fault was clear. Large rivers abruptly turned through 90°, their courses deflected by motion along the fault. At the village of Tangtse I found the most spectacular outcrops of mylonites exposed right beneath the village *gompa*. Here,

Figure 5.15 The Zanskar Range and Indus Valley from the Khardung-la.

Figure 5.16 Aerial view of the Zanskar Range in winter. The two large mountains in the distance are Nun (7,135 m) and Kun (7,086 m). Folded rocks in the foreground are the Palaeocene limestones around the Spontang ophiolite.

high metamorphic grade gneisses and marbles were sheared along the strike of the fault along with some intrusive leucogranite sills. In a few places I found some later leucogranite dykes cutting across the mylonite fabrics, but they were subsequently cut through again by the bounding brittle fault.[4] This was a magnificent example of a strike-slip fault at deep levels in the crust—exposure levels that are very rarely attained at the surface for geologists to study anywhere around the world. If we could date these early concordant leucogranites and the younger cross-cutting dykes using U-Pb isotopes on zircon or monazite crystals, we could place tight time constraints for initiation and motion along the Karakoram fault. It was an ideal site to test the great model of continental extrusion.

Now that both the Nubra Valley and Pangong Lake areas were opened to foreigners I thought it would be a good idea to get a D.Phil student to work on this Karakoram fault problem for a future field season. I persuaded my friend Randall Parrish, who was now head of the Isotope Geoscience unit at the British Geological Survey in Nottingham, and his colleague Steve Noble, to supervise the dating programme. Back in Oxford we ended up taking two D. Phil. students—Richard Phillips from Birkbeck College, London University, and Andy Thow from Edinburgh. Richard was given the task of mapping out the Karakoram fault in

Ladakh, and Andy was given the granites along the Baltoro Glacier in Pakistan. I had already mapped large tracts of the Baltoro Karakoram, but we really needed more tightly constrained U-Pb age data on the granites and metamorphic rocks.

Karakoram Fault

In 1996 the northern borders of India were still poorly defined. As soon as the Nubra and Pangong Valleys were first opened to outsiders I made a reconnaissance of the entire region by jeep, horse, and foot. Fida and I drove along the southern shore of Pangong Lake and into Chinese territory. In those days there was nothing to mark the border and it was only when we came across a lone Tibetan horseman that we realized we had inadvertently crossed into China. Two rivers carved deep gorges right across the strike of the two faults that splayed out along this sector. One was the Darbuk Gorge connecting north to the Shyok River and the other the Tangste Gorge connecting east to the great drowned river valley of Pangong Tso (*Tso* means lake in Tibetan). The dirt track descended from the Chang-la down to the Tangste Valley and then wound along to the tiny hamlet of Muglib and thence eastwards to the ice-blue waters of Pangong Tso. The lake had been formed when a series of large earthquakes had dammed the Pangong River. In one small sector successive earthquakes had resulted in an uplifted segment of the valley and forced the river to flow in the opposite direction for a small distance. The Pangong Tso was dammed and then filled up, since it had no outlet. Spectacular outcrops of gneisses, mylonites, and a series of leucogranite dykes were beautifully exposed along the valley walls. In the Nubra Valley we found evidence that frictional heating along the fault during past earthquakes had actually melted the rocks, resulting in some amazing examples of *pseudotachylytes*, glassy black rocks with tiny melt veins streaking off into the surrounding country rocks. To our great surprise, however, looking back at the earthquake records, we found there were very few earthquakes along the fault. How could this most prominent geomorphological feature that apparently offset major rivers and cut across glaciers not be actively moving?

During 1999 and 2000 Richard Phillips spent the summers mapping the structures along the Karakoram fault in the Nubra and Tangste Valleys in intricate detail, and the winters dating the rocks in Randy Parrish and Steve Noble's laboratory in Nottingham. The results were amazing. We found that the mylonites were formed in a very narrow time span between the two sets of leucogranite intrusions, so were able to deduce that ductile shearing at least along this section of the fault, had occurred between 15.7 and 13.7 million years ago.[5] Using several geological markers and the offset course of the Indus River we then showed that motion along the fault could not have been more than 120 kilometres, and could have been as little as 35 kilometres. These results showed that the Karakoram fault, the largest and most prominent strike-slip fault in Tibet, could not have been responsible for 1,000 kilometres of eastward extrusion of the Tibetan Plateau, as the continental extrusion model required. As Thomas Huxley so elegantly put it: 'There is nothing so ugly in science as the slaying of a beautiful theory by an ugly fact.'

The political triple junction where the borders of India, Pakistan, and China all meet at a point is somewhere on the remote icefields of the upper Siachen Glacier. All the borders are strongly disputed and there was absolutely no chance of us getting there from any of the three countries. Across the border in China the Karakoram fault runs along the Shaksgam Valley in the south-westermost part of Xinjiang Province. Along this sector a line of 7–8,000 metres of high mountains parallels the fault from K2 all the way along the Broad Peak and Gasherbrum Ranges to Baltoro Kangri. K2 lies immediately south-west of the fault in Pakistan. We could use my earlier mapping along the Baltoro Glacier and around K2,[6] together with the dating results from the K2 region across the border to interpret the exhumation and age relationships along the adjoining parts of the Karakoram fault.[7]

Meanwhile, Paul Tapponnier, the great exponent of continental extrusion, and his colleagues had started a project of his own along the Karakoram fault, but in the south-western Tibet sector, south-east of where we were working in Ladakh. They made similar detailed structural maps, and dated various granitic rocks along the fault. The relationships appeared to be quite similar to the structures we were mapping in Ladakh, but their interpretations were vastly different. The French group assumed that all metamorphic rocks and granites exhumed along the fault must have been formed by shear heating and therefore the ages from the granites would date movement along the fault. They claimed that their work showed that large-scale offsets and high slip rates were apparent, supporting the original model of continental extrusion of Tibet.[8] Our group, on the other hand, claimed that the metamorphism along the fault was much older than shearing, and we could tightly constrain the ductile shearing along the fault using Richard Phillips' U-Pb ages from the earlier deformed and the later cross-cutting granite dykes. Our U-Pb ages from metamorphic rocks collected right on the fault in the Pangong Range turned out to be much older—108 million years old[9]—proving that the metamorphism and shearing were unrelated.

The two camps on the continental extrusion question staked out their distinctive points of view. In one corner were the proponents of large-scale extrusion of Tibet. These geologists wanted large offsets, hundreds if not a thousand kilometres of motion, rapid slip rates, synchronous metamorphism and melting caused by shear heating, and faults extending right through the crust into the upper mantle. In the other corner were the sceptics who wanted only upper crustal faults showing smaller offsets (100 kilometres or less), restricted timing of shearing, and no connection between older metamorphism and shearing. It was an argument that would rage on and on, not only with the Karakoram fault but with many of the other great strike-slip faults of Asia, with neither side willing to concede.[10]

Return to the Baltoro

Although the Pakistan side of the Karakoram had now been mapped geologically we still had little idea of the age of the whole range of metamorphic rocks and the great granite batholith that many of the high peaks were made of. It was important to get this age data so that we could compare the timing of crustal thickening, metamorphism, and melting along the Asian side of the collision. This would form the basis of Andy Thow's D.Phil project, so to start we returned to the Baltoro Glacier area in 2001 to collect all the dating samples needed. My old friend from Hushe village, Anwar Ali, came with us as guide and we hired several of his village compatriots as porters. By now the 'road' had been constructed all the way up to Askole village, although it was always sliding down into the raging torrent of the Braldu River. Some sections along parts of the Braldu Gorge were the most dangerous tracks I have ever travelled on. In one place below Askole, the track traversed a 40 degree slope of unconsolidated scree under cliffs that overhung the thundering waters of the Braldu. We camped in the apricot orchards of Askole and next day set off once again for the Baltoro Glacier. Unlike the mass army of nearly a hundred Balti porters we had in 1985, this time without all the climbing paraphernalia, we managed with only about ten porters.

Trekking along the northern ablation valley of the main Baltoro Glacier we explored several side glaciers leading up to the incredible rock spires of the Uli Biaho, Shipton Spire, and numerous others. This was magnificent mountain scenery and mapping out the geological contacts at first glance appeared easy. The two- or three-kilometre-high cliff faces along the flanks of the glaciers were like a slice cut through the crust with the geology laid bare. We traversed the southern flanks of the Cathedral and stopped off at our old 1985 Biale Base Campsite at the junction of the two glaciers. After this we were forced back onto the main glacier to get to Concordia—the site where four major glaciers converged. Concordia is one of the most impressive places anywhere on the planet. To the north are the huge southern ramparts of K2 with the great icy peaks of Skyang Kangri and watershed, with the Shaksgam Valley in China beyond. Swinging around towards the east, enormous icefalls pour down the Godwin Austen Glacier dividing K2 from the massive twin peaks of Broad Peak. Directly above Concordia is the beautiful, shapely peak Gasherbrum IV, then further on a range of 8,000-metre-high giants, Gasherbrum II, Hidden Peak, and the wall of ice and rock that divides the upper Baltoro Glacier on the Pakistan side from the upper Siachen Glacier on the Indian side. At the head of the Gasherbrum Glacier rises another massive giant mountain massif, Chogolisa, and then the whole array of mountains around Biachedi and Masherbrum bordering the southern flanks of the upper Baltoro Glacier. A single shapely steep icy arrow of a peak, Mitre Peak, rises directly above Concordia, a lonely Matterhorn surrounded by much larger giants. This truly was the 'Throne Room of the Mountain Gods' from ancient mythology.

Andy and I collected nearly a hundred samples across the metamorphic complex and through the granite batholith. We made a rapid trek up to K2 and Broad Peak Base Camps, following the northern intrusive vertical contact of the granite along the mountains to the south. On the Landsat satellite photographs this contact was very clear: pale-coloured granites to the south intruding black shales to the north. On Mitre Peak the contact was beautifully exposed together with the area which had been metamorphosed by the heat of the intrusion (called the thermal metamorphic aureole). This proved that the granites had intruded into the sedimentary rocks, and were not enclosed within the melt zone like the Himalayan granites. It was this same clear vertical contact that we could trace eastwards across the inaccessible cease-fire line to the upper Siachen Glacier on the Indian side of the watershed. To our amazement, the offsets matching this contact in the Pakistan Baltoro to the same contact on the Indian Siachen side across the Karakoram fault indicated that the total geological offset across the fault could be as little as 35–40 kilometres. Here was the largest, apparently most active fault in Tibet, supposedly responsible for hundreds or even a thousand kilometres of offset, yet the rocks were telling us that actually the offset was quite small. Once more it seemed that the extrusion model for Tibet was invalid as soon you looked at the rocks (Figures 5.17 and 5.18).

Trekkers going up the Baltoro Glacier to K2 frequently make a diversion north-east to the Gasherbrum–Hidden Peak region along the furthest border of Pakistan with the Chinese province of Xinjiang. The mountains here get even larger and more impressive, with wide, heavily crevassed glaciers cascading down from the 8,000-metre-high peaks of the Gasherbrum Range and Hidden Peak. Beyond this impenetrable mountain barrier lie the desolate wastes of the Shaksgam Valley and the Rimo Range. Rather than retrace our route along the Baltoro Glacier we decided to cross the Gondogoro-la, a 5,585-metre-high pass cutting across the Chogolisa massif (Figure 5.19). It was a tough trek with a final climb up a steep 800-metre-high icy cliff section. The views from the Gondogoro-la were spectacular, with the

Figure 5.17 Space Shuttle photo looking obliquely south-east at the western Himalaya and western Tibet (courtesy of NASA). The long prominent crack is the Karakoram fault which cuts obliquely across the Pamir and Karakoram Ranges. Tibet (left) is moving east relative to the Himalaya. The drowned river valley of Pangong Lake can be seen.

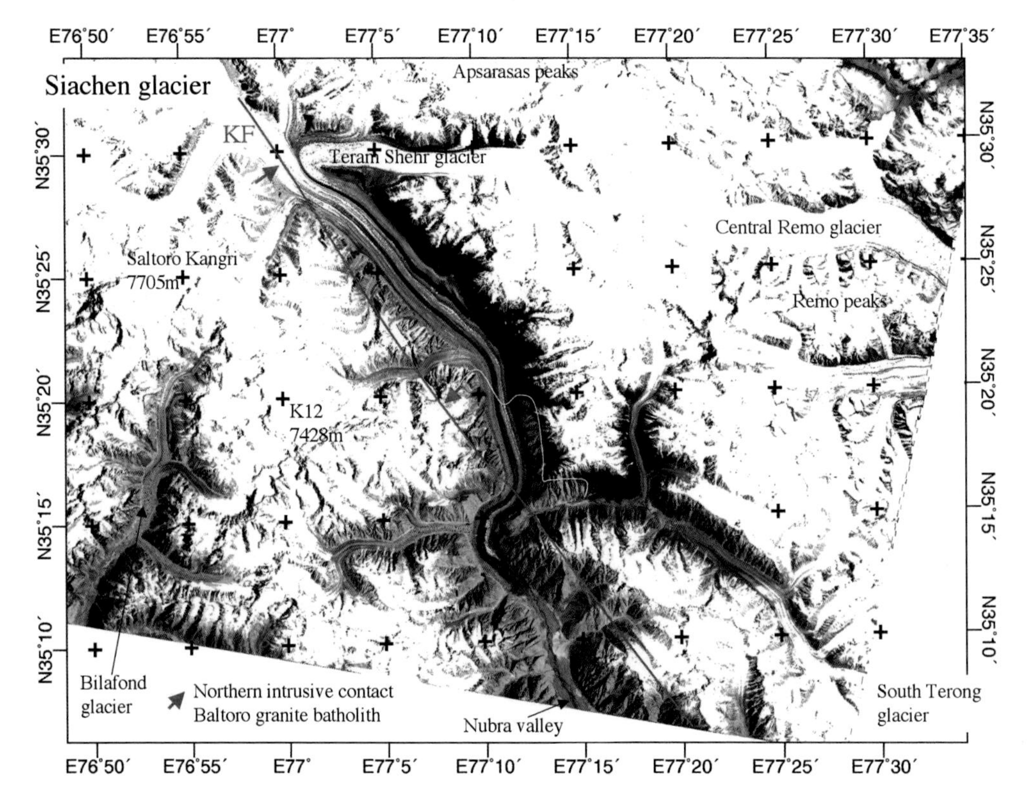

Figure 5.18 NASA space shuttle image of the Siachen Glacier region in the eastern Karakoram showing the offset of the Baltoro granite batholith. Red arrows mark the offset northern contact of the batholith showing that only about 20 km of offset has occurred in total along the fault.

Figure 5.19 Views of K2, Broad Peak, and the Gasherbrum Range from the Gondogoro-la pass.

giants of K2, Broad Peak, and all the Gasherbrums standing head and shoulders above a sea of mere 6,000- and 7,000-metre-high peaks. The Gondogoro-la trek also traversed the roof zone of the Baltoro granites, making it possible to map out the three-dimensional geometry of the granite batholith. Descending the southern side of the Gondoro-la we returned to the familiar peaks around Layla and the Trinity group and finally back to Anwar's home village of Hushe.

It was wonderful to enjoy the luxury of stretching out on a grassy meadow and drinking clear water from a bubbling stream after the rigours of living on the Baltoro ice. I wanted to make one side trip up the South Masherbrum Glacier to map out the source region of the younger phase of intrusive granites. I had seen Masherbrum from every angle and had attempted and failed to climb the long east ridge in 1985. It was certainly one of the most shapely and beautiful of all the Karakoram peaks and also one of the most difficult to climb. The upper tiers of Masherbrum showed pale-pinkish-coloured granite with streaks of black gneisses enclosed right up near the summit. Here along the southern flanks of the mountain the deeper structural levels revealed some extraordinary rocks: a massive thickness of layered leucogranite sills intruding almost horizontally through the older, dark gneisses. In some respects the structures appeared similar to the deepest levels of the smaller Himalayan granites. These horizontal sill complexes did not appear in any textbooks or papers I had read on granites, so I was intrigued to map out in as much detail as I could the structures exposed around Masherbrum. Granites were usually thought to be vertical intrusions yet here the structures clearly showed that the granitic melts were transported laterally for considerable distances. Only occasionally could you find vertical dykes feeding magma upwards through the section. Indeed, the upper three-kilometre thickness exposed on the upper level of the mountain showed that the granite melts were intruded almost horizontally along even thicker sills. The gneissic rocks recorded pressures up to 10 kilobars, equivalent to about 35 or 40 kilometres' depth, so this amount of overburden must have been eroded off the top of Masherbrum in order to expose these rocks at the surface.

We could now use my mapping of the Baltoro granites, combined with the pressure–temperature determinations and the U-Pb dating carried out by Andy Thow and Randy Parrish, to interpret the structural evolution of the Karakoram. The ages of monazite crystals extracted from the gneisses showed that high temperatures and high pressures resulting in metamorphism had been peaking sporadically for the last 60 million years along the southern Karakoram. The fact that such high-grade metamorphism could last such a long time was a major new discovery. The age dating also showed that most of the Baltoro granites had crystallized between 20 and 15 million years ago, remarkably similar to the ages of many of the Himalayan granites on the Indian side of the collision. The youngest granites in the Baltoro, those forming the Trango Towers, Nameless Spire, and Biale Peak, were intruded as recently as 13 million years ago. Since all the Baltoro granites were undeformed and cut by the Karakoram fault, clearly the fault had to have initiated later, after 13 million years ago.[11]

Our work along the Karakoram fault in Ladakh and the Baltoro region of the Pakistan Karakoram had turned up some very exciting new data. Despite being such a large and impressive fault our data proved that the Karakoram fault shows only very limited geological offsets and was initiated long after the India–Asia collision. Our pressure–temperature data from across the Karakoram showed that the crustal thickening processes began prior to the India–Asia collision and continued on for a long time, as long as 65 million years. The ages from across the Baltoro granite batholith also showed that the huge amount of crustal melting following the thickening process culminated in intrusion of the granites between 20 and 13 million years ago. These data appeared to show that crustal thickening, regional metamorphism, and melting to form granites propagated both south to the Greater Himalaya and north to the Baltoro Karakoram after the closure of the Tethys Ocean and collision of the two great continental plates. The plan now was to look further west in the Hindu Kush mountains along the Pakistan–Afghanistan border where these impressive mountains were made up of similar metamorphic rocks and granites. Were these rocks formed as a result of the collision of India and Asia or were they older, related to earlier plate collisions in central Asia?

CHAPTER 6

North-West Frontier

Kohistan, Hindu Kush, Pamirs

The Arabs called the Pamirs the Bam-i-Dunya or 'Roof of the World', a fitting name for a region where the valleys are not cosy little clefts but open steppes as cold and windswept and almost as high as the peaks that bound them.

John Keay, *When Men and Mountains Meet*

The Hindu Kush Mountains run along the Afghan border with the North-West Frontier Province of Pakistan. Following the First Anglo-Afghan war of 1839–42 the British government in Simla decided that the North-West Frontier of British India had to have an accurate delineation. Sir Mortimer Durand mapped the border between what is now Pakistan and Afghanistan in 1893 and this frontier is known as the Durand Line. Unfortunately it is a political frontier and one that splits the Pathan or Pushtun-speaking lands into two, with the North-West Frontier Province and Waziristan in Pakistan to the east and the Afghan provinces of Kunar, Nangahar, Khost, Paktiya, and Kandahar to the west. The border regions north of Baluchistan in Quetta and Waziristan are strong tribal areas and ones that have never come under the direct rule of the Pakistani government. Warlords run their drug and arms businesses from well-fortified mud-walled hilltop fortresses. During the period that Lord Curzon was Viceroy of India from 1899 to 1905 the entire border regions of British India were mapped out along the Karakoram, Kashmir, Ladakh, and south Tibetan Ranges. During Partition, in 1947, once again an artificial border was established separating mostly Muslim Pakistan from India. Lord Mountbatten, the last Viceroy, gave Sir Cyril Radcliffe the invidious task of delineating the border in haste to avoid a civil war that would surely have come, and on 17 August

1947 Pakistan inherited all the territory between the Durand Line and the new Indian frontier, the Radcliffe Line. In the north, the disputed Kashmir region still remained unresolved and the northern boundary of Pakistan ran north to the main watershed along the Hindu Kush, Hindu Raj, and Karakoram Ranges. To the west, Afghanistan was a completely artificial country created by the amalgamation of the Pathans of the east, Hazaras of the central region, the Uzbeks in the Mazar-i-Sharif area, and the Tadjiks of the Panjshir Valley along the border with Pakistan's North-West Frontier Province. The British lost three wars trying to invade this mountainous land between 1839 and 1919, and the Soviet Union which occupied Afghanistan for ten years from 1979 also withdrew across the Oxus River in failure in February 1989. History and geography clearly tell us that Afghanistan is an unconquerable land.

Following the withdrawal of the occupying Soviet forces, Afghanistan spiralled out of control with the Western-backed *mujahadeen* fighting on all fronts. After an epic battle at Jalalabad in 1989, the Communist forces under Dr. Najibullah finally defeated the *mujahedeen*. The Najibullah government was supported by the old Soviet Union, but when the USSR collapsed in 1992 so did the Afghan government. The *mujahedeen* ransacked Kabul, hung Najibullah, and attempted to form a government. A bloody civil war followed as Afghanistan once again disintegrated along tribal lines into warlord-controlled pockets, the Uzbeks under the control of General Dostum, the Tadjiks in the north-east under Ahmad Shah Masood, and the Pushtun tribes along the south-east and south. On 4 February 1998, at the height of the winter snows, a massive earthquake hit north-east Afghanistan killing at least 4,000 people and making a further 15,000 people homeless. Sixteen days later a second major quake shook the same region, flattening the few remaining buildings. A third earthquake hit on 30 May killing another 5,000 people. Undeterred, the Taliban, the *mujahedeen*, the Uzbeks, Tadjiks, and Hazaras all continued fighting in the northern provinces.

The Western-backed *mujahadeen* were a thoroughly incompetent, corrupt, and ruthless government and carried out many atrocities. The Taliban movement arose in direct opposition to the *mujahadeen* and started in Kandahar under the leadership of Mullah Omar.[1] The Taliban represent a fundamentalist branch of Islam resembling the puritanical Wahabi sect of central south Arabia. By 1996 the Taliban controlled the entire country except for the Uzbek stronghold of Mazar-i-sharif in the north, the Tadjik stronghold of Badakhshan, and the Panjshir Valley along the Pakistani border under the control of a charismatic leader, Ahmed Shah Masood. Pakistan was the only government to recognize the Taliban, and the frontier city of Peshawar was overwhelmed by Afghan refugees. Weapons, drugs, and black money flooded the city and lawlessness was rife. In October 1999 Prime Minister Nawaz Sharif was ousted by a coup and a military government headed by General Pervez Musharaf was installed in Islamabad. It was into this maelstrom of danger and intrigue along the North-West Frontier that I decided to extend my mapping of the Karakoram west to the Hindu Kush Range that borders Afghanistan in 1995.

Hindu Kush

By 1990 I had mapped large tracts of the Karakoram from the Hushe region along the disputed border with India, across Baltistan in the Baltoro and Biafo Glacier valleys and west as far as Hunza.[2] The Karakoram Mountains continue westwards into various mountain ranges along the northernmost borders of Pakistan, the Hindu Raj, the Wakhan, and the Hindu Kush. Several spectacular peaks rise abruptly above the Chitral Valley and seem to parallel some of the major faults that radiate south from the Pamir knot. Highest of these peaks is Tirich Mir, a magnificent 7,690-metre-high mountain that dominates the skyline above Chitral bazaar. I was keen to extend our Karakoram geological maps westwards towards Afghanistan.

My old friend Asif Khan from Peshawar University was looking after our ancient Leicester University jeep so we first decided to drive all around the northern areas, making a reconnaissance around the northern margin of the Kohistan island arc, sandwiched between the Indian and Asian plates, and along the old suture zone that divided Kohistan from the Asian plate. We drove across from Gilgit west to Chitral, exploring the beautiful valleys of the Kunar, Ishkuman, and Lutkho. I had given a new Oxford D.Phil student, Peter Hildebrand from Zimbabwe, the Hindu Kush Ranges to work on. Peter had been born and brought up in the remote Burma valley of south-eastern Zimbabwe and nothing fazed him, which was just as well given the project area I had given him. Some days after a particularly long hike up in the mountains we would take an hour or so off to go fishing in the clear waters of the rivers. Nearly one hundred years ago the British had stocked these rivers and lakes with brown trout. Peter would cut a long bamboo fishing line and cast into the eddies. In only a few minutes we would have caught a handful of delicious trout for our supper.

The mineralized granite above the little hamlet of Gharam Chasma was our main target of research. Initially I thought this granite was very similar to those I had seen along the main Himalaya, full of black tourmaline, red garnet, and white flaky muscovite crystals. I wondered if it might turn out the same Miocene age, about 20 million years ago, and how much of the metamorphism and granite melting along the Hindu Kush correlated in age with the Karakoram or even across the suture zone to the Himalaya. Peter spent two years dating all these rocks together with Steve Noble back in England. The results were very surprising.[3] Most of the metamorphic rocks and the granites along the main Hindu Kush Range turned out to be old—Jurassic or Cretaceous in age, formed long before the India–Asia collision. This metamorphism and melting must have been related to tectonic events along the Asian plate above the Tethyan, northward-dipping subduction zone in a similar way to granites exposed along the Andes. Much of the mineralization in the Hindu Kush could now be directly correlated to Andean-type processes. The Andes are rich in copper, gold, silver, and other ores so it seemed as though the Hindu Kush might be a good place to look for similar mineral deposits.

The one exception to the generally old metamorphic and granitic rocks was the Gharam Chasma granite that had initially aroused my curiosity.[4] Peter and Steve showed through U-Pb dating of monazites and zircons extracted from the granite that it was indeed very young, 21 million years old. We could now prove that some 30 million years after India and Asia initially collided, crustal melting events occurred along the Indian plate in the Himalaya, along the southern margin of the Asian plate in the Baltoro Karakoram Ranges, and now sporadically also along the Hindu Kush.

For two summers we explored and mapped along the Hindu Kush right up to the passes looking westwards into Afghanistan. Suspicious donkey and pony caravans laden with cloth sacks were crossing some of these remote passes, bringing supplies and aid money over to Ahmad Shah Masood's forces in the Panjsher Valley. Occasionally those same donkey caravans returned to Chitral laden with spectacular minerals and gems, fabulous slabs of turquoise and blue lapis lazuli from the mines in Badakhshan, sack loads of fist-sized garnets, topaz, aquamarine, tourmaline, and rubies. Rifling through some of the mineral shops in Chitral bazaar was a wonderful way to spend an afternoon.

Before we left Chitral we made one last memorable trip with Asif, Zafar Ali Khan, Shakirullah, and other geologist friends working from Peshawar. We all drove up to the Shandur Pass to watch the three-day annual polo tournament between Chitral and Gilgit. Polo is the national sport of the northern areas of Pakistan and the teams were all kitted out in their respective valley colours. Although there was a marked-out pitch with goals, the horses and riders frequently jumped the barriers and went careering through the crowds. It was a spectacular show of brilliant horsemanship against a backdrop of the magnificent snow-covered peaks of the Hindu Raj.

Kafiristan, Land of the Unbelievers

I was keen to trace the structures and rocks we had mapped in Gharam Chasma and Chitral southwards along the spine of the Hindu Kush. South of Chitral there are three valleys that lead up to the Afghan border along a spiny ridge of 3,000–4,000-metre-high mountains—the Rumbur, Bumburet, and Birir Valleys. These valleys are inhabited by the Kalash Kafir, the 'black infidels', the only non-Muslim enclave in the whole of North Pakistan. Kafiristan, the 'Land of the unbelievers', is a tiny enclave of pagans, who practise animism and ancenstor-worship and remain unconverted to Islam. Across the border similar Kalash Kafir villages had been conquered by Amir Abd el Rahman of Afghanistan in 1897 and forcibly converted to Islam. Their country had been renamed Nuristan, the 'Land of light'. In Balanguru, the main Kalash village in the Rumbur Valley, we stayed with Saifullah Jan, a fluent English speaker who represented the Kalash valleys in the Pakistani government. Saifullah was a fascinating companion who was very knowledgeable about the local forests, wildlife, and trekking routes. We made some day treks tracing geological contacts up to the border ridges and were able to peer down into the remote Nuristan valleys on the Afghan side of the frontier. The countryside was magnificent with walnut and apricot trees in lush and verdant valleys giving way higher up to firs and pine and Alpine pastures.

The Hindu Kush Mountains on both the Pakistani and the Afghan sides were majestic and idyllic. It was ironic that up in the mountains all was peace and tranquillity whilst down in the Taliban-controlled valleys of eastern Afghanistan turmoil and treachery reigned. A few days before the 9/11 atrocities in New York and Washington in 2001, a suicide bomber assassinated Ahmad Shah Masood in Panjshir, wiping out the only resistance to the Taliban government in Afghanistan. By then the country was awash with foreign Arab fighters of the shadowy Al-Qaeda network, many of whom freely crossed the border from Peshawar to Jalalabad. A month after the September 11 attacks, American and Allied Forces began bombing Al-Qaeda strongholds in Afghanistan, and the border region from Waziristan and Tora Bora through the Khyber agency north to Chitral became a high-altitude escape route across the Hindu Kush for every terrorist and innocent civilian caught up in the fighting. The Hindu Kush became a very dangerous place to do geology: it was time to move back east to Kohistan.

Island Arcs in the Mountains: Kohistan

The mountains of Kohistan have long been renowned for showing a spectacular section through a complete Cretaceous island arc. As we noted earlier, island arcs are best known from the western Pacific (such as the Philippines, or Tonga-Kermadec arc) and the Caribbean (Lesser Antilles arc), but these active island arcs only show the upper volcanic parts of an edifice that is of crustal proportions. In Kohistan, the island arc is sandwiched between the colliding continental plates of India to the south and the Karakoram to the north. The rocks have been folded, up-ended, and tilted to the north. The Karakoram Highway, the main trunk road that runs from Rawalpindi north to the Chinese border, crosses all the major rock units of the Kohistan island arc, and these sections have been the subject of numerous studies over the last fifty years. Rashid Khan Tahirkheli, the Professor of Geology at Peshawar University, was the pioneering geologist who carried out the entire initial mapping and exploration of Kohistan. He was followed by two of his best students, Qasim Jan and Asif Khan, both of whom followed Tahirkheli to become heads of the Geology department. Peshawar University has a splendid campus, laid out a hundred years ago by the British. The old buildings are a wonderful fusion of Oxbridge college and Lutyens-type colonial architecture, and the extensive grounds include over twenty cricket pitches and acres of gardens and forested areas. It was a haven in what was a sea of craziness outside the gates.

Peshawar had by this time become full of refugees fleeing first from *mujahedeen* atrocities and later from Taliban atrocities in Afghanistan. Our small Daihatsu jeep, still permanently bent over to one side after the accident on the Grand Trunk Road, was now in its last throes of life. One of my new D.Phil students, Simon Gough, was working on the rocks along the base of the Kohistan arc and we drove up to Kohistan along the Karakoram Highway. The engine finally suffered terminal failure at the village of Jijal, where we had to abandon it in a roadside garage. Simon and I trekked up several of the side valleys branching off from the main Indus Valley. We did not find the Kohistan villages friendly; some sported the welcoming sign, 'Proud to be a Muslim fundamentalist'. High above the Indus, however, the country is magnificent, with Alpine forests and meadows full of flowers. The rocks here are also spectacular, with impressive eclogites and garnet-bearing granulites, high-pressure and high-temperature rocks that formed along the deepest levels of the Kohistan island arc.

The Kohistan arc consists of a series of volcanic rocks in the higher units along the north overlaying a great thickness of amphibolite—metamorphosed ocean basalts that lie above the deep-level granulites and mantle rocks.[5] The deepest roots of the Kohistan arc are exposed in beautiful desert-varnished outcrops, an impressive array of peridotites and pyroxenites composed entirely of olivines and pyroxenes. Scattered along the southern boundary of the Kohistan arc are remnant ophiolite-type rocks indicative of the oceanic crust that lies in between the arc and India to the south. One of these ophiolite fragments crops out near the village of Sapat. These rocks are similar to the Oman or Spontang ophiolites I had worked on previously but the main difference was the phenomenal preservation of gem-quality pale green olivine or peridote crystals. Usually olivines in ophiolite rocks are strongly altered by hydrothermal processes to serpentine but these Sapat olivines were perfectly pristine and very valuable. Simon ended up getting chased out of the region by the peridote miners, afraid that he was coming to steal their gemstones. The 'gold rush' of peridote miners was soon to be replaced by another frenzy when a few sites with gem-quality pale blue aquamarine crystals were found high in the mountains above Nagar in the Hunza Valley (Figure 6.1).

Figure 6.1 Gemstones of Northern Pakistan, clockwise from top left: Emeralds, Rubies, Lapis lazuli, topaz, green peridote (olivene), andalusite, blue kyanite, red garnet, pale blue aquamarine (beryl), and black tourmalines.

Ancient Subduction Zones

Eclogites, rocks composed almost entirely of red garnet and green pyroxene crystals, are indicative of high-pressure metamorphism that occurs along subduction zones. In the 1980s eclogites were first discovered in northern Pakistan, but not along the suture zone. Instead huge boudins or lozenge-shaped nodules of black eclogites were found within the continental crustal rocks along the northern margin of the Indian plate. This was a surprise because geologists normally expect eclogites in oceanic subduction zones, yet here their presence was telling us that the leading margin of the Indian continental crust had been subducted to great depths. Since continental crust is lighter (~2.8 grams/cm^{-3}) and more buoyant than dense mantle (~3.3 grams/cm^{-3}) this was an enigma. How could lighter granitic-type crust possibly be subducted deep into the mantle?

Similar ultra-high-pressure eclogites had recently been discovered in a few older mountain ranges around the world: the western gneiss region of Norway, the Sulu and Dabie Shan mountain ranges of eastern China, and the Kochetav massif in Kazakhstan.[6] By definition, ultra-high-pressure rocks have two distinct minerals:

coesite, a high-pressure polymorph of quartz, and diamond, the high-pressure polymorph of graphite or carbon. The presence of coesite indicates pressures above about 27 kilobars and diamond indicates even higher pressures, above about 30–40 kilobars, equivalent to depths of about 150 kilometres.[7] In all these areas tiny micro-diamonds had been discovered, micron-scale grains armoured and encased in dense minerals such as garnet, pyroxene, or zircon.

The new findings of very high-pressure eclogites along the northern margin of Indian plate gneiss in Pakistan were very exciting. An expert on these rocks, Paddy O'Brien from the University of Bayreuth in Germany, first discovered coesite from the Pakistan eclogites and determined pressures of 27 kilobars, equivalent to over 100 kilometres depth of burial.[8] This was the first real indication that the Indian continental crust had been dragged down a subduction zone at some stage during the India-Kohistan arc-Asia collision. The question was when: during the earlier emplacement of ophiolites onto the Indian plate margin, emplacement of the Kohistan arc onto India, or during the early stages of the main continental collision between India and Asia?

Simon Gough and I collected numerous samples of these eclogites along the Kaghan Valley area in southern Kohistan (Figure 6.2). Simon also confirmed the pressures of formation of 27 kilobars and temperatures of around 750°C from the mineral assemblages in the rocks, and Randy Parrish dated tiny zircon and

Figure 6.2 A thin section of an eclogite from the Kaghan Valley, North Pakistan. The pink, rounded crystals are garnet and the pale green crystals are clinopyroxene.

allanite crystals using U-Pb techniques precisely at 46.4 million years ago.[9] He even managed to extract U-Pb ages from lower-pressure mineral phases like titanite, and with these data we were able to plot a very accurate pressure–temperature–time path. The results showed that extremely tight constraints on temperature, pressure, depth, and time could be extracted from a single rock sample. It was exactly this sort of data that was required to make meaningful models about the tectonic evolution of the Himalaya. Our new ages from the Kaghan eclogites opened up a whole new series of questions: 46 million years ago was very shortly after the initial 50-million-year closure of the Tethys Ocean and the first meeting of the two continental plates, India and Asia. In the geological time frame 3 or 4 million years is regarded as a short period of time. The new age data suggested that the leading edge of the Indian plate was subducted either at the end of the Kohistan arc emplacement process, similar to the eclogites I had seen in Oman, or during the first effects of the continental collision process.

A Diamond Factory Beneath the Hindu Kush?

Another fascinating aspect about the geology of the Hindu Kush region was the earthquake story. The greatest concentration of continental earthquakes anywhere in the world lies along the border between Pakistan and Afghanistan, a zone known as the Hindu Kush seismic zone (Figure 6.3a, b). Here, a narrow zone of earthquakes dips very steeply to the north-west and extends to over 250 kilometres depth. A second parallel zone of earthquakes curves around the Pamir arc to the north, dipping steeply to the south-east, the Pamir seismic zone.[10] Earthquakes at this depth in the

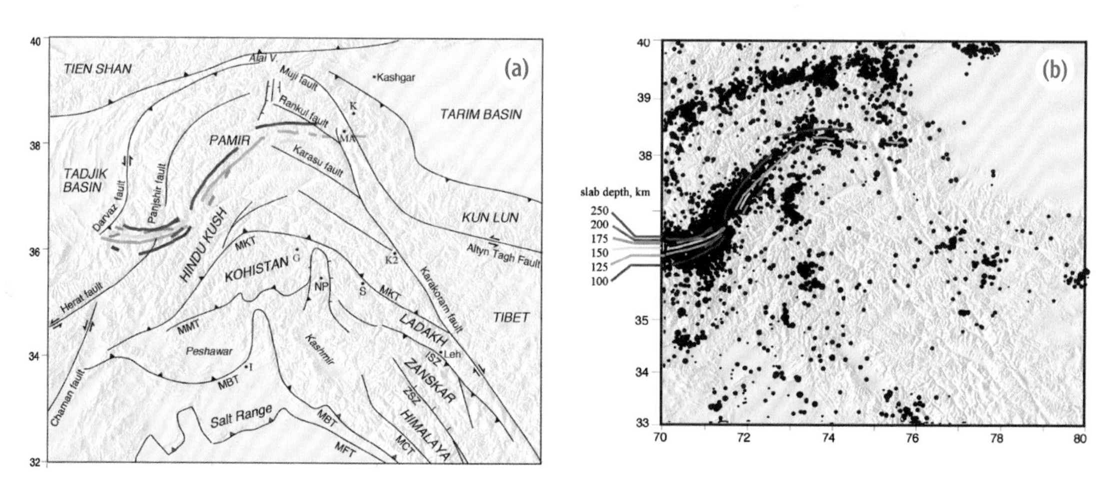

Figure 6.3 (a, b) Digital elevation model of the western Himalaya, Pamir, Hindu Kush, and Tadjik basin area showing the major faults. The coloured lines are depth contours of the Hindu Kush seismic zone, black spots indicate earthquake locations, and depths in kilometres are shown on the left.

continental lithosphere are extremely unusual. Earthquakes record the brittle fracturing of rocks and can only occur at temperatures below about 450°C. At temperatures higher than this the rocks would deform by plastic, viscous, or ductile flow, and hence would not set off an earthquake. Normally earthquakes only occur in the top 20 kilometres or so of the continents, above the magic line, the brittle-ductile transition. The fact that earthquakes occurred with hypocentres as deep as 250 kilometres beneath the Hindu Kush was quite extraordinary. It could only mean that along this narrow, contorted zone temperatures were extremely low compared to the hot mantle either side, a structure compatible with a subduction zone. During the time we were working in the Hindu Kush and in Kafiristan we felt earthquakes rattling off at least once a week; the locals have got used to them and build wooden houses that sway with the shock waves. Occasionally a big earthquake sends parts of mountains tumbling down, but to the Chitralis and Kafirs of the Hindu Kush discussing earthquakes is like the English discussing weather.

In 2000 Roger Bilham, a seismologist from the University of Colorado, was on sabbatical in Oxford and we used to frequently meet to discuss Himalayan matters over coffee. Roger is an expert on Himalayan earthquakes and the history of the Great Trigonometrical Survey of India. One day, a significant earthquake was recorded on our department seismometer along the Pakistan–Afghanistan border and we got around to discussing the Hindu Kush seismic zone. I told Roger about the coesite-bearing eclogites we had worked on from Pakistan and how these rocks could only have formed in a narrow and deep subduction zone. On our third cup of coffee we finally made the connection: maybe the Hindu Kush earthquakes were recording an active continental subduction zone where the thinned lower crust was being dragged down to the north. At these depths, pressures would be sufficient to form first coesite, then diamond. Maybe there was a diamond factory deep down below the Hindu Kush! We recruited Brad Hacker from the University of California, Santa Barbara, a leading expert on ultra-high-pressure rocks and wrote up a speculative paper presenting a model of how we thought these unusual diamond-bearing eclogites could have formed, using the Hindu Kush seismic zone as a paradigm (Figure 6.4).[11]

Tso Morari and the Subduction of the Indian Continental Crust

There was one other area along the Himalaya where similar ultra-high-pressure rocks had been found—in the Tso Morari region of eastern Ladakh. All of the Greater Himalayan metamorphic rocks were located along the main Himalayan Mountain Range and were bounded by the north-dipping Main Central Thrust fault zone along the south and by the north-dipping low-angle Zanskar shear zone, or normal fault, along the north. In the few areas where metamorphic rocks were

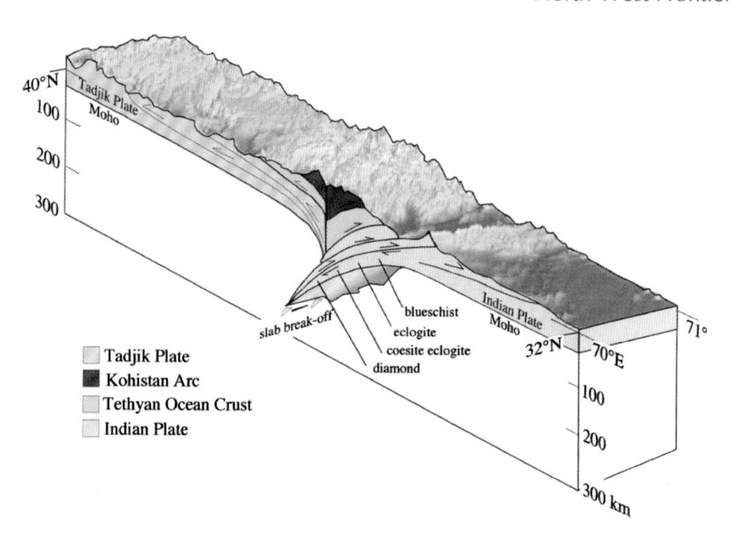

Figure 6.4 Block diagram showing the north-dipping Hindu Kush seismic zone with depths at which progressively deeper blueschists, eclogites, and diamond-bearing eclogites are formed.

exposed in the northern belt they appeared to be in the core of huge folded dome structures with low-angle ductile shear zones and normal faults surrounding them, as if the deep crustal rocks had been punched up through their sedimentary cover.

The Tso Morari region in eastern Ladakh was typical of one of these North Himalayan domes. In the 1980s much of this country around the salt lakes of Tso Morari and Tso Kar in far eastern Ladakh and Rupshu was in the 'Inner Line' and off-limits to foreigners but later with political problems in the Kashmir Valley area where the only major road from Ladakh to the outside world ran, a new road from Leh to Manali was constructed and we were able to get to this fascinating area. Rupshu showed more Tibetan type landscape than Ladakh or Zanskar, with wide open rolling hills, deep azure lakes, and arid snow-capped mountain ranges.

Hidden amongst the metamorphic rocks of the Tso Morari dome were several outcrops of black, dense eclogites, very similar to the Pakistan examples (Figure 6.5). These eclogites were composed mainly of red garnet and green pyroxenes, but also contained rare blue amphibole minerals called glaucophane, all minerals characteristic of high-pressure metamorphism. The eclogites occurred as small lenses and blocks enclosed within marbles and metamorphosed shales or pelites. The eclogites were originally basaltic sills and dykes that intruded into the passive margin sediments of the north Indian plate margin. They were subsequently buried and metamorphosed, highly deformed, flattened, and sheared. Thermobarometry studies soon revealed that these eclogites had been formed at great depths, more than 15–20 kilobars pressure, indicating a depth of nearly 100 kilometres, but these rocks only seemed to occur along the northernmost part of the Indian plate immediately south of the suture zone itself. How had these original surface

Figure 6.5 A thin section of an eclogite from Tso Morari, Ladakh. Apart from garnet and pyroxene, this rock also shows pale blue glaucophane amphibole and white phengite mica.

sediments, injected throughout by basalt sills and dykes, become buried so deep? By using our reconstructed structural cross-sections it was possible to link the formation of the eclogites to intense ductile shearing when the leading margin of India was buried to depths of around 100 kilometres.

The density contrast between continental crust and mantle makes it very difficult for even fragments of continental crust to sink deep into the mantle. This process only occurs along subduction zones where the continental crust is physically pulled down to these depths by a dense eclogitic root, rather like pulling a more buoyant cork down to the bottom of a swimming pool. When the dense eclogite root detaches and sinks into the mantle, the lighter continental crust (the cork) that was attached to it rises rapidly as a result of buoyancy forces and is returned to the surface by the easiest route—along the same subduction zone that was responsible for its burial. So these subduction zones are really large-scale conveyor belts, carrying surface rocks along continental margins down to mantle depths and then returning them back up the same way. The discovery of the eclogites, both along the Kaghan Valley in Pakistan and at Tso Morari in Ladakh, was a crucial piece of evidence that during the collision process the leading edge of the Indian continent was dragged down to great depths, as much as 100 kilometres beneath the southern margin of Asia, and then rapidly exhumed back up the surface.

CHAPTER 7

Faces of Everest

There was no mistaking the two great peaks in the west: that to the left must be Makalu, grey, severe and yet distinctly graceful, and the other away to the right—who could doubt its identity? It was a prodigious fang excrescent from the jaw of the world. We saw Mount Everest not quite sharply defined on account of a slight haze in that direction. This circumstance added a touch of mystery and grandeur; we were satisfied that the highest of mountains would not disappoint us.

George Mallory, 1921

The Khumbu South and South-West Faces

Trekking to Everest from the Sola Khumbu in Nepal (Figure 7.1) is most definitely one of life's great treats. When Nepal first opened up to foreigners in 1950 there was only one road from India to Kathmandu via the border town of Raxaul.[1] Early expeditions to Everest had to trek from the plains of India either from Jogbani or Jaynagar in south-eastern Nepal. For the purist, the trail nowadays starts in the Kathmandu Valley, whilst the road head at the village of Jiri is the normal starting point for overlanders. The first week's walking goes from west to east towards the village of Junbesi, against the grain of the land, crossing three passes and several rivers draining south from the Rolwaling and Khumbu Himalaya. Once across the Dudh Kosi River and up the hill to Lukla, the trail heads north up into the high country. Many trekkers nowadays fly directly into Lukla, where the plane lands at the impressive and frighteningly tilted airstrip built by Edmund Hillary and his Sherpa friends high on the side of the Dudh Kosi. From Lukla, the trail

Figure 7.1 View of the Everest-Lhotse massif from a hot-air balloon (courtesy of Leo Dickinson). Nepal is on the left, Tibet on the right, and the Kangshung Valley leading up to the east face of Everest below. The South Col lies to the left of Everest and the North Col is just visible to the right leading down to the Rongbuk Valley in the north.

winds through forests of blue pine, fir, silver birch, and the ubiquitous rhododendron. In spring the hills are a mass of red, pink, and white rhododendrons. Meadows are carpeted in wild flowers—gentians, primrose, edelweiss, and the magical Himalayan blue poppy. Small Sherpa villages with their sturdy homes built from slabs of schist and gneiss have expanded with new trekking lodges springing up annually.

The terraced rice paddies of the lowlands are soon left behind and apple orchards are a mass of blooms in the spring. Clouds well up and float quietly down into the valleys. The forests with their hanging mosses become eerily quiet. The senses dwell on the serene beauty of the forests and streams, all green and full of life and sound. Suddenly one's eye is caught by something higher up, way above the clouds. With amazement, one realizes that is no cloud up there: it is a mountain, five miles high, far above the peaceful green of the valley. The sheer scale, the vastness and steepness of the Himalaya is an affront to the senses. How can these magical mountains attain such incredible heights? How can these enormous precipices tower 4 or 5 kilometres vertically above us? What great forces are holding them up?

The trail to Everest begins to steepen on approaching Namche Bazar, the Sherpa capital, a spectacular village turned trekking metropolis built around an amphitheatre beneath the towering splendour of Ama Dablam. Abruptly the gorges become steeper; the rivers cascade down with increasing force. Namche Bazar is surrounded by some of the most splendid peaks—Kusum Kanguru, Kangteiga, Tramserku, Ama Dablam (Figure 7.2). Two great rivers meet just below Namche—the Bhote Kosi, the 'river from Tibet', gushes down from the north, cutting right

Figure 7.2 The north face of Ama Dablam (6,828 m).

through the Rolwaling and Khumbu Himal. The Dudh Kosi, the 'milky river', flows down from Everest itself; its name comes from the thick suspension of micaceous silt eroded off the metamorphosed rocks of the high Himalaya.

There is a point halfway up the final climb into Namche Bazaar from the gorge of the Dudh Kosi where one can get a first glimpse of the great peak itself. Everest is largely hidden from the south by the huge southern ramparts of Nuptse and Lhotse, its giant neighbours, but from this vantage point it rises just higher. The black summit pyramid appears almost to float above the hanging glaciers of Lhotse, jutting up into the indigo blue atmosphere. A massive plume of spindrift spews from the summit towards the east, blown by the jet stream winds that constantly pound the mountain. It is an awe-inspiring, mystical sight. Off to the east, Ama Dablam, almost too perfect a mountain, rises right above, with glaciers hanging improbably off every face. Thyangboche Monastery is set in a high meadow, beneath Ama Dablam, surrounded by fluttering prayer flags, ringing with the sounds of bells, conch horns, and the mesmeric chanting of the monks. To the west, two satellite peaks, Taweche and Cholatse, rise in splendid isolation above the rhododendron forests of the upper Khumbu.

Above Thyanboche Monastery beautiful rhododendron forests alive with bird song gradually thin out as dwarf rhododendron and juniper bushes are all that's left of the lush forests below. The landscape becomes more desolate and wild, the mountains steeper, and rivers deeper. Another great river, the Imja Khola, joins the Dudh Kosi from the east. The Imja Khola is fed by the huge glaciers cascading down from the south face of Lhotse and Nuptse, the north face of Ama Dablam, and the wild array of snowy peaks around the watershed with the Arun Valley to the east. The Dudh Kosi steepens here, pushed aside by the enormous bulk of Nuptse towering four kilometres higher. The trail crosses at the snout of the Khumbu Glacier, the sixteen-kilometre-long river of ice that flows from the south face of Everest and the Western Cwm. A few trekking lodges nestle into dusty ablation valleys at Lobuche and Gorak Shep. The lodges at Gorak Shep are the highest, and from here it is a few hours' further trekking along the ablation valley and moraines of the Khumbu Glacier to the site of Everest base camp. The early expeditions used the dry lake-bed at Gorak Shep as their site for base camp, but nowadays the site is scattered amongst the boulders and ice ridges at the foot of the Khumbu icefall. The small peak of Kala Patar is a few hours' scrambling above Gorak Shep. This is where many trekkers reach their personal summit, and from here can be seen some of the finest views of Everest, Nuptse, and the famed Khumbu icefall (Figure 7.3).

Everest Base Camp is a desolate place. Every spring and autumn, hundreds of tents appear as the annual charade of climbing the world's highest peak begins. The narrow trail to Everest can become packed with climbers and Sherpas, trekkers, and yaks, all jostling for space. Any of the side valleys west of the Khumbu Glacier are, however, usually empty and peaceful. During the summer monsoon and in the depths of winter, the place becomes deserted again, left only to those avian acrobats, the scavenging choughs and lammergeier vultures.

The summit of Everest cannot be seen from base camp itself, being hidden behind the peak's western shoulder. Above base camp, huge granite cliffs rise up to the Lho-la, the pass first ascended by Mallory and Bullock back in 1921. Across the Khumbu Glacier, three magnificent peaks, Khumbutse, Lingtren, and Pumori rise along the watershed, the outer bastions of Everest itself. Immediately to the south, the massive face of Nuptse shows glints of white granite beneath its icy carapace (Figure 7.4). The Khumbu icefall, right above base camp, is the most treacherous kilometre of ground anywhere in the Himalaya, where the river of ice pours over the lip of the Nuptse granite. It has claimed more lives than any other glacier in the Himalaya. Over a kilometre above base camp, the Khumbu Glacier tumbles down the icefall in a chaotic mass of towering ice seracs, yawning crevasses, and surreal icy architecture. The glacier creaks and grinds as the ice suddenly cracks. Seracs, hundreds of metres high, collapse in a thunderous cacophony of spindrift. Boulders

Figure 7.3 The Western Cwm and Khumbu icefall between the Everest massif (left) and the Nuptse-Lhotse wall (right). Scattered coloured tents in the moraine at the bottom is the site of Everest Base Camp.

released from their icy grip plummet into hidden caverns in the frozen underworld. Climbers reaching the top of the icefall breathe a huge sigh of relief. The fear of the unknown is replaced by wide-eyed wonder at the beauty of the Western Cwm.

The Western Cwm is the source of the Khumbu Glacier. The great glacial bowl is carved out of the south face of Everest, the north-west face of Lhotse, and the north face of Nuptse. Ice sheets have scoured out a perfect amphitheatre as the glacier flows west through the narrows, where sheer cliffs of gleaming granite guard the southern approaches to Everest. Above, the glacier flattens towards the foot of the climb up to the South Col. Avalanches sweep down from the steep north face of Nuptse, but Everest's south-west face has little snow. The thin-banded schists have a foreboding blackness to them. The snows have been blasted off by ferocious winds that continuously pummel its towering pyramid.

The climb to the South Col is a gut-wrenching plod through increasingly thin air, but the views across the Lhotse–Nuptse ridge to the peaks of the Khumbu appearing beyond make up for the effort. The South Col is as bleak and desolate a place as any. Sometimes the debris of expedition paraphernalia litters the col, for it

Figure 7.4 Everest (8,850 m) and Nuptse (7,861 m) rising above the Kumbu Glacier in Nepal.

is from here that climbers leave for the final push to the summit. Earlier expeditions placed one, or even two, more camps along the final south-east ridge, but the less time spent at these extreme altitudes the better, so recent summit attempts usually leave from the South Col.

The final south-east ridge from the South Col up to the summit weaves a line through the black schists. The ridge itself is exposed and buffeted by all the winds of Asia. Up here the views become intense. The air is pure: no dust or pollution rises up here to the rarified atmosphere, eight kilometres high. Towering now above the Lhotse–Nuptse ridge, the Khumbu peaks appear far below—spiky Ama Dablam, elegant Pumori, shapely Lingtren (Figure 7.5). Away to the east, a sea of snowy mountains around Baruntse and the upper Arun Valley culminate in the stunning, pale orangey-pink pyramid that is Makalu, the world's fourth highest mountain.

Three hundred kilometres away to the east, the huge mass of Kangchenjunga, the world's third highest peak, breaches the skyline on the Nepal–Sikkim border. To the west, the Himalayan crest continues with the great ramparts of Gyachung Kang and Cho Oyu. A sea of 6,000- and 7,000-metre-high peaks spill over into Tibet around Menglungtse, north of the twin peaks of Gauri-Shankar, once thought to be the highest mountain in the world.

The South Summit appears tantalizingly close to the true summit, but the difficulties of the Hillary Step, a small limestone cliff, must be surmounted. If the snows are not banked up, this can make tricky climbing, especially with a full down suit and oxygen equipment. Finally, the seemingly endless plod to the real summit, and quite suddenly there is no more mountain to climb. The vast, barren, brown landscape of Tibet is spread out way below, the icy streaks of the Rongbuk Glacier snaking off towards the northern wastes of Tibet. The Himalaya forms the most spectacular watershed on Earth. To the south are high mountains,

Figure 7.5 The south face of Lingtren (6,749 m) showing a thick massive granite sill overlying a zone of black gneisses intruded by numerous thinner granite sills.

steep cliffs, deep valleys, jungles, waterfalls, rain. To the north are the high, but relatively flat, plateau-lands, brown, dusty, and arid. Glaciers flowing north to the deserts of Tibet feed gentle streams, many of which simply die out on the Tibetan Plateau.

The Rongbuk Valley: Northern Approaches

The northern approaches to Everest from Tibet could not be more different from the southern, Nepalese side. Flying over southern Tibet into Lhasa, either from Kathmandu to the south, or from Chengdu, in Sichuan to the east, at least prepares one for the vast, barren landscape of Tibet. From Kathmandu, the plane skips over the Himalaya in less than an hour, but it is peering out of the window on the Chengdu to Lhasa flight that one realizes the sheer scale of this incredible, untravelled, largely unexplored land of Tibet. Eastern Tibet is not a plateau at all. It is a series of stupendous mountain ranges, swinging around from the eastern highlands all the way into Yunnan, Burma, northern Thailand, and Laos. At least three of these mountain ranges are larger than the European Alps. The rivers draining this land carve deep gorges in between the mountains—the Tsangpo-Brahmaputra, the Salween, the Mekong, and the Jinsha rivers. The Yangtse-Kiang, the greatest river of Asia, collects all the drainage from north and east Tibet and flows over 6,000 kilometres east into the South China Sea. The flight from Chengdu to Lhasa crosses right over the huge massifs of Minya Konka (Gongga Shan), and then to the south directly over the great Yarlung Tsangpo Gorge, which cuts right through the twin peaks of Namche Barwa and Gyala Peri, the two giants of the far eastern Himalaya. After another hour's flight west over the rolling brown lands of southern Tibet, the plane descends to Gonggar, the airport of Lhasa in the Yarlung Tsangpo Valley.

Lhasa, the capital of Tibet, is situated north of the Yarlung Tsangpo Valley, but the road to Everest heads off west through Shigatse, and then south from Shekar over the Pang-la towards the Himalaya. The Rongbuk Valley widens out onto the plateau, with small Tibetan villages scattered along its flanks. The rock faces above show outrageous folds where the rocks have been twisted and turned in all directions with the compressive forces of colliding continents (Figure 7.6). The track fords rivers and crosses mudflows. Small Tibetan horse-drawn chariots joust for room on the single dirt track with Chinese trucks and Land Cruisers. Every year in the early spring and late monsoon, this remote Tibetan valley sees an influx of affluent Western climbers, eyes all set on the big hill at the head of the valley. The track gets progressively worse, crossing great boulder screes and fording rivers before finally reaching Rongbuk Monastery. The monastery was almost completely destroyed during the Cultural Revolution, but the Chinese, now realizing the tourist potential, have been slowly rebuilding it, and now a few ancient and bemused lamas have returned to live there.

The track from Rongbuk Monastery up to base camp winds through the boulder fields and terminal moraines of the Rongbuk Glacier. A few wild Himalayan tahr, mountain goats, inhabit this desolate tract of land, and apart from the piercing whistle of a startled marmot, or the mournful cry of a Himalayan snowcock, there appears to be little sign of life. Vegetation is almost completely lacking up here in this high-altitude desert, where there is little protection from the freezing temperatures that plummet to minus forty degrees in the depths of winter.

Every year several hundred mountaineers make the pilgrimage up to Rongbuk Base Camp.[2] East and west of Mount Everest, the Himalaya continue with a whole array of sublime mountains, yet trekkers continue to follow the hordes along the same path reconnoitered by Mallory and Bullock in 1921. Base camp is situated at the terminus of the Rongbuk Glacier around the wide, flat ablation valley where the river disgorges from the melting ice. Ahead to the south, Everest forms a

Figure 7.6 Folded Tethyan sedimentary rocks north of the Rongbuk Glacier, Everest area.

mighty monolith soaring up into the blue. The satellite peak of Changtse lies right in front, dividing the North and East Rongbuk Glaciers. The main route follows the true right bank of the glacier then branches up following the East Rongbuk Glacier.

The East Rongbuk Glacier is a spectacular river of ice. The glacier widens up here, with an avenue of ice pinnacles and weirdly shaped ice penitents adding to the surreal landscape. Further glaciers join, draining down from the peaks of Khartaphu and Khartse. One branch leads up to the Raphu-la, the pass that heads over towards the Kangshung Valley. Gradually as one climbs higher, the altitude begins to take its toll, lungs working overtime to inhale the increasingly thin air, head pounding. Skirting around the eastern flank of Changtse, the site of advanced base camp is a human oasis of colour and life, set amidst the granite boulders beneath the North Col (Figure 7.7).

The climbing route up to the North Col is a long and exhausting snow plod. One can imagine the excitement of George Mallory when he first set foot on the North Col in 1921 and scanned upwards along the north-east ridge towards the summit. The distance from the North Col to the summit is actually about three kilometres, but from the North Col the whole route looks foreshortened. With altitude hallucinations, it might seem like an easy walk, with just three small steps of jutting rock to surmount en route. Those three kilometres are the hardest of all, not because of the difficulties of climbing, for it is easy angled, but from the sheer extremity of the situation. The ridge is very exposed and winds, touching the jet stream, can be ferocious at these altitudes. The rocks are thin slabs of shales and limestones, sloping gently to the north, rather like a series of overlapping roofing slates. One slip, and then the great surf ride all the way down the north face of Everest. Somewhere on these upper slopes of the north-east ridge, George Mallory and Sandy Irvine did just that in 1924. Mallory's body was found frozen into the mountain he both loved and hated in 1999 by the American climber Conrad Anker. Unfortunately there was no sign of the small camera he carried, so the mystery of whether Mallory and Irvine actually made it to the summit remains to this day.

The three small cliffs of the north-east ridge are thick limestone units that stand proud of the intervening black shales. During their early ascent, the Chinese bolted an aluminium ladder up the Second Step, and many climbers nowadays simply climb up the ladder before continuing the plod up to the summit. It is only after the final steps are taken to tread on the summit itself that the great panorama of the Khumbu Peaks in Nepal is seen to the south. It is like looking into another world.

Figure 7.7 The north face of Everest from the Rongbuk Glacier in Tibet, with the satellite peak of Changtse in the foreground.

The Kangshung Valley: Eastern Approaches

The eastern approach to Mount Everest is the remotest of the three main access routes.[3] The Kangshung Glacier drains east into Kama Chu and eventually into the upper Arun River, which rises in southern Tibet and carves a deep chasm through the Nepal Himalaya, east of Makalu. The road from Lhasa crosses the Pang-la, where some of the classic distant views of the Everest massif can be seen, then heads east into the Kharta Valley. At Kharta village the road ends, and expeditions must hire yaks and local porters from here. There are two main passes crossing over from the Kharta Valley to the Kama Chu, the Langma-la at 5,500 metres and the Sho-la at 5,400 metres' altitude. Both these passes are high and can be difficult and dangerous to cross. Early in the spring, the yaks will be floundering about in two metres of soft snow. In the summer months, the snow may have melted but the lashing monsoon rains and thick cloud can make any crossing of the passes a traumatic experience. For autumn expeditions, there is always the fear of being trapped in the Kama Valley by the first winter snowstorms.

The Kama Valley, once accessed, is a hidden Himalayan gem of unparalleled beauty, surrounded by some of the most spectacular Himalayan scenery anywhere (Figure 7.8). The Kama Chu drains east into the Arun, and flows through a deep and mainly inaccessible gorge south into Nepal. In only 30 kilometres the river plunges from an altitude of 4,000 metres down to less than 1,000 metres. The Kama Valley

(a)

Figure 7.8 (a) (*opposite*) The Kangshung face of Everest showing Lhotse, the South Col, and the east face of Everest, taken from a balloon. Courtesy of Leo Dickinson. (b) (*above*) Peak 38 (7,589 m) and Shartse (7,381 m) east of Lhotse, and Everest rising above the Kangshung Glacier in Tibet.

is uninhabited except for occasional yak herders bringing their yaks over from Kharta for grazing. In spring the valley is a mass of colourful flowers. Monsoon rains sweep up the Arun Valley, and some of the most pristine and beautiful forests remaining are hidden in the inaccessible cliffy regions of the lower Kama and upper Arun Valleys. The desert landscapes around Kharta are soon replaced downstream by forests of dwarf rhododendron, silver birch, willow, and juniper, sometimes laced with the yellowy-green hanging mosses of 'old man's beard'.

Three-thousand-metre-high cliffs of flawless pale granite, forming the unclimbable blank walls of the north face of Chomolonzo (7,790 m), line the southern side of the valley. The ridge leads west to the beautiful peak of Pethangtse (6,738 m) rising right above the site of the Kangshung Valley Base Camp. At dawn, the snows on the hanging glaciers turn delicate shades of red, orange, and pink. Away to the south, the north face of Makalu (8,475 m), the world's fourth highest mountain, forms the perfect backdrop, towering above everything in the vertical world of the Kangshung. The north wall of Chomolonzo has never been climbed. Avalanches, set off from the summit snowfields between Chomolonzo and Makalu, free-fall over the north face precipices. Waterfalls of powder snow end up blowing in the wind, swirling back up into the atmosphere (Figure 7.9).

At the head of the valley, the Kangshung face of Everest is the headwall, with its myriad hanging glaciers and rock buttresses. It is the steepest and most impressive of all the faces of Everest, and has two of the hardest routes yet climbed on the mountain. The bowl carved out by the Kangshung Glacier is one of the most spectacular—its 6,500–7,000-metre-high wall includes the north faces of Shartse (7,502 m), the unnamed Peak 38 (7,589 m), Lhotse Shar (8,383 m), and Lhotse (8,501 m) (Figure 7.8a and b). From here the ridge drops down to the South Col at 7,986 metres, then rises to Everest's summit. The long north-east ridge descends all

Figure 7.9 The north face of Makalu looking south across the Kangshung Valley, with an avalanche falling down the sheer granite cliffs.

the way to the Raphu-la at 6,510 metres, then to the lower 'foothill' peaks east of the Lhakpa-la, extending off north into the Tibetan distance.

Everest Mountaineering

Everest has always been a magnet for mountaineers since it was first discovered. When George Mallory was asked why he wanted to climb Everest so much, he replied off-handedly 'because it's there'. There is no good reason to climb Everest, and Mallory, like many after him, have wondered exactly what it is that drives people to do it. The Tibetans and Nepalese Sherpas who live beneath the mountain never thought of climbing it until the 'mad English' came along in the 1920s. Their only reason for climbing high was to find good pastures and grazing for their yaks or occasionally to cross the high passes over to Tibet for trade. Many Nepalese porters to this day shake their heads in bewilderment at what drives these crazy people to do something so pointless, dangerous, expensive, and frankly ridiculous. Many want to do it so badly that they are quite prepared to lose fingers, toes, brain cells, and even their lives in the process.

The sense of mystery, inquisitiveness, and wonder, touched by that required degree of madness, so prevalent in the early explorations, has been replaced by the egotism, nationalism, and commercialization of the modern era. The Chinese climbers in 1960 were ordered to climb Everest by the Communist Party, some plucked from obscurity to attain glory for the party and the country. Numerous Japanese climbers have attempted the mountain: a postcard from one Japanese expedition on a hotel wall in Kathmandu in the 1970s announced boldly over the picture: 'Success climb! Four on summit. Only two dead.' Jon Krakauer wrote in his book *Into Thin Air*: 'I'd always known that climbing mountains was a high-risk pursuit. I accepted that danger was an essential component of the game—without it climbing would be little different from a hundred other trifling diversions. It was titillating to brush up against the enigma of mortality, to steal a glimpse across its forbidden frontier. Climbing was a magnificent activity, I firmly believed, not in spite of the inherent perils, but precisely because of them.'[4]

In the early 1990s the Nepalese authorities allowed only one expedition at a time on Everest and the peak fee was $2,300. Soon, they realized that demand far outstripped opportunity, with a long list of applicants willing to pay over the odds to secure the coveted permission. By 2000, the fee had risen to $70,000 for an expedition of up to seven members, with $10,000 extra for every additional climber. The authorities removed the limit to the number of simultaneous expeditions. Commercial expeditions used to only take clients who had some experience of Alpine or Himalayan mountaineering, but these conditions were soon relaxed. Despite the enormous cost, anywhere between $35,000 and $70,000 per person, there was no shortage of clients waiting to sign up for a guided trip to Everest. These clients included a few experienced mountaineers who had no time to organize the expedition themselves, but were willing to pay someone else to do it. But most were not experienced climbers, and some indeed had never put on a pair of crampons or wielded an ice-axe before they set out for Everest. There are, however, a few real travellers and characters still attracted to Everest. Göran Kropp, for example, bicycled from his home in Sweden to base camp, climbed Everest alone and then cycled all the way back to Sweden. Tim McCartney-Snape, a true Australian pioneer, walked 620 miles from sea level in the Bay of Bengal, swam across the Ganges River, and then climbed Everest, reaching the summit on 11 May 1990 without using bottled oxygen.[5] This remains the purest, and only real ascent of Everest, ever!

Everest now holds little of interest to mountaineers. All the ridges and faces have been climbed, the huge expense of the peak fees and permits make it out of the reach of most climbers and the reality is that climbing Everest is a rather unpleasant pastime. There are many dead bodies still lying encrusted in the ice, and stories of climbers having to literally step over frozen corpses are not exaggerated. In the past few years, on the few days when conditions permit an ascent, there can be a queue of climbers waiting to stand at the summit. Many clients spend so much money and energy to attain their goal that they never stop to think about the layers of support and infrastructure that make their visit possible. Exploitation of Nepalese porters has been regrettably common, and only a very small percentage of the large sums of money paid in peak fees finds its way to their homes and villages. After his first ascent of the south-west face of Everest in 1975, Doug Scott, the greatest of British Himalayan climbers, started his own charity—Community Action Nepal—that builds schools, hospitals, and porter shelters for the most needy communities throughout Nepal (see page 427).

Everest can bring out the best and the worst in people. There are heroic stories of climbers who have given up their chance for a summit bid to rescue others in danger and get them safely down the mountain. There are also stories of extreme selfishness: of some climbing right past dying climbers on their blinkered way to the summit; of climbers refusing to lend radios to help in emergencies; and of people stealing gear from tents high on the mountain. What is so extraordinary is that 10 kilometres away from Everest in every direction is a sea of unclimbed, unexplored routes on beautiful mountains, some of which do not even have a name. In one valley to the west of Everest, a line of four magnificent mountains between 6,000 and 7,000 metres high is within sight of Everest. Only one of these has been climbed and very few people ever trek there.

When all is said and done, though, the mountains of the Sola Khumbu, and particularly Everest, cast a spell that is difficult to assuage. The ethereal beauty of watching the clouds well up and float down above the Khumbu Glacier, the sight of four- or five-kilometre-high walls of rock, of avalanches screaming down the south face, of the jet stream trailing a plume of spindrift off the summit, is truly hard to beat. Despite all the human craziness that goes on around Everest, the mountains hold a power over people that will never be diminished, as long as man retains some inquisitiveness and wonder.

Mountain Madness

The first year I started my geology studies around Everest, in 1996, the climbing season was the pinnacle of mountain madness. Commercial exploitation of Everest reached its zenith, and fate was to take a heavy toll. Altogether eleven climbers died, eight of them in one storm on 10–11 May, five on the south side and three on the north. The 1996 story forms one of the classic sagas of Everest mountaineering.[4] At least five commercial expeditions were on each side of the mountain, including two led by the most experienced Western Everest climbers of all, New Zealander Rob Hall's Adventure Consultants and American Scott Fischer's Mountain Madness.

On 9 May 1996 about forty climbers from five expeditions left the South Col for their summit bid in what appeared to be perfect, calm conditions. By midday, when most of the climbers were either at the summit or close to it, a ferocious storm swept across the mountain and the climbers were fighting for survival. The eight fatalities occurred in the epic struggle to get down the mountain that day, including Rob Hall and Scott Fischer, in the worst single tragedy Everest has seen. It is clear that the one overriding cause of the tragedy, apart from the weather, was hypoxia—altitude sickness. Hall and Fischer, probably the two strongest and most experienced of all Himalayan climbers, broke all their own rules, continuing upwards to the summit way past the turn-around time (1 a.m.), leaving the South Col too late to make it up in time, and failing to turn their clients around below the summit when it was obvious that they would not make it. Rob Hall stayed with one of his struggling clients, and in the process both died near the South Summit. Before he died he was in radio contact with base camp and they even managed to hook up a radio link to his wife back in New Zealand.

The survivors were left permanently scarred, both physically and mentally, from the trauma of their experience. In the confusion of retreat that afternoon, Beck Weathers was given up for dead, snow-blinded and suffering severe frostbite below the South Summit. He somehow staggered down alone and survived a night out above the South Col by some miracle. After sixteen hours spent comatose in the snow he got up and staggered into Camp Four the following morning. Todd Burleson and Pete Athans described 'his ghostly appearance, grotesquely frostbitten, his right hand bare to the wind, outstretched in a kind of frigid salute, as reminiscent of some kind of horror movie'. Back in America, he had one arm amputated below the elbow, all the fingers on his left hand were amputated, and his nose had to be cut off and reconstructed. The price of Everest was indeed very high.

Anatoli Boukreev, a strong Kazakh climber and friend of Scott Fischer's, was the main guide on Fischer's team.[6] On 10 May, once the storm had died down, having witnessed Beck Weather's miraculous survival after a night out, Boukreev made an heroic ascent back up the mountain from the South Col in the slender chance of rescuing his friend, but to no avail. In truth, both Hall and Fischer were probably suffering from hypoxia long before things started to go wrong, possibly even when they initially left the South Col. In the final analysis, Jon Krakauer wrote: 'It can't be stressed strongly enough that Hall, Fischer and the rest of us were forced to make critical decisions while severely impaired with hypoxia. In pondering how this disaster could have occurred, it is imperative to remember that lucid thought is all but impossible at 29,000 feet.'[4]

David Breashers was also on the South Col route in 1996 making the IMAX wide-screen film of Everest with a $5.5 million budget.[7] Breashers' team included Ed Viesteurs, Robert Schauer, Araceli Segarra, and Jamling Norgay Sherpa, son of Tenzing Sherpa, the man who made the first ascent in 1953. They were by far the most competent and strongest team on the mountain, and had given up half of their oxygen supply to help the others in the rescue efforts. They abandoned their filming to evacuate the survivors of the 9 May storms, and without their rescue efforts several more climbers would certainly not have survived. They managed to get Weathers down to the Western Cwm and a Nepalese pilot, Lt. Col. Madan Khatri Chhetri, made an heroic landing above Camp 1 on the ice of the Western Cwm, to evacuate him. This was an altitude record for the helicopter, nearly 21,000 feet above sea level. Breashers' team went on to make the IMAX film, a stunning tribute to the mountain and the climbers that take on Everest.

There are a number of factors that conspired to bring about the 1996 tragedy on Everest. There can be no doubt that commercial expeditions were the source of many problems. Anyone who had $50,000 could sign up for the trip of a lifetime, guided to the highest summit in the world by a team of professional guides. Rob Hall claimed in 1995 that he thought he could get almost anybody up Everest, given that they were reasonably fit. City bankers, socialite climbers, and people who had no climbing background at all signed up for all sorts of reasons, mainly to do with egotism and nationalism. Rob Hall had climbed Everest five times already, and had a superb team of dedicated Sherpas helping him. The clients would have all their gear carried up for them, their tents erected, and their food cooked. The Sherpas would keep the icefall open, stock the high camps, and tight rope them on steep sections. Climbing Everest bore no resemblance to mountaineering in most other places. The publicity of the 1996 disaster, however, had no effect in halting the flow of potential Everesters. Quite the opposite: it generated more and more people keen to have a slice of the action. Mountain madness certainly prevailed, and was there to stay.

Extracts from an Everest Diary, Nepal, 1997

The following year, 1997, there were even more expeditions lining up to have a crack at Everest. More than 500 climbers attempted the mountain from both the Nepal and Tibet sides. It seemed that no lessons from the 1996 tragedy had been learned. Commercial expeditions were thriving, with increasing numbers of clients. For the cash-strapped Nepalese government, Everest was a pot of gold.

In the spring of 1997, trekking up the side moraines of the Khumbu Glacier en route to Everest Base Camp, Ian Brewer and I saw a helicopter droning up the valley, skimming just above the ice walls and pinnacles on the glacier, another ferry service up to the roof of the world. It had been summoned by radio to evacuate a climber with frostbitten fingers. After it rounded the moraine and came in to land on a flattened patch of ground just outside base camp, a vicious updraft of wind sweeping down from the Lho-la picked it up and tossed it down again. A rotor blade clipped some boulders and the helicopter catapulted out of control, crashing down onto the ice. The rotors were twisted but the helicopter amazingly just rolled onto its side and a very sheepish-looking pilot clambered out. It was Lt. Col. Madan Khatri Chhetri of the Nepalese Air Force, the hero of the 1996 rescue of Beck Weathers from the Western Cwm.

* * * * *

Diary extract, May 1997: 'We staggered into Base Camp, a sprawling city of about three hundred tents scattered amongst the boulders and moraines of the Khumbu glacier. The first thing we saw was the crashed helicopter, lying on its side like a stranded whale. It seemed amazing that no-one had been killed, given that there were thirty or so people milling around waiting to load the stricken climber on board. On the rubble-strewn glacier, every flat piece of ground had a tent on. After weaving around the chaos of Base Camp we finally found my friend Jon Tinker's Base Campsite. That evening we were sitting around the table sipping mulled wine, eating excellent food when the satellite telephone rang. I happened to be sitting next to it so lifted the receiver to hear a strange Nordic voice enquiring if that was Everest. "Yes," I replied, "this is Everest speaking, where are you?"

"This is the office of the Prime Minister of Iceland in Reykjavik and we would like to speak to the Icelandic climbers please."

"Sure, hold the line." Jon took the phone and explained that the first Icelanders to reach the summit were now safely back at the south Col and would they like him to try to connect the phone link with them? Everest it seems was just a phone call away from anywhere in the world.

The following day news came on the radio that the first Malaysian climber finally reached the summit of Everest, guided by two burly Russian professional guides. The Malaysians were camped next door to us just across the glacier. Theirs was a huge expedition with at least 20 Malaysians, a tent full of video and audio equipment, and a tame American to run the show. As the Malaysian and Russian climbers approached the summit, the video man yelled into the microphone that they should wait five minutes before treading the final few steps, then he could make a live video-link all the way back to Kuala Lumpur, so that then they would hit prime-time TV news. The summit moment came, and the team erupted into a roar of cheers and back-slapping before the whole team turned towards Mecca to pray thanks to Allah.

Just up the hill at the American camp, Freddy Blume, a geodesy PhD student of Roger Bilham's from Colorado was manically running the camp, whilst the climbers were trying to fix GPS permanent stations at the South Col and Hillary Step. Freddy was running a couple of Internet sites with daily reports from base camp and any visiting scientists, climbers, or just about anyone was interviewed for the 'live' Internet chat station. Mountain zone.com was sending live news and pictures over the Internet worldwide. The radio crackled and the climbers on the South Col, preparing for their summit day the following morning were asking us where we wanted our rocks collected. "I need at least one from as close to the summit as you see rock please, one or two from the Yellow Band, and as many as you can carry at regular intervals back down to the South Col. Just try and note the exact locations, and heights, you picked them up from, then we can put them on the map correctly. Thanks!" I said. This was a very different way to do geology from what I was used to. I sat back in my chair, coffee in hand, and looked up at the plume of snow streaming off the summit of Everest. Just another day at the office really for a geology Research Fellow from Oxford. Unfortunately, the reality of climbing Everest in poor weather, high winds, and driving snow made rock collecting a low-priority, and only a few small samples made it down.

On our final night at base camp, we had a big party at Jon's base camp mess tent. Nick Kekus had just arrived with the Icelanders fresh from the summit, my old friend Simon Yates strolled in, down from near the top of Lhotse. All of Jon's team of climbers was down now and safe. Only a few late expeditions were still retreating down the Western Cwm and icefall. There was a strange mixture of peoples in the tent: Mexicans, Icelanders, Europeans from at least 5 different countries, Russians, Kazakhs, Nepalese Sherpas, and a Tibetan. Everyone was euphoric, knocking back the wine and the whiskey, the cold air cut by Jamaican reggae music and loud stories of Everest climbing. I felt that all we needed was for Mallory and Irvine to stroll in to cap a surreal evening.'

* * * * *

Tibetan Climbs

Although Tibet was opened up to Western visitors in the late 1970s, the Chinese kept a strict control on numbers and movements of people. Only a few areas were open, and it was not until 1979 that the Chinese authorities allowed foreign expeditions to climb mountains. Instead of the month-long trek from Sikkim and across the plateau, routinely done by the British expeditions of the 1920s and 1930s, it was now possible to fly into Lhasa or drive by road from Kathmandu along the Friendship Highway, and all the way to base camp. The Chinese had constructed a dirt road to Rongbuk in order to supply their massive 1975 expedition. The monastery at Rongbuk had been almost completely destroyed, vandalized, and looted during the Cultural Revolution, but now, the Chinese started to restore it and even imported a few old lamas.

The Rongbuk Glacier approach to Everest was the same route used by all the early British expeditions in the 1920s and 1930s. The Chinese Mountaineering Federation controlled the access closely though and set up a similar application and fees system as the Nepalese government. Base camp was situated on the flats at the snout of the Rongbuk Glacier, where every year now hundreds of tents would suddenly appear with the Everest climbing season. Advanced base camp is a long day's walk up the East Rongbuk Glacier, from where the assault on the North Col and north-east ridge starts.

The Japanese had paid one-and-a-half million dollars to be the first expedition allowed into the Rongbuk, setting the trend of upwardly spiralling greed that has dominated Everest mountaineering ever since. The Japanese and Chinese climbed both the north-east ridge and a new route on the north face in the spring of 1980. In complete contrast to the million dollar and multi-crew ascent of the Japanese and Chinese, the following year saw Reinhold Messner breaking new barriers yet again with his amazing solo climb of Everest's north face, without using bottled oxygen. Messner went on to become the first person to climb all fourteen 8,000 metre peaks, a truly incredible feat of mountaineering.[8]

Several impressive new routes were established on the north side of Everest soon afterwards, perhaps the most spectacular being the Swiss ascent of the Hornbein Couloir in 1986 without oxygen, camps, or even ropes. Our friends from K2 in 1985, Jean Troillet and Erhard Loretan, then glissaded all the way down in only three and a half hours. The Australians made an impressive ascent of the Great Couloir in 1984, and several British expeditions attempted the long north-east ridge in the 1980s, but it was not climbed in its entirety until 1995 when a large Japanese expedition finally succeeded, but only by using thirty-one Sherpa climbers in support and a vast number of fixed ropes.[9]

The east, or Kangshung face of Everest is the most remote and the steepest face of Everest. It has two of the hardest routes on the mountain; the first, climbed by Americans in 1983, goes directly up the buttress beneath the South Summit, and the second, climbed by an Anglo-American team, ascends directly up the Neverest buttress to the South Col. Stephen Venables completed this climb, reaching the summit alone on 12 May 1988.[3]

Extracts from an Everest Diary, Rongbuk, 2000

Rick Law, a British expat professor from Virginia Tech in the US, and I had obtained a three-year grant from the National Science Foundation to study the South Tibetan Detachment and the structures associated with this low-angle normal fault. This was the same low-angle normal fault that I had mapped along the Zanskar Valley and climbed up to on the Shivling expedition in Garhwal. The fault appeared to clip the tops of many of the Himalayan mountain giants all the way along the chain. Everest was the biggest and the best of all three-dimensional exposures of this unusual structure. In the spring of 2000 I organized a 'tourist trek' to Rongbuk together with Rob Simpson, a new D.Phil student of mine from Oxford, and two old climbing friends, Nick Groves and Paul Harris. We spent several weeks geologizing along the Phung Chu and Kharta Valleys to the north-east of Everest. We then hired six yaks from the village of Kharta and trekked along the Kharta Chu with the intention of crossing the Sho-la at 5,400 metres over into the Kama Chu, the valley leading up to the Kangshung face of Everest. The winter snows that year had been unusually heavy and by May there were still huge snowdrifts and banks along the Kharta Chu. Our yaks floundered shoulder-deep in the drifts and there was no way we could get them up to the pass. Eventually we unpacked, sent the yaks back, and laboriously carried the gear ourselves up to the pass. We spent a wonderful few days camping right on the Sho-la itself, with

Figure 7.10 The golden glow of the setting sun shining on Makalu's north face.

its spectacular views looking south to Makalu (Figures 7.9 and 7.10) and the great granite walls of Chomolonzo, and north across the desolate brown hills of Tibet. Occasionally avalanches would thunder down the whole 4,000-metre-high face of Makalu and fill the Kangshung Valley with powder snow.

After our failure to get into the Kama Valley, we decided to drive around to the Rongbuk Valley, the northern approach to Everest and the route taken by all the pre-war British expeditions. Here we spent a month mapping the geology, and collecting rock samples to the north of Everest. At Rongbuk Base Camp we met Russell Brice, a New Zealand mountaineer who spent several months every year at Everest, guiding commercial expeditions of climbers. Russell managed to get us special permission from the Chinese liaison officers to climb to the advanced base campsite at 6,400 metres below the North Col.

* * * * *

Diary extract, 17 May 2000: 'Spent the whole day climbing wearily up the Rongbuk and East Rongbuk glaciers towards ABC (Advanced Base Camp). It was fascinating to weave through the great ice penitents of the East Rongbuk Glacier. I remembered the historic photos in the books by Ruttledge and Mallory showing the pinnacles of gleaming white ice pointing vertically out of the brown rubble of the glacier. The ice has moved on but the "avenue of the penitents" still looks exactly the same. It was extremely hard going, with altitude, exhaustion and dehydration, and not a moment too soon, the prayer flags strung across the granite cliffs above ABC appeared.

We staggered into Russell's camp and immediately a friendly Sherpa thrust a mug of hot tea into my hands. There was quite a village up here, with about fifty or so tents. At least ten different expeditions were trying this route. The North Col looked tantalizingly close and we could see along the famous NE ridge to the lower steps. Massive glaciers stretched east to the Raphu-la, where Mallory had climbed up to in 1928 and Lawrence Wager had collected geological samples in 1933. A gong sounded and we went into another tent for lunch, an excellent chicken curry followed by a chocolate pudding, coffee and biscuits. This was five-star living at 6,000 metres. Everest lifestyle had gone a full circle, from the opulence of the General Bruce era to the frugality of Bill Tilman, Eric Shipton, Ed Hillary and George Lowe, and now back to luxury, with good food, whiskey, telephone and Internet. The real explorers would be turning in their graves!

The radio crackled and an English voice was reading out an Everest weather forecast, direct from Berkshire. A satellite telephone stood on a table next to a laptop computer and there was a hi-fi playing CDs. Next door, another tent was bristling with radio, TV and computer equipment. Internet connection gave us instant communication to the outside world. I sent a bunch of e-mails to family, friends and work. "Bet this is the highest e-mail you've ever had! I'm sitting at 6,400 meters on the North Face of Everest, the North Col is right above us, and the view is stunning. . . . ." I bragged to colleagues back in Oxford. This was a far cry from the Mallory days of the 1920s. How the world has shrunk!'

* * * * *

Everest 2001–11: More Madness

By the end of 2001 the total number of ascents of Everest had reached 1,501, including 85 ascents without the use of bottled oxygen. One-hundred-and-seventy-one climbers had died on the mountain, and most of their bodies are still somewhere up there. Before 1970 over 50 per cent of fatalities occurred in the Khumbu icefall, whereas nowadays more fatalities occur in the 'death zone' above 8,000 metres. This is probably because technical advances in equipment have made the icefall increasingly safe, and the fact that summit attempts now leave routinely from the South Col, making the summit day extremely long and arduous. Hillary and Tenzing left for the summit from Camp Nine, halfway up from the South Col, which left them sufficient time to top out and return to the South Col the same day.

In 2001 there were nearly fifty expeditions on Everest, all except one of which were attempting the standard routes of the South Col (now sometimes derogatorily referred to as the 'Yak route') and north-east ridge. Climbers seemed to be completely fixated on the summit. There was no exploration or innovation in their thinking at all. Over one thousand people were at the two base camps. A total of 183 climbers summited in five days (between 19 and 25 May) that year, over five times the number that summited in 1978. On one day alone, 23 May, forty-seven climbers from the Nepalese side and forty-two from the Tibet side reached the summit. The summit of Everest that day was indeed a crowded place with eighty-nine climbers, all jostling for the top, another ghastly record. As of 2011, the Nepalese mountaineer, Appa Sherpa, has now climbed Everest twenty-one times. The record time for an oxygen-assisted ascent is Pemba Dorjey's 2004 climb that took only 8 hours, 20 minutes on the south-east ridge. Frenchman Jean-Marc Boiven paraglided from the summit in September 1988 and Slovenian Davo Kamicar made the first complete descent by ski in 2000.

With all these people on Everest one has to wonder about the ecological state of the mountain. An American Environmental Expedition the previous year collected a phenomenal amount of rubbish. The team put thirteen climbers, including ten Sherpas on the summit and they brought down 509 oxygen cylinders, 12 kg of dead batteries, 168 kg of burnable rubbish and 35 kg of non-biodegradable rubbish. They also took 832 kg of human excrement and 750 kg of waste food back to Gorak Shep where it was buried. That was the year before one thousand people showed up at base camp in 2001, and another thousand or so in 2002, and another in 2003. Where is it all going?

Everest Anniversary, 2003: Fifty Years On

In the spring climbing season of 2003, the fiftieth anniversary of the first ascent of Everest, Everest records tumbled like a pack of cards. On 22 May, Lhakpa Sherpani climbed Everest for the third time, a record for a woman. She was accompanied by her 15-year-old sister, Ming Kipa, and her brother, Mingma Gelu, a family outing to remember. The following day, Pemba Dorje Sherpa climbed from base camp to the summit in 12 hours 45 minutes, the speed record. This was broken three days later by Lhakpa Gelu Sherpa who made the trip in an incredible 10 hours 57 minutes. On 1 June Yuichiro Miura, who filmed *The Man Who Skied Down Everest* in 1971, became the oldest man to climb Everest at 70 years old. (This record was broken in 2008 by 76-year-old Min Bahadur Sherpa who climbed the Nepalese side.) The youngest person to climb Everest is 13-year-old Jordan Romero, who summited in 2010. Over 400 people are at base camp on the Nepal side of the mountain and a similar number camped on the Rongbuk Glacier on the Tibetan side every season. Live telecasts from base camp and the summit are de rigeur, and there has even been an Internet café opened in a tent amongst the boulders at the Khumbu Base Camp. The Everest fest was rudely interrupted when a Russian helicopter crashed coming into land at base camp and killed two people, the second helicopter crash at this site.

On 29 May 2003, the exact anniversary of the first ascent of Everest, a large gathering of Everest veterans, Sherpas, and mountaineers met in London in front of the Queen and several members of the royal family for a celebration evening. Fifty years earlier the news had reached London, after some remarkable scheming and trickery by James Morris, *The Times*' correspondent on the expedition, on the very day that the young Queen Elizabeth II was crowned.[10] 'Everest—The Crowning Glory', 'All This and Everest Too!' shouted the banner headlines, celebrating the triumphal ascent by Hillary and Tenzing. The surviving members of the 1953 expedition (George Lowe, George Band, Mike Westmacott, Charles Wyllie, Griff Pugh, Michael Ward) reminisced nostalgically about their historic trip. Stephen Venables, Chris Bonnington, and Doug Scott gave their take on the Everest experience, and Jan (formerly James) Morris gave a witty account of the way he managed to get the news relayed back to London. There was a live video link to Edmund Hillary in Kathmandu and another live link to base camp on the Khumbu Glacier, where climbers just back from the summit were interviewed. That beautiful, sunny summer's afternoon in London, the British Raj was once again alive and kicking.

Human Impact and the Future

The early attempts in the 1920s and 1930s used to last anything from six months to years. They involved steamship voyages from England east to Bombay or Calcutta, followed by a train ride north to Siliguri and Darjeeling, then an arduous month's trekking through Sikkim and Tibet just to reach the Rongbuk Glacier. Nowadays, Everest expeditions can last as little as three or four weeks. The well-known trail to Everest in Nepal has tea-shops every few hundred metres along the way. The trail is crowded during the season, and campsites are usually full. Sleepy little Sherpa villages became busy watering holes, with new lodges springing up every year. Porters carry canned beer up from Lukla for the thirsty trekkers. Namche Bazaar boasts several bakeries, Internet cafés, and modern lodges. It has been estimated that 27,000 trekkers pass through Namche Bazaar each year. This trade has made some Sherpas very wealthy. Many own large houses in the Sola Khumbu and also in Kathmandu. Less wealth has trickled down to the Tamang and Rai villages further south, where the majority of the porters come from.

This disproportion in the wealth generated from Everest has contributed greatly in creating the 'Maoist' insurgency in Nepal. Most 'Maoists' are simple village folk who have no idea who Mao was, or what he stood for. They are people in need of a job, an income, and some hope for the future. They live in rather poor villages and farms in the lower parts of the Nepal Himalaya and every year they see a steady stream of wealthy Western mountaineers walking past, on their way to climb a mountain that many Nepalese regard as the abode of gods. The meagre wages paid to porters goes only a little of the way towards their needs for looking after their families and dependants. Often, the porters stagger under huge loads, some as much as 35 or 40 kgs, sometimes wearing only flip-flops on their feet. They are a hardy, happy, and hard-working people, and they rarely complain.

Several prominent Everest climbers have devoted a huge amount of time and effort to helping their Sherpa friends and families. Edmund Hillary is the prime example, and has been an inspiration to many. Perhaps his greatest achievement was setting up his Himalayan Trust, which builds schools and hospitals, and raises money for the general improvement of the Sherpas. The policy of the Trust is to respond to requests from the Sherpa community, never to impose projects, and always to involve the local community. In 1961 Edmund Hillary and his Sherpa friends started to build the first school at Khumjung. Later, many other schools were built in villages all over the Sola Khumbu, and in December 1966 the hospital at Khunde was opened. Edmund Hillary was always very wary of increasing the commercialization of Everest and before he died in January 2008 called for a ban on commercial expeditions to the mountain. Doug Scott is another who has worked tirelessly to get fair employment pay for porters, and the charity he set up, Community Action Nepal, builds schools, hospitals, and porter shelters all over Nepal. This is a wonderful way for Western trekkers and climbers to put something back into the country.

Tenzing Norgay was also an inspiration.[11] He was a natural, gifted mountaineer, had a wonderful flashing smile, and by all accounts was a perfect gentleman. After Everest he was practically deified by the Nepalese and Indian press, who concocted various untrue stories about how he hauled Hillary up to the top. However, he had the strength of character to survive the media onslaught, and his relationship with all the British climbers of the 1953 expedition, and particularly John Hunt and Edmund Hillary, was always strong. He went on to found the Himalayan Mountaineering Institute in Darjeeling, which trains young climbers, and he lived in that beautiful town looking out over to the snows of Kangchenjunga until he died there on 9 May 1986. His son, Jamling, and his grandson Tashi Tenzing have both climbed Everest (another record: the first time three generations of one family have summited Everest), while his nephew Gombu was the first person to climb Everest twice. Tenzing is affectionately remembered all over Darjeeling and beyond, to this day.

The Sherpas of the Sola Khumbu nowadays are relatively well off by Nepalese standards. It is the more remote villages of the Tamang and Rai communities who now need most help. The impoverished Terai is the recruiting ground for the Maoist rebels. But it is not really some hallucinatory Communist paradise they seek, it is merely the simple chance to have a job and an income to help feed and house their families. For a while, the Maoists took to holding some trekking groups to ransom. The more polite gangs even handed out receipts after extorting money from the group, so that others would know who had paid up. Some of these modern-day Robin Hoods actually serve the poor communities well, whilst others are simple highway robbers.

It is difficult to see where the future of Everest trekking and climbing is going. As long as the supply of mountaineers willing to risk life and limb to climb the world's highest peak remains, so the Everest business will continue to thrive. Perhaps the most needed visitors to Everest and the Himalaya should be the scientists, for problems thrown up by the natural world could cause the human problems to pale into insignificance. The chance of a massive earthquake flattening Kathmandu, with hundreds of thousands of casualties, is very high; we do not yet know exactly where it will be, or when, but happen it certainly will. The monsoon system, which controls the lives of billions of people in the entire Indian subcontinent, is governed by geological factors, such as the uplift of Tibet and the Himalaya. If the monsoon fails, millions could die; if the monsoon is strong, then Bangladesh and the plains of Pakistan drown. Deforestation is a major problem around the overcrowded cities and towns of Nepal. Pollution, waste disposal, corruption, and politics are all major problems.

The hope for the future is that, since Everest is such an icon to millions of people all over the world, the world will not allow this wonderful and unique place to be destroyed. Preserving the environment must be an absolute necessity, come what may. If that involves restricting commercial trips or banning large-scale expeditions as Edmund Hillary suggested, so much the better. The sheer numbers of people going to Everest will have to be controlled somehow in the not-too-distant future. Is it ethically right that rich, foreign expeditions fly in huge amounts of luxury goods, employ porters at meagre wages to carry their gear up, rely on Sherpas to lead them all the way to the summit, and leave so much rubbish on the mountain, when Nepalese villagers live in squalor in the shanty suburbs of Kathmandu and the poverty-ridden villages on the Terai? How true today are the words of Howard Somervell, written in 1924: 'I verily believe his death (Mallory's), as that of his well-loved and splendid companion (Irvine), is a clarion call to our materialistic age, which so terribly needs the true unselfish spirit typified by George Mallory alike in his life and in its ending.'

Of all the money spent on Everest in the last fifty years, less than 1 per cent has been on the science of Everest. Ultimately it is science that will make the world a better place to live in, it is science that will relieve poverty, and it is scientific knowledge that will enable us to understand how the Earth works. Maybe it is time that, if we really want to understand this planet we live on, what makes mountains, what drives climate, and how the Earth works, we devote our efforts and resources into funding the science that will ultimately tell us. What better place to start than Mount Everest?

CHAPTER 8

Mapping the Geology of Everest and Makalu

For the stone from the top for geologists, the knowledge of the limits of endurance for the doctors, but above all for the spirit of adventure to keep alive the soul of man.

George Mallory, 1923

There are few places in the world where a geologist can actually take a look at the rocks and structures 5 or 6 kilometres down beneath the Earth's surface (Figure 8.1). The opposing forces of nature—the uplift of rocks towards the Earth's surface and their erosion and removal—usually balance each other out, at least roughly. It is only where the rate of uplift of rocks greatly exceeds erosion that high mountains are built. This is precisely why the Himalaya are so unique to geologists studying mountain-building processes. The Himalaya is an active mountain range: the plate convergence rates are high, uplift rates are extremely high, and glacial and fluvial erosion has carved deep channels in between the mountains. By walking and climbing all around Everest we can actually map and study the rocks in three dimensions, which elsewhere, beneath the Tibetan Plateau for example, remain buried below the Earth's surface.

After the Survey of India discovered that Mount Everest was the highest mountain in the world, a pioneering expedition set out to fly across the summit and take photographs. On 3 April 1933 a Houston-Weston biplane piloted by Lord Clydesdale flew across the summit and took the first photos of the mountain (Figure 8.2). Clydesdale wrote: 'We were in a serious position. The great bulk of Everest was towering above us to the left, Makalu down-wind to the right and the connecting range dead ahead, with a hurricane wind doing its best to carry

Figure 8.1 Aerial view of Mount Everest rising behind the Nuptse-Lhotse wall, taken from the south. Ama Dablam rises in the foreground. Photo courtesy of Paul Harris.

us over and dash us on the knife-edge side of Makalu.'[1] The earliest geologists to study the structure of Mount Everest, A. M. Heron and Noel Odell, both noted the apparent conformity of strata with sedimentary rocks on top of the mountain lying above the more metamorphosed rocks around the base[2] In his 1965 paper on the structure of Everest, Lawrence Wager wrote: 'It never ceases to surprise the writer that the highest point of the Earth's surface is composed of sedimentary rocks which are relatively flat-lying and but little metamorphosed.'[3] Wager thought that 'the relatively simple tectonic pattern so far found in the eastern Himalaya was presumably due to the recent date of the Himalayan orogeny, so that the high level and the structurally simpler, upper part is still partly preserved. Beneath what can now be seen there is probably a more complex pattern of tectonic, metamorphic and igneous phenomena, and on a scale commensurate with the grandeur of the range.' Writing before the advent of plate tectonics, Wager implied that the sedimentary rocks on the summit pyramid of Everest had been simply pushed upwards with the uplift of the range. He was, after all, used to working in the

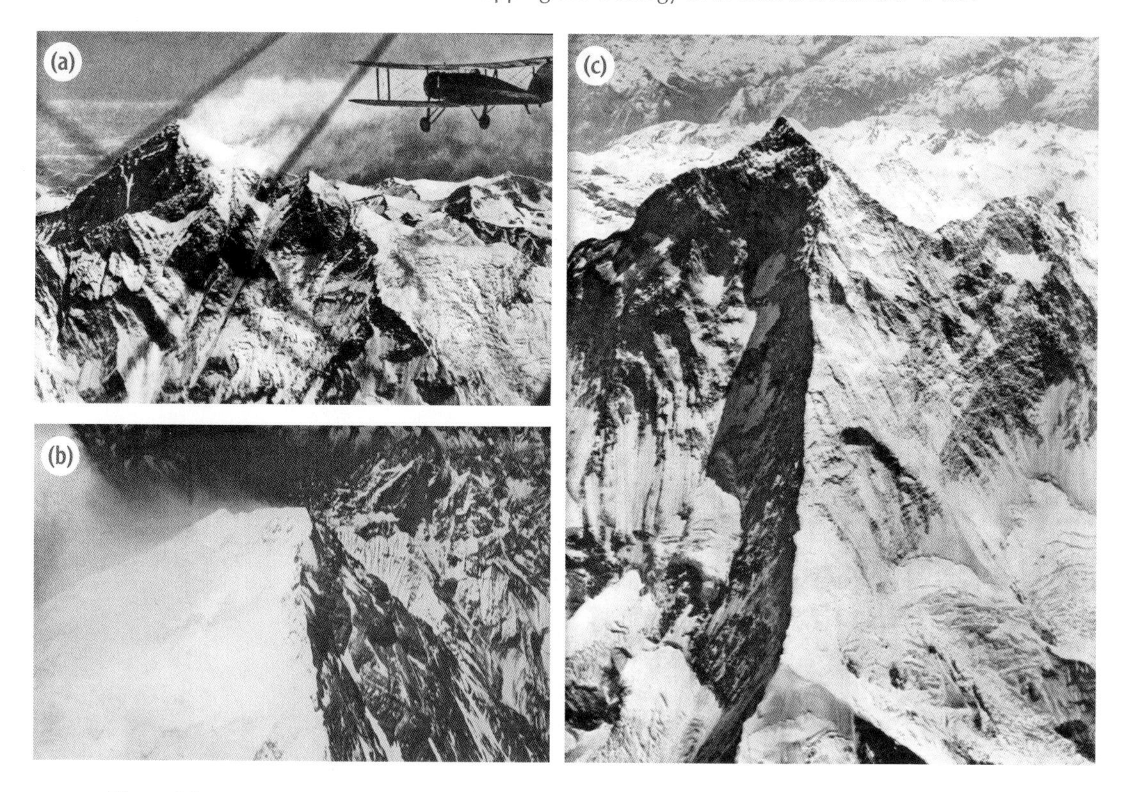

Figure 8.2 The first photos of the summit of Everest (a, b) and Makalu (c) taken in 1933 during the first flight over Everest by the Houston-Westland expedition. The spectacular north ridge of Makalu was initially mistaken for Everest.

deeply eroded, old mountain ranges of Scotland and Greenland, where the upper 20 or so kilometres had long been eroded away.

Everest is a giant pyramid situated right on the main continental divide. Deep glacial valleys have carved out 4,000–5,000-metre cliff faces along three flanks. The main watershed lies along the west ridge, which continues along the Himalayan divide across the summits of Lingtren, Pumori, Gyachung Kang, and Cho Oyu to the west, and along Lhotse, Chomolonzo, and Makalu to the east. The Everest–Lhotse–Nuptse horseshoe ridge is the southern catchment for the Khumbu Glacier which has scoured a hole 4 kilometres deep right into the Everest massif. The structures of Everest are beautifully exposed along these flanks. The only problem for the geologist is accessing them. Mapping Everest geology is a strange game, very different from conventional geological mapping, where one can walk over the ground taking measurements, and collecting samples across most of the area. On Everest, one has to rely on cunning, binoculars, and inference to support the small acreage over which one can actually get one's face onto the rock.

Aerial photographs provide the greatest asset for geological mapping (Figure 8.3). During the making of the National Geographic map[4] in 1988, Brad Washburn's team flew across Everest and took a series of superb colour aerial photographs, which can be used for accurately mapping the geology. Likewise, oblique aerial photographs of the Everest massif have been used for plotting the geological structures. Climbers have been helpful; some have managed to collect samples

Figure 8.3 Aerial photograph looking directly down onto the summit of Everest taken during the making of the National Geographic Society map of Everest. The SW face is in shadow below bounded by the west ridge (left) and southeast ridge (right). The North face is sunlit (upper left) and the Kangshung, or East face, is top right. Photo courtesy of Bradford Washburn and Swiss photos.

from high on Everest and bring them back for analysis. Mostly, though, climbers have other things on their minds at altitude, like simple survival, and collecting geological rock samples is not high on their agenda. I did, however, persuade three of my climbing friends, all guides, Jon Tinker, David Hamilton, and Kenton Cool, to collect rock samples from between the South Col and the summit of Everest.

One unique source of base map information came from an unexpected source. Leo Dickinson is a professional cameraman–adventurer, who specializes in extreme sports in wild places. He has filmed base-jumping off the Trango Towers, canoeing down the Dudh Kosi River in Nepal, and sky diving with peregrine falcons, among numerous other adventures. In 1991 he hatched a wild scheme to float a hot-air balloon over the summit of Everest.[5] After years of raising funds, he and his old climbing mate Eric Jones teamed up with Australian balloon pilots Chris Dewhurst and Andy Elson. They took off from Gokyo Lake, south of Cho Oyu, at dawn one fine day in October and floated serenely up to 8,000 metres above the Khumbu Himalaya. Suddenly they hit the jet stream, winds that blew them at speeds up to 90 miles per hour right across the summit of Everest on a perfect, cloudless blue-sky day. The basket was full of gas canisters and other ballooning paraphernalia, so Leo had to hang on to a tiny platform strapped to the outside of the balloon. From here he shot some of the most sublime photo images of Everest ever taken. As they rocketed over the summit, he shot panoramas of the whole of the Khumbu, close-up shots of the Kangshung face, and amazing views south towards Makalu and eastern Nepal. After a flight lasting only 95 minutes, they crash-landed in a remote valley in southern Tibet, much to the amazement of a couple of passing yak herders.

Mapping Everest's geology has been a journey of discovery (Figures 8.4 and 8.5). Altogether I spent seven field seasons mapping the Everest–Lhotse–Nuptse massifs around the Khumbu south side in Nepal and the Rongbuk and Kangshung sides in Tibet between 1997 and 2003. The map was first published in 2003, the fiftieth anniversary of the first ascent, and a second edition with the addition of the Makalu massif, Kangshung Glacier, and Barun Glacier region was published in 2007.[6]

Flowing Rocks and Non-Rigid Tectonics

As more and more geological and geophysical information came out of the Himalaya and Tibet, it seemed likely that the theory of plate tectonics, with its rigid, undeforming plates, was not really applicable to the structure and tectonics of great continental masses like Tibet and parts of some mountain ranges like the Himalaya. The oceans, with their large plates of dense basaltic crust, did appear to

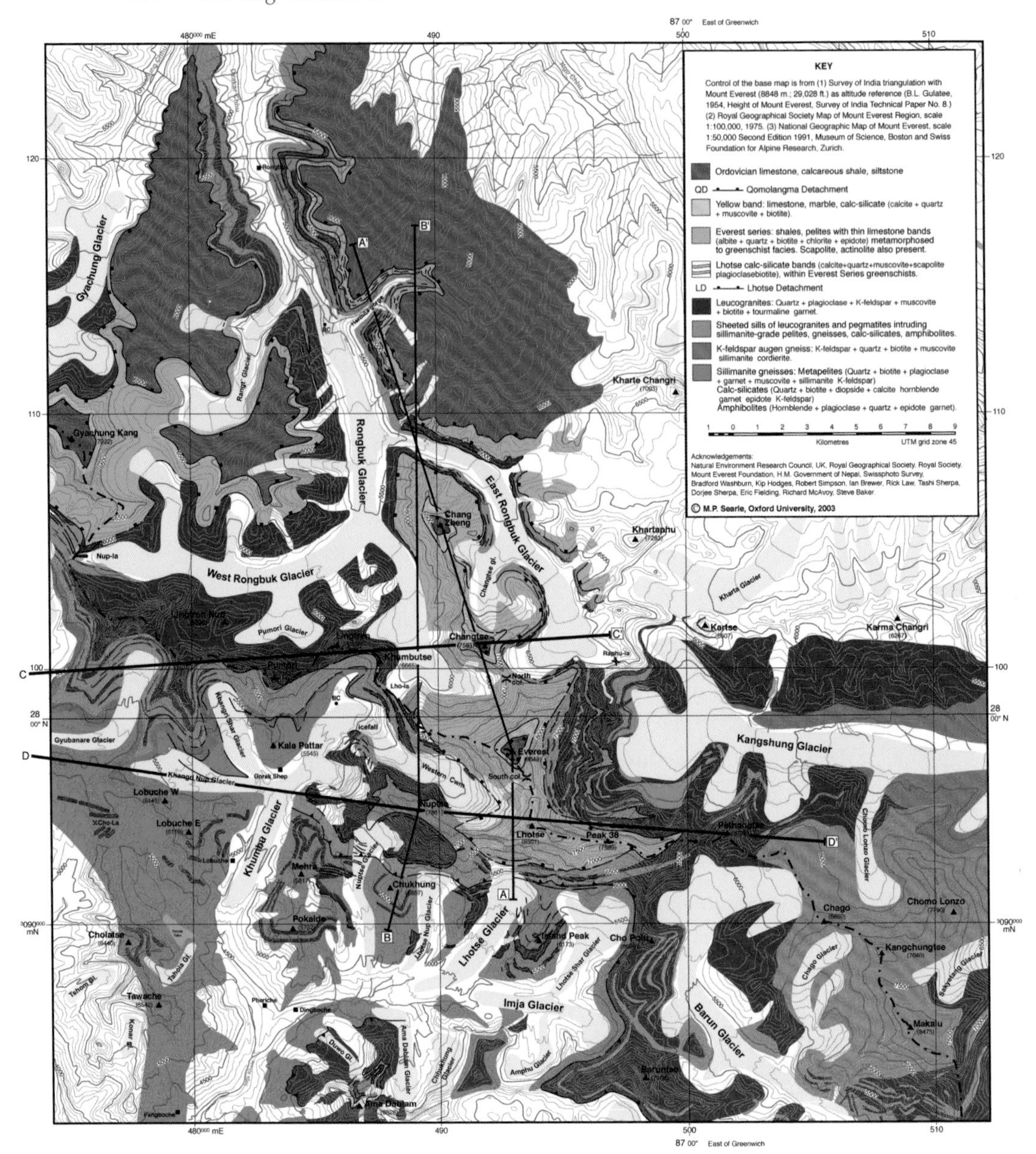

Figure 8.4 Geological map of the Mount Everest and Makalu massifs in Nepal and South Tibet. Published by the author first in 2003, and subsequently in 2007.

be largely rigid, deforming by brittle faulting and fracturing, or gentle buckling, but the pattern of deformation on the continents was not like this at all. Even the pattern of earthquake distribution showed that earthquakes were scattered all over the Tibetan Plateau, not just restricted to its deforming margins. Earthquakes only occur in brittle deforming rocks; when the temperatures become greater than

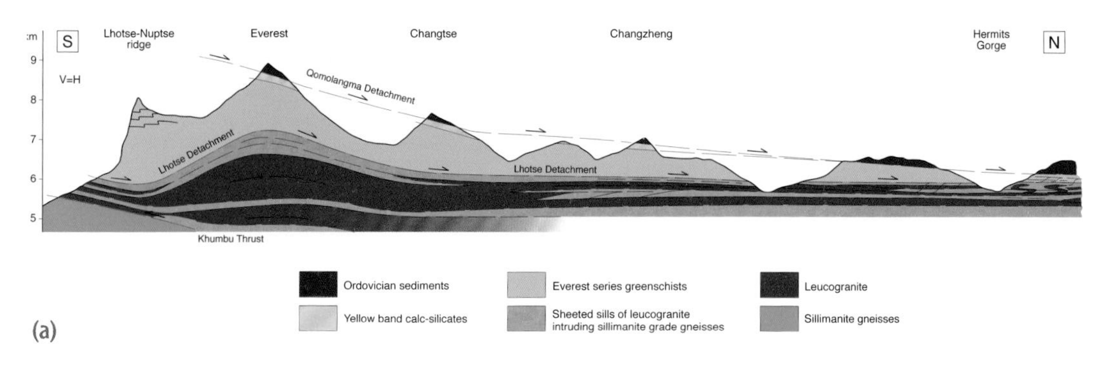

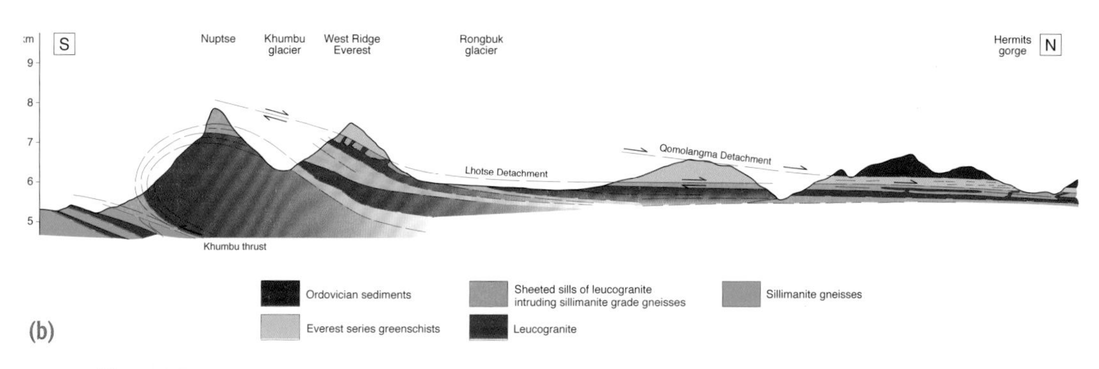

Figure 8.5 Geological cross-sections of the Everest massif.

around 400–500°C the rocks deform by ductile flow rather than sudden, brittle fracture. Where lower crustal rocks have been exposed along the high Himalaya these gneisses show intense deformation in the form of ductile folds which behave almost randomly, like treacle flowing out of a can.

Large parts of the higher Himalaya, including much of the Everest region, show that the rocks were once partially molten, with pockets of granitic liquid intercalated within the original gneisses. In the highest peaks of the Everest Himalaya, sills of molten granite magma intrude as a dense network into the surrounding metamorphic rocks. The deep structure of the Himalaya and Tibet was turning out to be nothing at all like the stable continental masses of peninsular India. It was crucial then to map out in detail the geology of these mountains and valleys, to date the rocks to know exactly when they formed, and to obtain all the other parameters of pressure, temperature, depth, and time needed, before we would really begin to know how these mountains evolved.

Where better to look than the Everest massif, where the largest vertical relief in the Himalaya exists? Here, there are whole networks of granites that could be mapped out in great detail in three-dimensional space (Figure 8.6). The valleys

Figure 8.6 Magnificent exposures of layered granites (white) intruding into gneisses in an unnamed peak to the west of Pumori, Everest region of Nepal.

that drain south of the Nepal–Tibet watershed provide ideal access routes into the high mountains, and passes connecting these valleys could be crossed to correlate the geology from one valley to the next. The accessible geology in the Everest Himalaya still only provides us with direct access to the upper 10 to 20 kilometres or so of crust. To look deeper into the crust, we would have to rely on geophysical means.

Geological Structure of Everest

During the continental collision, rocks of the Indian plate were thrust down to the north beneath Asia, heated and metamorphosed, and then returned to the surface by thrusting towards the south. Everest lies right at the top of the Greater Himalayan slab, the layer of high-grade metamorphic and granitic rocks that characterize many of the highest peaks of the Himalayan Range. These rocks are part of the giant mid-crustal slab that has been thrust out towards the south, bounded by major shear zones. A ductile shear zone is the deep equivalent of a brittle fault in the upper crust. The shear zones at the top of this slab of deep rocks are exposed on the Everest massif itself. Two such normal faults have been mapped across the Everest–Lhotse–Nuptse massif and also further west across Gyachung Kang and Cho Oyu,[7] and even further west to Shisha Pangma. These normal faults always put shallower-level, lower grade metamorphic rocks on top of deep-level, higher grade (higher temperatures and pressures) metamorphic rocks.

The low-angle normal faults on top of the Greater Himalayan sequence slab extend along the entire length of the Himalaya, and in the Everest region show two major faults. The upper one, called the Qomolangma Detachment, is the highest normal fault in the world. It cuts right through the summit pyramid of Everest above the Yellow Band, a prominent marble band that encircles the summit, placing very sheared but generally un-metamorphosed shaley limestone rocks on top of the metamorphic marbles of the Yellow Band, rocks that have been heated up to higher temperatures and buried to deeper levels than the summit rocks. The lower normal fault, termed the Lhotse Detachment, places these intermediate grade rocks of the Everest Series above very high-grade metamorphic rocks containing the high-temperature minerals sillimanite, cordierite, and garnet. Between 30 and 50 per cent of the rocks beneath the Lhotse Detachment are granites, whereas above this fault, very few granites occur.

Mapping on Everest and along the Rongbuk Glacier to the north has shown that the Qomolangma Detachment is a shallow north-dipping planar feature that clips the summit cone of Everest and also encircles the summit of Changtse Peak to the north. The fault is beautifully exposed along the entire length of the east and central Rongbuk Glaciers showing black shales above and layered granites and metamorphic rocks beneath. It dips north about 10° on Everest and flattens out to the north along the Rongbuk Glacier to become almost horizontal in places. The lower Lhotse Detachment is more complex, a deeper, hotter ductile shear zone showing more distributed deformation. The detachment is also folded around the top of the bulbous Nuptse granite, and can be traced west along the south face of Cho Oyu and all around the mountain of Gyachung Kang. The Lhotse Detachment has truncated and cut all the huge granite sheets, the inflated sills of Nuptse, and the massive Makalu granite beneath. If we could date the granites cut by the fault, the age would give us a maximum age of movement along the fault, since it must be younger than the granites. The next step in the puzzle then was to collect samples of these crucial granites and get them dated.

Sola Khumbu Field Campaigns

Although I had made a rapid trek up to Everest Base Camp before our Langtang Lirung expedition in 1980, the first year I started working on the geology of the Everest region was in 1996 when I made two trips to Nepal. The first was in the spring, with Ian Brewer, a seven-week trek up to Everest Base Camp, then over the Cho-la Pass to Gokyo and up the glacier to the south face of Cho Oyu. During this trip I realized that the major granites around Everest Base Camp were actually horizontal sheets or sills and they extended for tens of kilometres both along the strike and down strike towards the north. These granites were completely different from the usual Andean-type granites that intruded vertically upwards in bulbous plutons. The Himalayan granites were also intruded horizontally into their metamorphic host rocks, from which they had been derived by partial melting. After the monsoon I returned with a BBC crew to film the Himalayan parts of *Earth Story*, a documentary directed by Simon Lamb dealing with all geological aspects of the Earth from volcanoes to earthquakes to the making of the Himalaya and the Andes.[8] We trekked up to Ama Dablam north face from where we could get a magnificent panorama of the entire southern aspect of the Everest massif. We had bad weather most of the way but on the very last morning at dawn the clouds parted and revealed Everest in all its majesty.

The following year I took a new D.Phil student from Oxford, Rob Simpson, with me. Our *sirdar* was a young Sherpa, Tashi, from a village south of Lukla. We hired three porters to carry our rock samples and trekked up from Lukla to Everest where I had arranged to call in and stay at Jon Tinker's base camp. We sampled the granites and gneisses all around the southern flank of Everest and ascended up to the big cliffs on the Lho-la and Lingtren. Jon managed to collect a few summit samples for me.

Tashi Sherpa

At the end of my work I left Rob up at Lobuche to give him time to become immersed in his project and trekked back to Lukla. Tashi volunteered to come as far as Lukla with me and carry my big rucksack. We left Lobuche one cold, snowy day and trekked rapidly down as far as Namche Bazar in one long day. That day we over-nighted in a trekking lodge where the rooms were so small we had to leave our rucksacks piled up outside the door of the room. At 5 a.m. we woke to a commotion outside when our neighbours left to ascend to their base camp. When we emerged at a more leisurely 7 a.m. Tashi discovered that they had accidentally taken his small rucksack with all his possessions in their rush to leave. Instead of the typical Western reaction of rage and swearing, I was amazed to see Tashi grinning broadly when he told me the story, then laughing uproariously. Sherpas certainly have a splendid sense of humour, even when the joke is on themselves. There is so much, it seems to me, that Westerners can learn from this culture and attitude.

The next day as we trekked downhill through the magnificent pine forests of the Khumbu towards Lukla, Tashi turned to me and said, 'Dr. Mike I want to come to England.' 'No problem Tashi, you are welcome; you can stay my house any time,' I replied casually. Back in Kathmandu he went to the British embassy to obtain the information on precisely what was required in my letter of sponsorship. Two months later, back in Oxford, I received an email from Tashi: 'Namaste Dr. Mike; thank you very very much for embassy letter—they gave me three month visa and I bought ticket for next Saturday—I coming Royal Nepal airline to Gatwick airport—I think that place nearby you in Oxford yes. I very much looking forward seeing England. See you soon, Tashi.'

Rob Simpson went down to Gatwick airport to meet Tashi, who duly appeared loaded with bags of pashmina shawls and carpets. He had $50 in his pocket and only a vague idea where Oxford was in relation to London. Tashi appeared at my house, complete with broad grin and infectious enthusiasm for just about everything. Luckily I had advice from our local pub landlady, Debbie Hall at the Rose and Crown on North Parade, about an English language school on the Banbury Road, so we went there and enrolled him in a six-week English language course. His teacher turned out to be an English girl who could speak fluent Nepali, and Tashi took an immediate fancy to her and the school. We called in at the Rose and Crown for a quick pint and to thank Debbie for her excellent tip. She was running around doing three jobs at once in the busy pub, because one of her bar staff had suddenly quit. She sat down exhausted and suddenly said, 'Tashi I don't suppose you want a job do you?' Of course, Tashi leapt at the chance. In Nepal such a job would be almost impossible to get and he was keen to earn money to take home and support his family. In one day we had fixed Tashi up with an English language course and a job at the Rose and Crown.

Tashi's stay in Oxford was turning into a great success. He loved his part-time job at the pub, learnt how to pull pints, discuss the weather with learned locals propping up the bar, and how to cook 'sausage 'n mash' and Sunday roast lunches. He made friends with everyone in the pub from local academics, BBC directors, and journalists to Lithuanian gap-year girls on temporary jobs in the pub. Debbie said that he worked twice as hard as anyone else and nothing was ever too much trouble. Tashi set up a sideline selling Nepali pashmina shawls with Debbie and another friend Alison Bennet-Jones. Tashi would get his sister, Lami, to send pashminas over from Kathmandu and they sold them in Oxford and around the Sherpa community in Reading. Every time I returned to Nepal, Lami would sweetly ask 'Dr Mike, please could you take small parcel back for Tashi?' Then a large package of pashminas would surreptitiously appear next to my rucksack. After three months it was time for Tashi to return to Nepal. On his last evening we had a great farewell party in the Rose and Crown and lots of the pub regulars turned up. Tashi was rapidly becoming a local institution.

The following year he was back in England again, this time with a six-month visa obtained through the Buddhist Association. He had found a Nepalese monastery on the Cowley Road in Oxford, with a single resident monk from the Sola Khumbu district. There was also a network of expatriate Nepalese in Reading, many staying with retired ex-Gurkhas who were part of the British army. Whilst Tashi was in Oxford I was attending an HKT geological conference in Gangtok in Sikkim. I took the opportunity after the meeting to trek up to Kangchenjunga Base Camp from the east with a friend, Adam Szulc, from Edinburgh University, who was working towards a PhD on the detrital sediments eroded off the Himalaya and deposited by rivers into the Siwalik basin to the south. Tashi had told me that when we returned to Darjeeling we should go to a special hotel there, the North Star Hotel on K.D.Lama road. After three weeks of trekking through the magnificent forests of Sikkim up to the icefalls of Kangchenjunga, Adam and I arrived back in Darjeeling, tired and dripping with mud and Himalayan dirt. At the North Star hotel the owner, Sonam, turned out to be Tashi's uncle and Doma, his wife, the Chief Magistrate, both pillars of the Sherpa society. Adam and I were given a magnificent suite on the top floor with stunning views across to the entire Kangchenjunga massif.

For four days in Darjeeling we were treated like royalty, driven around by Sonam in his car, wined and dined by Doma and Sonam in their penthouse apartment, and visited various relatives of Tashi's all over Darjeeling. Sonam refused payment for anything from us. Coming down to breakfast on the first day I had noticed an elderly man who looked vaguely familiar sitting quietly in the corner. Sonam beckoned him over and introduced him to us. 'Dr Mike, this is Tashi's father; he has just arrived from their village in Sola Khumbu.' It turned out that Gyaltzen had walked two days from their village to the airstrip at Phaplu, flown to Kathmandu, then travelled for two days by bus from Kathmandu to the Sikkim border and up to Darjeeling, simply to say, as Sonam translated: 'Namaste Dr. Mike and thank you for what you have done for Tashi.' I was touched and amazed at this wonderful gesture, and replied that I had really done nothing except write a letter of support to the embassy and offer him a room in my house. Gyaltzen could not speak any English and my Nepalese was only basic, so on the two-day bus ride back to

Kathmandu we spent a long time grinning and clinking beer bottles to each other. The bus went through the night and, as almost all of the Terai lowlands of Nepal were under the control of Maoist rebels, we had to get out and unload at numerous check-posts every few hours in the jungle.

After six months Tashi's visa ran out and after another leaving party at the Rose and Crown we piled him onto the Heathrow bus laden with Oxford T-shirts, and presents for everyone back in Kathmandu. A few months later I had an e-mail from Tashi saying that he had bought some land on the edge of the Kathmandu Valley and was planning to build a house. He also got married to a delightful Sherpani lady, Kalpana, from the Sola Khumbu. On our next visit to Kathmandu we were all invited to a wonderful Sunday lunch, with copious amounts of Everest beer, on the roof of his new house overlooking the Kathmandu Valley, with the snows of the Langtang Range glinting above.

Kangshung Face of Everest

In the autumn of 2003 Rick Law, an expatriate Brit who was a professor at Virginia Tech in the US, and I managed to get a three-year grant to work on the structural evolution of the Everest massif. We had two new PhD students, Micah Jessup, an enthusiastic American climber and geologist based at Virginia Tech, and the following year John Cottle, a New Zealander who obtained a scholarship from New Zealand to come to Oxford to work for a D.Phil with me. In 2003 we made another trek up to Everest from the Nepal side and collected samples for both thermobarometry and strain analysis for Micah to work on. The following year, after meeting John in Lhasa, we headed to the Kangshung Glacier on the remote east side of Everest in Tibet. Flying over the Himalaya from Kathmandu to Lhasa, we had a spectacular view of the Everest massif and its eastern neighbour Makalu towering high above a sea of cloud.

Lhasa was a completely different city from when I first visited twenty years earlier. The only old part of town left was the Barkhor area surrounding the Jhokang Temple where old mud-brick, yak-dung, and straw houses radiated around the central Jhokang, with its golden roofs and gilded statues. Elsewhere all the old houses had been torn down and replaced with monstrous new blue glass and concrete Chinese constructions appearing just like any other modern Chinese city. The Potala Palace, the home of the Dalai Lamas, of course remained, but it was a dead building, devoid of any normal life and unlike all the active living *gompas* in Ladakh. There is no doubt that the life of many Tibetans in the cities has improved under the Chinese, but at the cost of a devastating cultural decapitation. Many of the nomadic pastoralists in the remote Chang Tang region had been housed in appalling concrete box houses in new cities along the truck routes. Chinese colonialism has, however, not dimmed Buddhist fervour, or the devotion to the now exiled Dalai Lama.

In Lhasa we met up with Réne Schrama, a Dutchman who was something of a legend in Lhasa and was running a travel agency in partnership with my friend in Kathmandu, Shiva Dhakal. Réne and his wife Kris, together with another Dutchman, Fred Madern, were also running a bar, the Dunya, attached to the Yak Hotel. Fred was another Lhasa legend, standing head and shoulders taller than most Tibetans with his huge handlebar moustache and chef's hat. The Dunya was an oasis for tourists and travellers who wanted a decent cup of real coffee, a beer, or a good meal. Fred helped us shop for a six weeks' supply of food from the amazing new market in Lhasa near the Potala Palace, where you could buy almost anything trucked in from China. We set off in a couple of Land Cruisers once again for the remote Kharta Valley to the north of Everest. We hired a few local porters and a team of sturdy yaks trekked in across the Sho-la to gain access to the Kangshung Valley, and reached base camp for the Kangshung face.

The Kangshung Valley is one of the most beautiful hidden gems of the Himalaya. The glacier drains from the hanging icefields of the east face of Everest and Lhotse and flows east before plunging down a series of dramatic waterfalls into the Arun Valley of Nepal. In spring, the Kangshung Valley is carpeted in colourful flowers, its rhododendron forests laced with hanging mosses. Above, the 3,500-metre high north faces of Makalu, Chomolonzo, and Pethangtse plunge straight down in sweeping walls of pale granite shrouded by hanging glaciers. Spectacular avalanches thunder down the slopes to erupt in clouds of powder snow above the forest (Figure 8.7).

Mapping the geology also proved to be fascinating. We first hiked up to the Raphu-la, the high pass to which Mallory and Bullock first climbed from the Rongbuk Glacier and spied the whole extent of the north-east ridge of Everest. Then we crossed the highly crevassed Kangshung Glacier to gain access to the ridge that connects Everest to Makalu. Here we found some extraordinary rocks—giant boulders of metamorphic rock completely enclosed in pale granite (Figure 8.8). The granite had tiny red garnets, black tourmalines, and shiny white muscovite micas, similar to most of the Himalayan granites. Both above and below this level were more sheets of granite. We were right in the middle of the Miocene molten core of the Himalaya, the southward extruding mid-crustal channel bounded by the two ductile shear zones below and above. We collected samples for Micah and John to do the thermobarometry and U-Pb dating work that made up their respective theses.

On the drive back we traced the low-angle normal fault known as the South Tibetan Detachment along the strike for nearly 50 kilometres. At one excellent

Figure 8.7 North face of Makalu and Chomolonzo from across the Kama Valley in Tibet showing five-kilometre-high cliffs of pristine white granite.

locality in the Dzacha Chu I spotted a cliff on the north side of the river that showed a wonderful section through the entire shear zone. It showed layered metamorphic rocks intruded by several sets of granite sills and cross-cutting dykes at the base, grading upwards to layered marble bands and then sedimentary rocks above. We managed to cross the river and clamber up to sample these rocks in great detail. John spent the winter analysing the pressure, temperature, and strain conditions of these rocks as part of his D.Phil project.[9]

We also returned to the Rongbuk Glacier where the most complete and spectacular section through this unique low-angle normal fault cropped out. Rick, Micah, and I trekked into the Hermit's Gorge above the Everest Base Camp and sampled multiple generations of leucogranite dykes, whose ages would give us very accurate time constraints on the formation of the metamorphic fabric. We climbed high on several sections in the lower Rongbuk Valley and discovered that the sedimentary rocks immediately above the ductile shear zone had tiny crinoid ossicles (stems of sea lilies) still preserved, similar to a few of the Everest summit rocks that Jon Tinker and David Hamilton had previously collected for me. These Ordovician-age sedimentary rocks that formed the uppermost summit cliffs of Everest itself dipped very gently to the north above the large-scale South Tibetan Detachment so that we could map the whole upper part of the Greater Himalayan slab in glorious three-dimensional detail.

Figure 8.8 Blocks of earlier-formed metamorphic rocks enclosed in young Miocene leucogranite, north slope of Pethangtse in the Kangshung Valley.

Summit Rocks

The north face of Everest shows an awe-inspiring 3,500-metre cliff falling directly down to the upper Rongbuk Glacier. The three climbing Steps on the north-east ridge are limestone bands that stand proud of the shaley rocks in between, and it is these limestone units that form the main difficulties encountered by climbers since the days of Mallory, Irvine, Norton, and Somervell. In 1924 Noel Odell mapped the summit pyramid of Everest as the 'Upper Calcareous Series' of presumed Permo-Triassic age. A sample from 25,500 feet appeared to have microfossils, but they turned out to be physical features resulting from dewatering, not fossil organisms. In 1933 Lawrence Wager thought that the summit rocks were a little older—Carboniferous–Permian age—by correlating them with the Lachi Series in northern Sikkim.[2] Augusto Gansser published the first thin section of a fossil sea lily or crinoid stem from a summit rock in his book *Geology of the Himalaya* in 1964—the first indication that the highest rocks on Earth were marine sediments.[10] As John McPhee said, 'When the climbers in 1953 planted their flags on the highest mountain they set them in snow over the skeletons of creatures that had lived in the warm, clear ocean that India, moving north, blanked out . . .This one fact is a treatise in itself on the movements of the surface of the Earth. If by some fiat I had to restrict all this writing to one sentence, this is the one I would choose: The summit of Mt. Everest is marine limestone.'[11]

Both Jon Tinker and David Hamilton had previously collected summit samples for me on their Everest climbs. Some of these samples from the summit of Mount Everest had revealed a rather dull sedimentary rock, a limey mudstone with no real distinguishing features, but a few thin sections showed some circular calcite crystals identified also as crinoid ossicles, stems from sea lilies. Detailed stratigraphic work on the better-exposed sections along the Friendship Highway by Paul Myrow and colleagues showed that the summit rocks were actually Ordovician in age, overlying Cambrian rocks of the Yellow Band (Figures 8.9 and 8.10).[12]

We now knew that Everest was cut through by two low-angle normal faults, each placing low-grade rocks on top of high-grade rocks. Rick Law and Micah Jessup had carried out some extremely detailed strain measurements using the long axes of quartz crystals on several of our transects in the Kharta and Rongbuk Valley profiles.[13] These showed that concentration of high strain occurred along the upper fault, the Qomolangma Detachment. We needed to try to constrain the upper structural levels of Everest in more detail.

During the making of a TV programme in 2009, *How the Earth was Made*,[14] I met up with a young climber from Gloucestershire, Kenton Cool, who had done a geology course at the University of Leeds as an undergraduate but had since made a career out of guiding clients to the summit of Everest. He made the ascent every year and by the time I met him he had already climbed Everest six times. The idea of the programme was that I would explain the geology, mainly filming in the Annapurna and Dhaulagiri region, whilst Kenton would heroically take a small camera to the summit of Everest and collect samples for me at specific locations. I asked Kenton to collect from rock outcrops rather than scree and at specific GPS-measured heights on the south ridge. Later back in England the crew filmed Kenton handing over the summit samples to me in Worcester College and talking over my geological map back in my office in the Earth Sciences Department. It was analysis of these samples that finally told us precisely where the fault line was on the summit pyramid of Everest, with a big jump in pressure-temperature conditions across the Qomolangma Detachment. The fault lies immediately beneath the summit Ordovician limestones with their fossil fragments still preserved and above the Yellow Band marble, which is dated as Middle Cambrian.

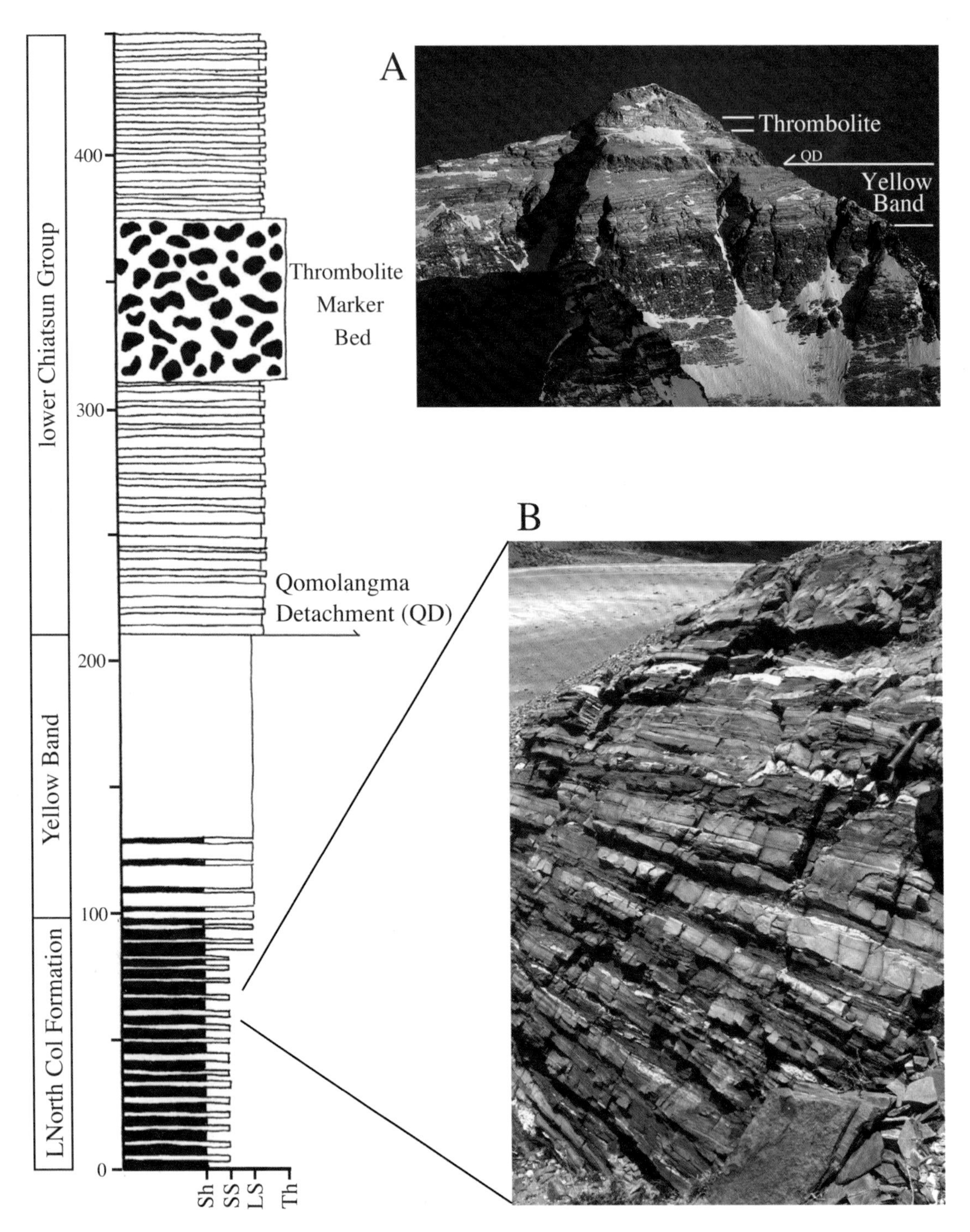

Figure 8.9 Stratigraphic section through the summit rocks of Everest. The Qomolangma Detachment cuts across the summit pyramid beneath the summit rocks which are Ordovician lime mudstones. Thrombolite is a reef-forming mound formed by cyano-bacteria and is indicative of a shallow littoral environment on a tropical shoreline. The Yellow Band is metamorphosed Cambrian limestone, now a marble.

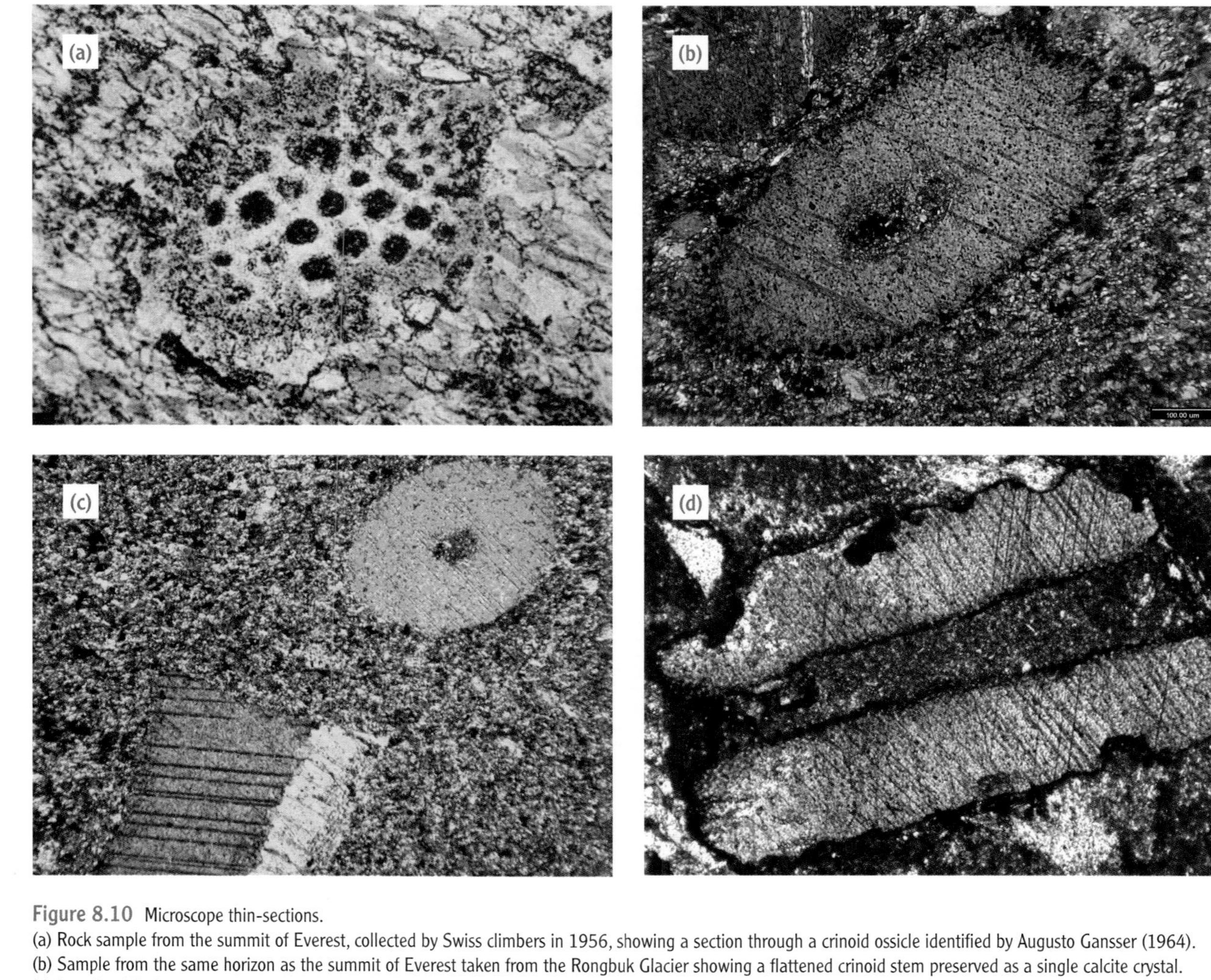

Figure 8.10 Microscope thin-sections.
(a) Rock sample from the summit of Everest, collected by Swiss climbers in 1956, showing a section through a crinoid ossicle identified by Augusto Gansser (1964).
(b) Sample from the same horizon as the summit of Everest taken from the Rongbuk Glacier showing a flattened crinoid stem preserved as a single calcite crystal.
(c) Sample of Ordovician lime mudstone with a crinoid ossicle and stem fragment enclosed within from the Rongbuk Valley.
(d) Longitudinal section through a crinoid stem infilled with calcite mud, from the Lower Ordovician section on Everest.

Yellow Band

The Yellow Band is a thick band of metamorphosed limestone or marble that sticks out prominently to form the South Summit of Everest and encircles the upper summit cone. Marbles from the Yellow Band were certainly metamorphosed, unlike the summit rocks immediately above, and the line of the low-angle South Tibetan Detachment fault separates the two rock types. This fault was called the Qomolangma Detachment as it was first defined along the Rongbuk Glacier profile to the north.[15] Below the Yellow Band marbles are thin-banded schists with biotite mica, but no minerals of particular use for determining pressure and temperature. These thin-banded schists continue all the way to the South Col and further down the Lhotse face. Within the schists are three or four thin white layers of marbles that show spectacular folds on the north face of Lhotse along the opposite side of the Western Cwm. Down in the Western Cwm below the Lhotse face the first granite masses come in, with some prominent thin dykes of granite shooting off the main Everest–Nuptse granite, intruding the surrounding schists. The entire south-west face is therefore the upper bounding zone of the mid-crustal channel, composed of granites and gneisses that were molten during the Lower Miocene period from about 23–14 million years ago.[16] It was the U-Pb ages of the Everest leucogranites that defined the time-scale of the partial melting and the ages of the shear zones above and below that determined the timing of the southward mid-crustal extrusion.

Everest Leucogranites

Many of the outlying high peaks of the Khumbu Himalaya in Nepal, such as Ama Dablam, Kangteiga, Tramserku, and Baruntse are composed of flat-lying sheets of granite which were once connected to the big granite intrusion seen most spectacularly on the south face of Nuptse. Here, a single granite sheet composed of multiple horizontal sills crops out all around the base of the Everest massif, and can be traced along the upper Rongbuk Glacier in Tibet on the north side of Everest. This giant granite sill originated from some 200 kilometres to the north of Everest, under the Tibetan Plateau, and was squeezed or ejected to the south between the large shear zones of the South Tibetan Detachment low-angle normal fault above, and the Khumbu Thrust, a south-vergent shear zone below. When the sill intruded upwards, it inflated like a balloon, folding the surrounding gneisses around it, and this spectacular oval-shaped granite can be seen on the southern slopes of Nuptse as one treks north towards Lobuche and Gorak Shep. As the granite cooled, the final puff of liquid was extracted from the melt zone, largely full of volatile and gaseous phases, and exploded as a network of dykes and veins of crystalline rock called pegmatites. These spidery white veins intrude the surrounding dark gneisses, and frequently contain very large crystals of black tourmaline, tiny red garnets, flaky white micas, and quartz, with the occasional beautiful blue, semi-precious mineral, beryl or aquamarine.

The Kangshung Valley in Tibet revealed the deepest eroded parts of the melting zone and here in the heart of Everest itself and in the great granite cliffs of Chomolonzo and Makalu that rise above the valley we found spectacular outcrops of the actual melt zone. Enormous rafters of metamorphic gneisses containing early sets of leucogranite sills were completely enclosed by a later phase of granite. These outcrops provided a superb opportunity for us to determine the entire melting history of the Himalaya. John Cottle, my D.Phil. student from New Zealand, spent three years systematically dating every phase of granitic melts, using U-Th-Pb isotopes across the Everest massif in the Kangshung Valley and the regions further to the north in the Ama Drime Range.[17] John's work showed that we could now pin down the precise timing of processes operating at deep levels of the crust during the Himalayan mountain-building event. The peak of burial metamorphism was precisely constrained at 39 million years ago with the highest temperature metamorphism at 28 million years ago. This metamorphism was followed by several sets of granitic melts injected at 20.8 and 16.7 million years ago (Figure 8.11). The ages from the granites could also be used to date the structures. Several outcrops we studied showed that one set of granite sills and veins was folded concordantly in with the metamorphic fabric, but that a later, undeformed set of dykes cross-cut all the earlier rocks. The ages of these two granites, 15.2 and 12.6 million years ago, could be confidently used to bracket the age of the structural fab-

Figure 8.11 Everest, Nuptse, and Lhotse viewed from the Renjo-la showing the distinctive white Nuptse granite beneath the black schists of the Nuptse-Lhotse ridge. Photo courtesy of Steve Razzetti.

rics. We were now able to show that the melting event along this part of the Himalaya reached its peak 23–20 million years ago but ended after 16 million years ago.

Makalu and the Barun Glacier

Makalu, at 8,475 metres, is the world's fourth highest mountain and lies only 20 kilometres east of Everest (Figure 8.12). It is connected to Everest along the watershed ridge that links Lhotse to a row of unnamed and unclimbed peaks. Makalu is a spectacular peak with immense rock walls made of gleaming white granite. The access to Makalu is a long and quite difficult trek along the Arun Valley, crossing the high Shipton-la, a pass named after the legendary explorer and climber Eric Shipton, and into the hidden recesses of the Barun Valley. I wanted to explore the Barun Glacier and the southern flank of Makalu, and so planned an expedition there in the autumn of 2005. I advertised for a new D.Phil student in Oxford and was lucky to get Mike Streule who was not only a top Oxford undergraduate but also an excellent rock and ice climber. Mike and I flew into the remote airstrip of Tumlingtar with Tashi Sherpa and Suka Ghale, our two *sirdar* friends, who hired the porters, cooked, carried rock samples, and negotiated with the Maoist rebels who were in almost complete control of the area. The village of Num, the last major settlement in the Arun Valley, was the headquarters of the Maoists, who by then had their sights set on taking over Kathmandu and toppling the Nepalese monarchy and the government.

The Maoists would suddenly appear in our camp out of the forest around dusk with their guns and sticks. We would invite them to sit around our campfire and have a cup of tea before they asked what we were doing and 'Would we care to make a donation?' After half an hour's talking we usually managed to negotiate the price of the 'donation' down a little. I always insisted that they spent our money on food not weapons, but I was under no illusions. It was effectively Nepalese-style highway robbery. Very few trekking groups came this way compared to the streams going up to Everest so we were a particular novelty and regarded as a potentially easy target. At the time kidnappings and extortion from Maoists were a common phenomenon throughout Nepal, although they rarely targeted Western trekkers, the main source of income for the country.

We managed to extricate ourselves from the Maoists and crossed the Shipton-la to the magnificent and remote Barun Valley. This was a botanical paradise with an incredible variety of trees and flowers and spectacular wildlife. The geology also proved to be very exciting. The high mountains around Baruntse and the other satellite peaks of Makalu were composed of the same horizontally layered leucogranite sills that occur beneath Everest and Cho Oyu, but on Makalu itself there was another type of granite, rarely seen, containing

Figure 8.12 Aerial view of Makalu (8,475 m) and Chamlang (7,319 m) above the Barun Valley east of Everest. Photo courtesy of Paul Harris.

large greenish crystals of cordierite, indicating higher temperatures of melting. The cordierite granites also contained garnet and tourmaline, but appeared to be intruding the earlier granites. Mike, Suka, and I clambered all around the cliffs on the south face of Makalu and climbed high enough to see the Everest and Lhotse massifs over to the west in the Sola Khumbu. We could even map out the vertical feeder dykes that channelled the liquid granite melts up to the higher levels on Makalu. Mike's project involved mapping out the various phases of granite on and around Makalu, obtaining precise pressure and temperature conditions of metamorphism, and dating the granites using U-Pb isotopes. The results of Mike's work showed the usual Himalayan metamorphic progression through time to sillimanite-grade migmatites and partial melting, to production of the 'standard' Himalayan garnet- and tourmaline-bearing granites, and finally to the cordierite-bearing granites. We thought this showed the decompression through time as the partially molten channel was flowing up towards the south during the Miocene period. The ages from Makalu were reassuringly similar to those we had obtained from Everest with the main phase of granite melting occurring from 24 to 21 million years ago, and the youngest melting events at 16 and 15.6 million years ago.[18]

Channel Flow

After ten years of research around the Everest and Makalu regions we had a large amount of data, both regional mapping of structures as well as data on temperatures, pressures, depths, and timing of metamorphism and melting of the rocks (Table 8.1 at the end of the chapter). Rick Law and I had managed to get a second three-year grant to study the lower part of the extruding channel and the Main Central Thrust zone, so we had complete coverage across the whole deep crustal core of the Himalayan orogen. We had suggested the overall geometry of a partially molten mid-crustal channel flowing towards the south, bounded by large-scale shear zones top and bottom, back when we started the project. Indeed our earlier work in Zanskar and Kashmir back in the late 1980s showing recumbently folded metamorphic isograds around the extruding channel implied exactly the same process.[19] Now we had obtained all the field structural, metamorphic, and geochronological data in the Nepalese Himalaya to back up our model and show that it was real. Chris Beaumont, a geophysicist at Dalhousie University in Canada, simulated the flow within a mid-crustal channel, and formulated a numerical model to explain the two dominant processes operating—channel flow within the middle crust and ductile extrusion of this weak layer—using all the known geological constraints from several different sources. Chris and his colleagues also suggested that channel flow could be coupled to rapid erosion and removal of rock material along the southern flank of Tibet, along the High Himalayan Range.[20] Once set in motion, there would be a positive feedback loop: the higher the rainfall and erosion along the Himalaya, the faster the channel extruded. It was fieldwork, intricate mapping combined with thermobarometry and geochronology, that provided all the starting parameters for the models (Figure 8.13).

A number of factors I thought argued against the idea that climate was the driving force. The highest exhumation rates seen anywhere on the planet occurred in the two corner areas, called syntaxes, of the Himalaya. In the west, Nanga Parbat is situated in the barren deserts of Kohistan in North Pakistan, with a very dry climate, whilst the eastern Himalayan syntaxis, centred on the peak of Namche Barwa in south-east Tibet, is in the wettest region on Earth. Both these areas were obvious targets for research to determine the amounts and timing of extrusion. We also knew that channel flow was operating during the Miocene from our extensive U-Pb ages of granite crystallization, but what was the climate doing back in Miocene times?

Figure 8.13 (*opposite*) The Channel Flow model showing the partially molten middle crust layer of the Greater Himalaya extruding southwards bounded by the Main Central Thrust below and the South Tibetan Detachment normal fault above during the Miocene. The upper photo shows the STD clipping the summit pyramid of Everest and the lower photo shows a detail of the outcrop 50 km north of Everest at Dzachaa chu. The lower figure is a restoration of the STD showing the depths from which the metamorphic rocks and granites were formed and their exhumation pathway.

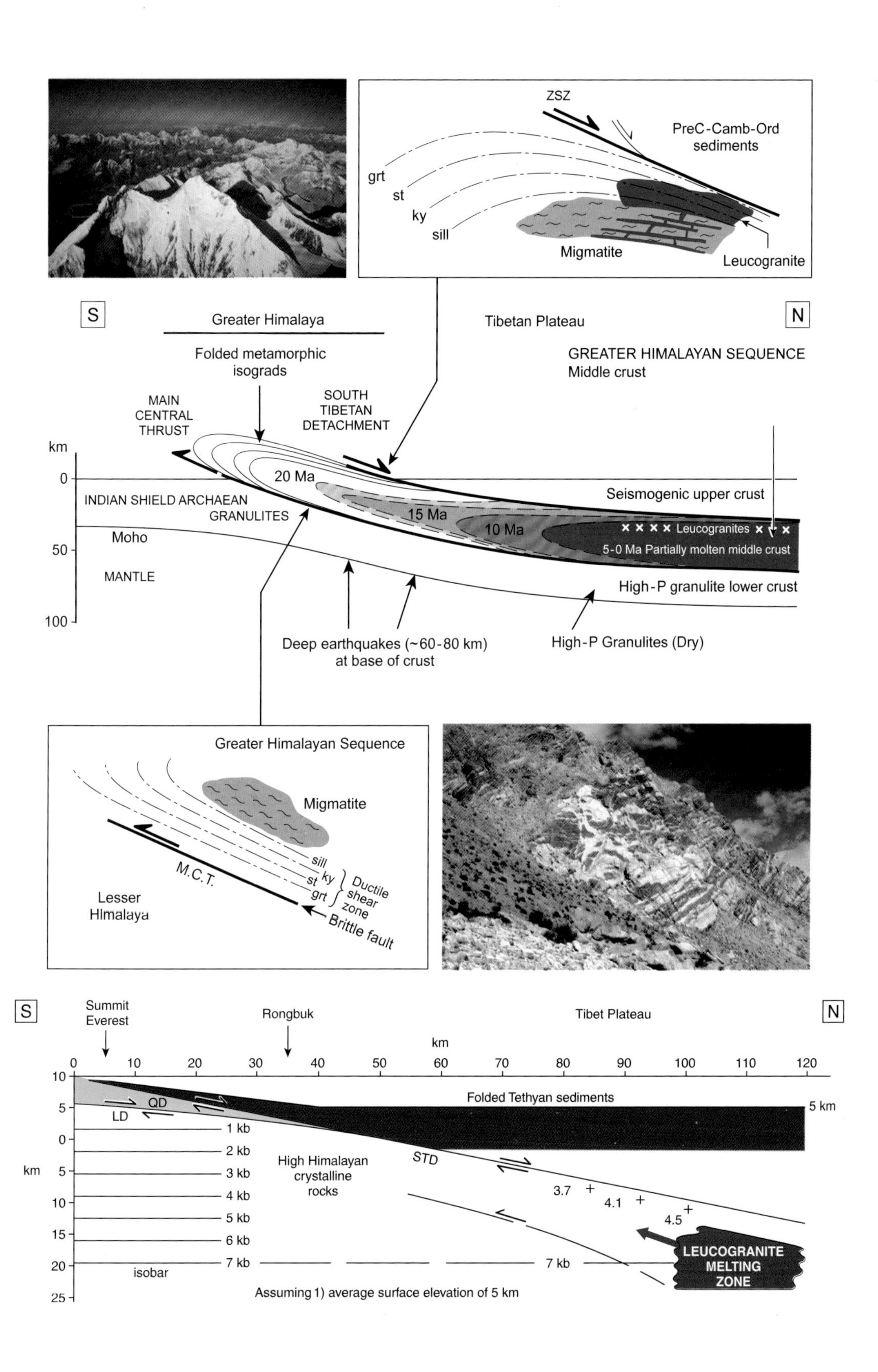

ZSZ
PreC-Camb-Ord sediments
grt
st
ky
sill
Migmatite
Leucogranite
S
Greater Himalaya
Tibetan Plateau
N
Folded metamorphic isograds
GREATER HIMALAYAN SEQUENCE
Middle crust
MAIN CENTRAL THRUST
SOUTH TIBETAN DETACHMENT
km
0
20 Ma
Seismogenic upper crust
INDIAN SHIELD ARCHAEAN GRANULITES
15 Ma
10 Ma
Leucogranites
Moho
5-0 Ma Partially molten middle crust
50
MANTLE
High-P granulite lower crust
100
Deep earthquakes (~60-80 km) at base of crust
High-P Granulites (Dry)
Greater Himalayan Sequence
Migmatite
sill
M.C.T.
st
ky
grt
Ductile shear zone
Lesser Himalaya
Brittle fault
S
Summit Everest
Rongbuk
Tibet Plateau
N
km
0
10
20
30
40
50
60
70
80
90
100
110
120
QD
LD
Folded Tethyan sediments
5 km
1 kb
2 kb
3 kb
4 kb
5 kb
6 kb
7 kb
High Himalayan crystalline rocks
STD
3.7
4.1
4.5
LEUCOGRANITE MELTING ZONE
7 kb
isobar
Assuming 1) average surface elevation of 5 km

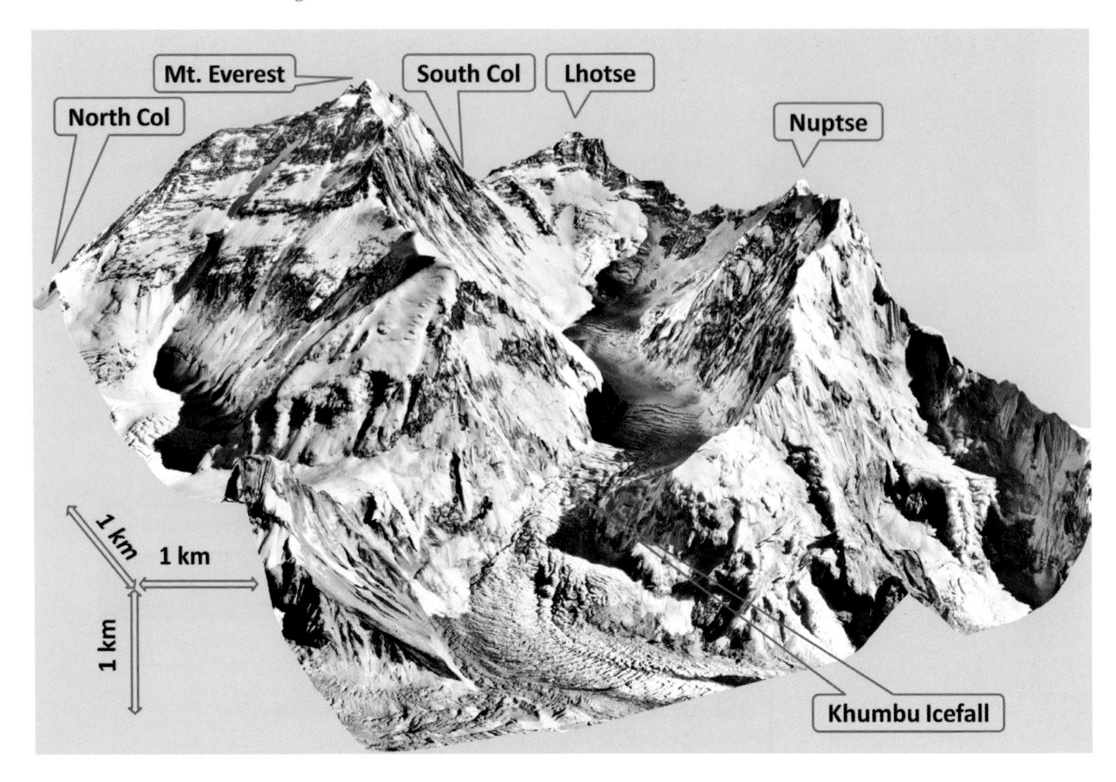

Figure. 8.14 Digital elevation model of the Everest massif. Courtesy of Gerhard Bax

Several questions remained. Was channel flow operating today, as suggested by deep crustal seismic experiments carried out across Tibet which showed the presence of blobs of liquid melt at relatively shallow depths beneath southern Tibet,[21] or was it restricted to the Miocene period, spanning all our U-Pb zircon and monazite age constraints? Were climate and high erosion rates along the Himalayan front responsible for channel flow, or was tectonics the driving force, with climate merely following along? It was clear that in the Everest–Makalu region of Nepal (Figure 8.14), channel flow was a viable model with several kilometre thickness of partially molten rocks and numerous granite sills. In other regions of western Nepal, such as the Dhaulagiri–Annapurna Himalaya, the highest peaks were composed of sedimentary rocks with a greatly reduced thickness of metamorphic and granitic rocks. Could similar geological processes still be operating along these parts of the Himalaya?

Table 8.1 Heights and rock composition of some Nepal–south Tibet peaks

Mountain	Height (metres)	Rock composition [m.y.= million years]
Dhaulagiri	8167	sedimentary rocks—marbles, limestones
Annapurna I	8091	marble, limestone (Cambrian–Ordovician)
Gangapurna	7455	marble, limestone (Ordovician–Silurian)
Annapurna III	7555	marble, limestone (Ordovician–Silurian)
Annapurna IV	7525	marble, limestone (Ordovician–Silurian)
Annapurna II	7939	marble, limestone (Ordovician–Silurian)
Machhapuchare	6993	marble, limestone (Cambrian)
Manaslu	8163	granite (24-19 m.y.)
Himalchuli	7540	gneisses, with granite sheet near summit
Ganesh	7429	gneisses with thin granite sills
Langtang Lirung	7234	gneisses with thin granite sills
Dorje Lakpa	6973	gneisses with thin granite sills
Shisha Pangma	8027	granite (20.2-17.3 m.y.)
Menlungtse	7181	granite (probably ca. 20 m.y.)
Gauri Shankar	7135	gneisses with thin granite sheets
Cho Oyu	8153	granite (probably ca. 20 m.y.)
Gyachung Kang	7922	granite sheets at base, greenschists high up
Chakung	7029	granite and gneiss interbanded
Pumori	7165	granite (17.5-18 m.y.)
Everest	8850	granite (21-17 m.y. at base); Everest series greenschists above, Ordovician limestone, shale on summit.
Lhotse	8501	granite (20.2 m.y.) at base, greenschists above
Nuptse	7861	granite (20.2 m.y.) bulbous sill with schists wrapped around. Pegmatite dykes on 'spider wall'
Ama Dablam	6856	granite (>17.7 m.y.) with metamorphic bands
Kangteiga	6685	granite sills above sillimanite gneisses
Tramserku	6608	granite sills above sillimanite gneisses
Kusum Kanguru	6367	gneisses with thin granite sills
Peak 38	7591	massive granite
Makalu	8463	massive granite (24 m.y.)
Chomolonzo	7790	massive granite (probably 24 m.y.)
Baruntse	7129	gneisses with granite sills
Chamlang	7319	gneisses with thin granite sills
Jannu	7710	granite with metamorphic inclusions
Kangbachen	7903	? granite with schists above
Kangchenjunga	8598	massive granite sheets, a few gneisses
Kabru	7353	massive granite sheets, a few gneisses

CHAPTER 9

Mountains and Maoists

Annapurna, Manaslu

When the imagination sleeps, words are emptied of their meaning.

Albert Camus 1913–1960

Sarangkhot village is perched on top of a hill overlooking the Phewa Lake and Nepal's second city of Pokhara. The views from here are some of the most wonderful in all the Himalaya. To the north-west, the huge snowy ramparts of Dhaulagiri rise above the Kali Gandaki Gorge (Figure 9.1), one of the world's deepest canyons. East of the Kali Gandaki the whole Annapurna Range (Figure 9.2) forms an icy rampart that blocks the entire northern horizon, with the majestic spire of Machapuchare, the 'Fish Tail', rearing up in the foreground (Figure 9.3). Hanging glaciers clinging to the side of Machapuchare glisten in the sunlight and cascade down to the green bamboo jungles of the Seti khola. The Annapurnas extend east into the Lamjung Himal, and then across another deep canyon, the Burhi Gandaki, the 8,000-metre-high massif Manaslu appears on the far eastern horizon (Figure 9.4). Winds funnel down the Pokhara Valley and rise up to the ridge at Saranghot, making this the perfect take-off place for hang-gliders. To jump off the ridge at Saranghot and see the whole Nepalese Himalaya unfurl to the north, ride the thermals with the black kites, eagles, and vultures cruising alongside, and land back at the lakeside is one of the new wave of extreme sports to have recently caught on in Nepal. It is difficult not to like Pokhara.

The monsoonal rains in Nepal can be dramatic. Clouds gradually form in the late morning, welling up the valley to hug the high peaks, with just the tips of the mountains standing clear and proud above a sea of cloud. Suddenly the skies

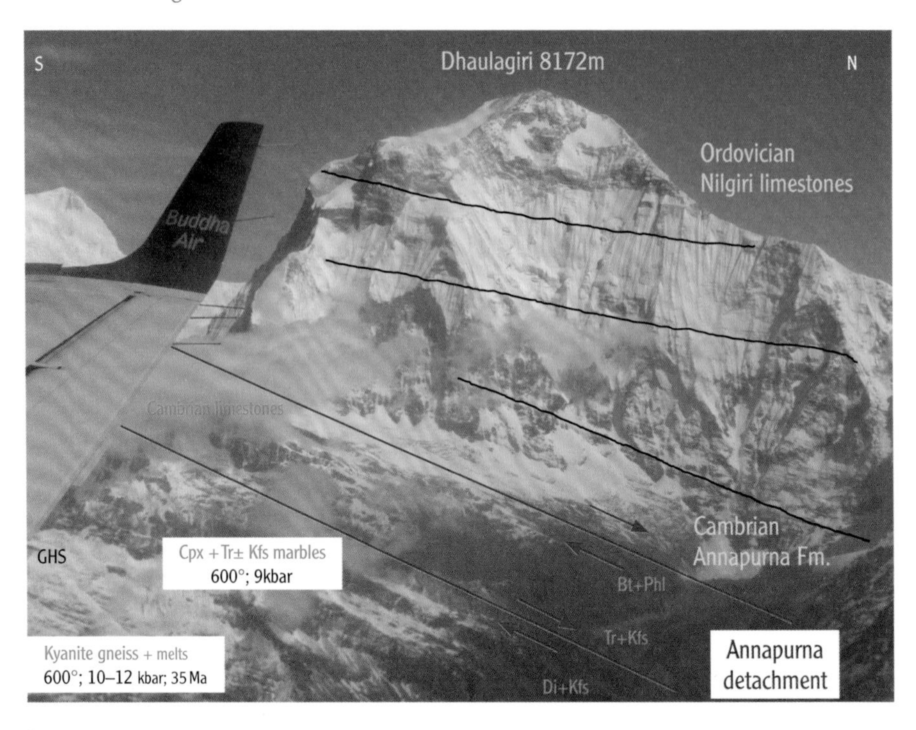

Figure 9.1 Aerial view of the south-east face of Dhaulagiri (8,167 m). Cambrian-Ordovician limestones lie above highly sheared marbles of the South Tibetan Detachment with metamorphic rocks of the Greater Himalaya beneath.

Figure 9.2 South face of Annapurna I (8,091 m) from the Annapurna Sanctuary.

Figure 9.3 The south faces of Gangapurna (7,454 m) and the 'Fish-Tail' Machapuchare (6,993 m) viewed from the air.

Figure 9.4 Manaslu (8,163 m), Peak 29 (7,871 m), and Himalchuli (7,893 m) rising above a sea of cloud.

darken, streaks of lightning appear out of nowhere, and thunderclaps ricochet around the mountain walls. One or two large raindrops herald a torrential monsoonal downpour, rain falling out of black clouds in sheets. Trickles of water in rivulets become raging torrents in minutes, and very soon waterfalls gush over every cliff. The rains finally ease as suddenly as they started, the skies gradually clearing as mists linger in the deep forested valleys. One by one the high mountains reappear above the rain clouds with their summit icefields glinting in the last rays of the sun. The landscape and weather of Nepal never cease to amaze and one can never tire of looking at these magnificent creations of geology.

Annapurna Sanctuary

I had wanted to study the rocks of the Annapurna Range for a long time, having seen spectacular photos of the huge fold structures on the Nilgiri peaks and northern ramparts of the Annapurnas. I got the chance in 1994 when, following the annual HKT conference in Kathmandu, Kip Hodges, a prominent geologist from the Massachusetts Institute of Technology, invited me along on a field trip to the Modi Khola. This deep valley cuts an incredible gorge in between the Annapurna–Hiunchuli massif and the Matterhorn-like peak of Machapuchare. The valley led all the way up to the Annapurna Sanctuary, a wide bowl-like area surrounded by some of the most beautiful peaks in Nepal. This deep canyon cuts south directly from the south face of Annapurna I and made an ideal profile along which we could map the structures. A large French geological project during the 1970s, largely the work of Michel Colchen, Arnaud Pêcher, and Patrick LeFort, had succeeded in mapping much of the Annapurna and Manaslu ranges[1] and their map provided an extremely useful starting point for our work in the Annapurna Range.

The folds exposed in the Annapurna Range were similar to those I had seen and mapped in Zanskar but here they were on a larger scale and beautifully exposed in three dimensions (Figure 9.5). We spent several weeks trekking up the deep gorge of the Modi Khola mapping along the way, taking structural measurements and collecting samples for thermobarometry and geochronology. We discovered a series of north-dipping normal faults along the profile similar to the Zanskar shear zone I had been working on in Ladakh (Figure 9.6). These normal faults were not on the French map and it was only very recently that they had been discovered along the northern side of the Greater Himalayan Range in southern Tibet, the Annapurnas, and in Zanskar.[2] In the Annapurnas, each one of these faults places higher, less metamorphosed rocks onto deeper more highly metamorphosed rocks. Randall Parrish was able to date our rock samples using U-Pb isotopes in his laboratory. By analysing the mineral assemblages we could

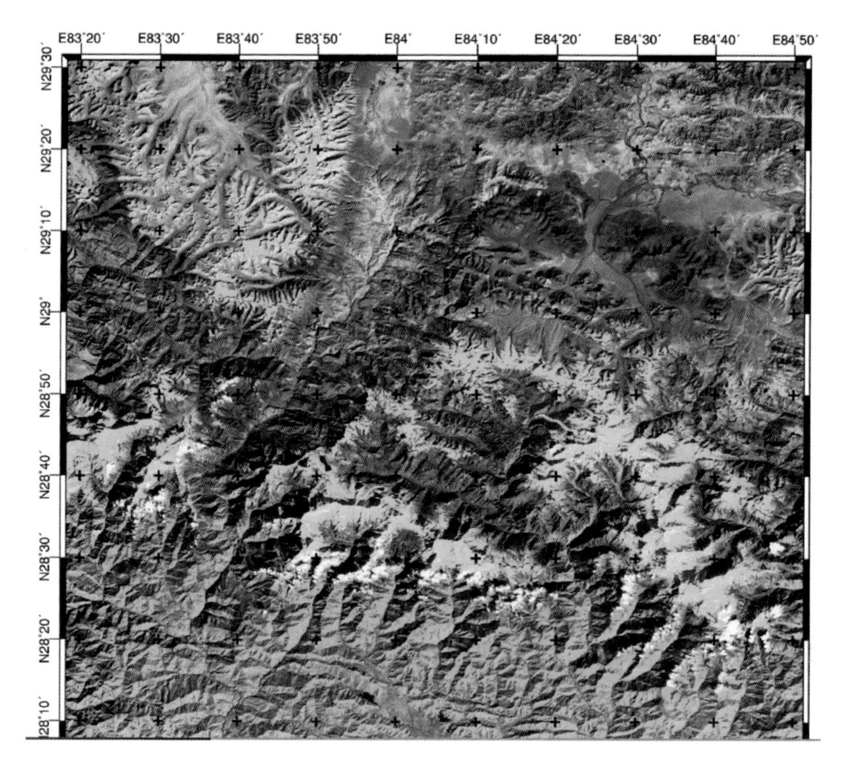

Figure 9.5 Landsat photograph of the Dhaulagiri (left) and Annapurna (right) massifs with the valley of the Kali Gandaki in between.

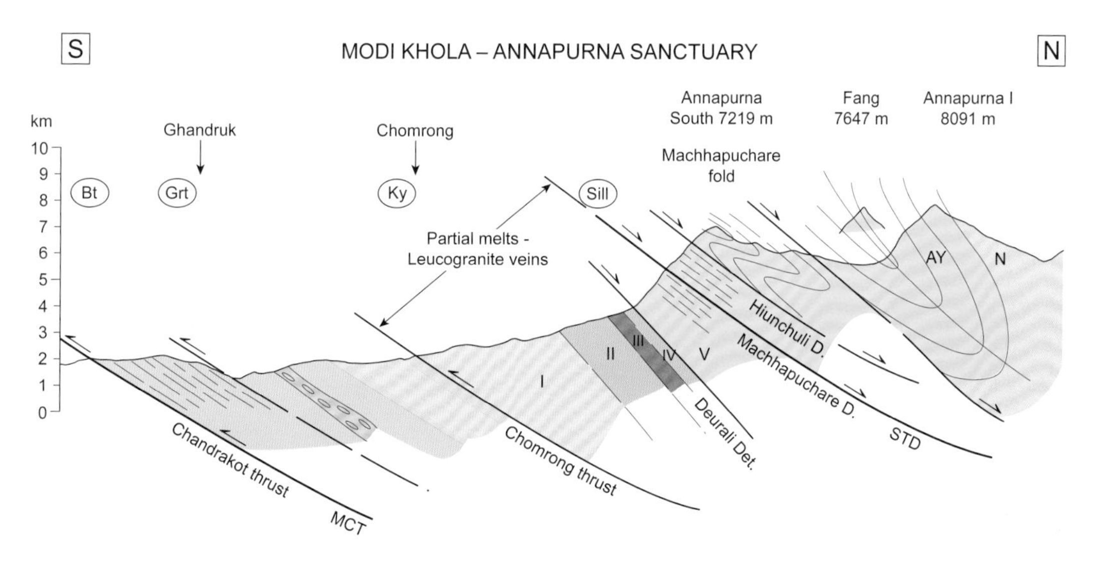

Figure 9.6 Geological cross-section along the Modi Khola profile across Machapuchare.

extract pressure, temperature, and depth constraints, after which we were able to add the most important fourth dimension, time, into the puzzle. We found that the normal faults were progressively younger upwards and outwards from the high-grade metamorphic core of the Himalaya to the north, but also that the thrust faults associated with the Main Central Thrust sheet along the base of the slab were progressively younger to the south, towards the Indian foreland.[3] These discoveries were to add one more piece to the puzzle that was later to become the Channel Flow model, whereby the once deeply buried, metamorphosed, and partially molten core of the Himalaya was extruded to the south, bounded by these low-angle normal faults along the top and low-angle thrust faults along the base.

Structurally above the low-angle normal faults, the rocks of the Dhaulagiri, Nilgiri, and Annapurna Ranges were made up mainly of unmetamorphosed sedimentary rocks of the northern or Tethyan Himalaya. In this part of Nepal they ranged from the earliest Cambrian, some 540 million years ago, up to the Eocene, some 50 million years ago, and they formed the northern passive sedimentary margin to the Indian plate before it collided with Asia 50 million years ago. Trekking along the Kali Gandaki Valley it is possible to see some of the largest and best-exposed fold structures anywhere on Earth. The Nilgiri Peaks show spectacular folds, but unlike most places along the Tethyan Himalaya, where folds are generally southward verging, these folds are backfolded to the north. They sit right above the low-angle north dipping faults of the Annapurna Detachment, and appear to have slid back to the north as the metamorphic rocks of the Greater Himalayan Series were extruded or thrust out to the south beneath the fault (Figure 9.7a, b).

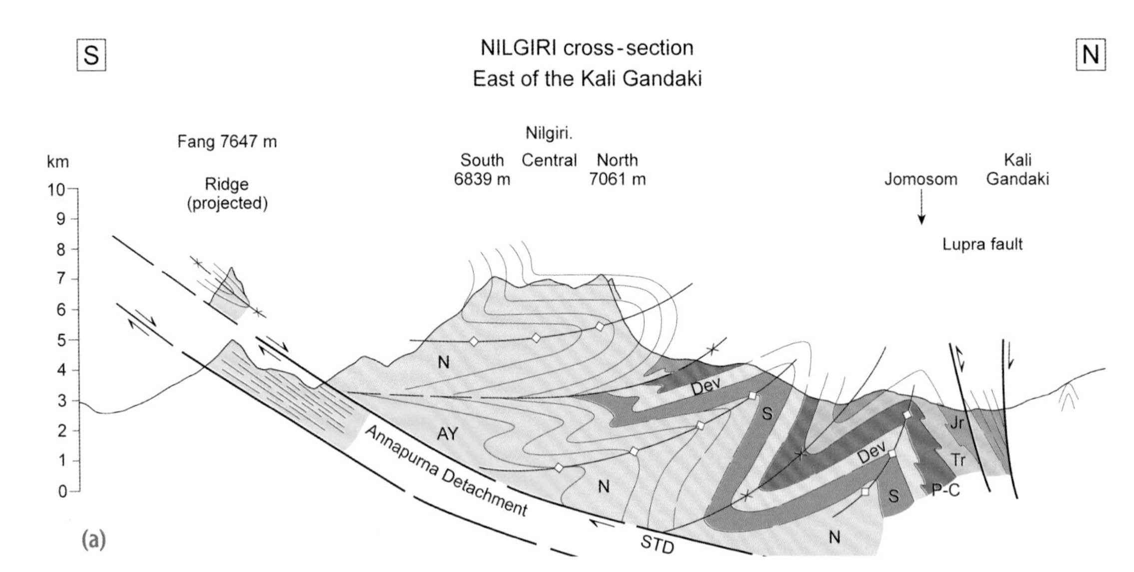

Figure 9.7 (a) (*opposite*) Geological cross-section of the Nilgiri Range east of the Kali Gandaki showing the relationship between the huge folds and the Annapurna Detachment. (b) (*above*) The large-scale folds in the Cambrian-Ordovician limestones exposed on the face of Nilgiri Peak (7,061 m)

Around the Annapurnas

A friend of mine from Queens University in Canada, Laurent Godin, had recently completed his PhD on these rocks in Nepal and had come to Oxford for a two-year post-doctoral fellowship. He showed me the results of his detailed mapping and structural analysis and we compared notes with my work in the Ladakh and Zanskar regions. In 2002 we planned a more extensive trip to the region, extending Laurent's mapping to the north in the remote Nar-Phu region, north of the Annapurna Range, along the border with Tibet. We planned a four-week-long trek around the Annapurnas starting off along the Marsyandi Valley, then heading north to Nar village, crossing a pass to descend to Manang, then rejoining the Annapurna circuit to cross the Thorung-la over to Kagbeni in the Kali Gandaki. One of our objectives was to map out the western margin of the Manaslu leucogranite, which had previously been described as intruding up across the low-angle normal faults into the Tethyan sedimentary sequence. No other leucogranites along the Himalaya had this structural geometry and it was even more of a puzzle because Manaslu had some the oldest crystallization ages of all the Himalayan granites, about 24 million years old.

The Nar-Phu Valley was a magnificent and remote trek and in those days required a special 'inner line' permit from the Nepalese government, being right on the sensitive Tibetan border. We mapped along the whole valley and along the remote Himlung Glacier leading up towards Manaslu in the east. Our results showed that rocks all around Manaslu were not sedimentary rocks of the Tethyan Himalaya at all, but highly metamorphosed marbles and pelitic rocks (metamorphosed mudstones), often intruded by a network of tourmaline-bearing, granitic dykes (Figure 9.8a, b). We discovered another large-scale ductile shear zone, or deeper-level equivalent to the Annapurna low-angle normal fault, that wrapped around the top of the Manaslu granite in the Himlung Range along the Tibetan border. Thus Manaslu was no different from any other Himalayan leucogranite, being entirely within the crystalline core of the Greater Himalayan Series.[4]

After exploring the valleys radiating up to the Tibetan border watershed and the rustic old Tibetan-type villages of Nar we finally crossed a new pass over to the upper Marysandi and descended towards the village of Braga near Manang. It had been a remarkable trek and had yielded important new geological results for our regional understanding of the structure of the Himalaya. We then crossed over the 5,416-metre-high Thorung-la, passed the Hindu pilgrimage temple of Muktinath, and the medieval walled bastion village of Jharkot, and descended to the archaic mud-brick village of Kagbeni, the gateway to Mustang and its spiritual capital, Lo Mantang. Laurent then planned to ascend the Dhampus Pass to the west and gain access to the Hidden Valley north of Dhaulagiri, but I decided to hike on down the Kali Gandaki and study this classic profile, then we would meet back in Pokhara. Pasang Tamang, our *sirdar,* collared the nearest porter and asked if he would carry my rucksack for the seven-day trek back to Pokhara. The porter, Suka Ghale, a young Gurkha from the village of Kashigaun near Manaslu, immediately agreed and in the next week we trekked down the Kali Gandaki, ascended to the Dhaulagiri icefall in the west and to Ghorepani in the east. Suka was a delightful companion and showed me several hidden gems of treks along the Kali Gandaki, notably up around the Dhaulagiri icefall above Larjung, high above the apple orchards of Marpha, and the rustic village of Tukuche. This trip reinforced my view that the best way to immerse oneself in the country was to do a solo trek with only a local porter or guide for company.

Figure 9.8 (*opposite*) (a) White granite dykes intruding gneiss in the Chako dome, along the Nar Valley in Nepal. (b) Water polished green and white marbles intruded by a granite dyke along the Nar Valley west of Manaslu.

(a)

(b)

The Maoist Insurgency

Nepal was undergoing major transformations during this time and a large-scale Maoist insurgency was under way with anarchy reigning across most of the country.[5] Gangs of youths from the Terai lowlands roamed the land, armed with sticks and guns demanding money and food from villagers and trekkers. The Maoists wanted to abandon the caste system and started to burn the elite Sanskrit schools that only allowed entry to the upper caste Brahmins. The 238-year-old Hindu monarchy in Kathmandu ended in 1990 when King Birendra transferred sovereignty to the people's parliament and Nepal set out on a democratic course. Nepal is a highly mountainous country inhabited mainly by Buddhists, with a significant proportion of Hindus maintaining a tribal structure dominated by the caste system. The caste system in Nepal is a more ancient one based on one's trade and tribe (Chhetris, Rais, Tamangs, Thakkalis, Gurkhas, etc.), rather than the strictly social hierarchy of the Indian caste system. Buddhism, Hinduism, and animism are all entwined together in the hill tribes. The one exception was the Newari caste of the Kathmandu Valley that included the royal family, an upper caste Brahmin family that ruled the Kathmandu Valley by some sort of 'divine' right.[5]

Parliament was becoming completely ineffectual with numerous parties and factions. The Communist Party of Nepal broke off into at least three factions, with the only real opposition coming from the Nepali Congress Party, backed by Western governments. The Maoist insurgency started in the Rapti region in the Terai in 1996, when disaffected youths with no job prospects and no money started burning government buildings and attacking army posts. Very soon the army entrenched, with less and less power in the countryside. Massacres of army and government officials spread across the country and the whole of Nepal descended into chaos. On 1 June 2001 in a drink and drug-filled craze, the Crown Prince Dipendra massacred his entire family, including his father King Bihendra, his mother the Queen, and several others in the royal palace in Kathmandu.[6] It was the nail in the coffin for the royal family. One of the few survivors of the 2001 royal massacre in Kathmandu was King Bihendra's brother Gyanendra, who then became King and tried to keep the show going. Gyanendra and his son, Prince Paras, were both extremely unpopular with Nepalis. In 2002 Nepal's economy was completely stifled and King Gyanendra dissolved parliament, sacked Prime Minister Deuba, and reinstated direct rule by the monarchy. Meanwhile the Maoists were becoming stronger and almost all the country was being terrorized by gangs of unemployed and anarchic youths roaming the villages, burning and pillaging.

The two most important trekking regions of Nepal are the Everest region and the Annapurna circuit. Since Nepal's income is almost completely dependent on the tourist trade, these regions were the most important ones for the Army to keep out of Maoist hands. The problem was that nobody could tell who was a Maoist. 'Maoist' became a byword for a *dacoit*, highway robber, young thug, or anybody with a grudge against the government. Certainly most Nepalese Maoists had no idea who Mao Tse-tung was, or what he stood for. As China and the Soviet bloc disowned Maoism, Marxism, and Leninism, Nepal alone seemed to embrace Communism, at least in name. By 2002 only Kathmandu, and the major trekking routes around Everest and Annapurna were not in Maoist control. Along the Kali Gandaki, heavily armed soldiers tried to keep order from their fortified bunkers, but the Maoists would creep in under the cover of darkness and burn schools and government buildings, and occasionally have full-scale pitched battles with the army. Lodge owners told me that gangs would approach the village from the southern lowlands, barge into the lodge, lay their guns and sticks on the table, and demand food. They would torch the nearest government buildings, extort money ('donations') from anyone who had some, and then disappear back into the forest. Villages were afraid of the Maoists, who often forced them to give food and money, and of army reprisals if they found out which villages harboured them.

Our geological work in Nepal throughout the 1990s and 2000s suffered from Maoist restrictions. Trekking routes were frequently blocked by Maoist 'checkpoints' where they demanded 'donations'; trekking permits issued in Kathmandu had little meaning. Usually the Maoists gave us receipts that were duly submitted to account for our research grants back in Oxford. In the box asking for what the amount was paid I wrote 'extortion' or 'bribery'. In April 2006 a seven-party alliance supported by the Maoists forced the King to return power to the parliament. The Maoist leader Prachandra emerged from the jungles for the first time and entered the democratic process. The Maoists decisively won the elections of 2008 and the jungle revolutionaries finally became the government.

Suka Ghale and Kashigaon Village

In the spring of 2006 I again joined up with Laurent Godin in an exploration of the Manaslu Himalaya to the north of Gorkha in central Nepal. Our goal was both to study the structures and metamorphism of the Burhi Gandaki Valley and to map around the margins of the largely inaccessible Manaslu leucogranite. Whilst Laurent and his new PhD student, Kyle Larson, spent two months mapping and sampling along the Burhi Gandaki, I left them to take a closer look at the Manaslu massif near the Tibetan border. I headed north with Suka Ghale and three porters from his village of Kashigaon, with the idea of crossing the Larke-la, a 5,098-metre-high pass near Tibet, where the northern contact of the Manaslu granite was exposed. This contact was critical to our theory that the Manaslu granite was contained entirely within the metamorphic core and did not intrude up across the South Tibetan Detachment into the Tethyan sedimentary cover, as was previously thought (Figure 9.9).

After two days of trekking, Suka and the porters invited me to spend some time in their village. Although it was a day's detour and strenuous hike up the steep eastern bank of the Burhi Gandaki Valley (Figure 9.10), it would give me a good vantage point to map in the structures, and I thought it would be great to see Nepalese Gurkha village life at first hand. Kashigaon village was extremely remote, perched high up above the Burhi Gandaki near the tree line, a row of rough stone and mud houses with wood smoke billowing out of the rafters. The forests outside the village stretched all the way up towards the Ganesh Himal. There was no electricity, and water was collected from a single standpipe. Inside, the houses were as simple and basic as you can get, with the odd water buffalo or bleating goats sharing the floor space. Suka's family gave me a wonderful reception and soon the entire village gathered to witness this first visit to the village from a foreigner. It was a humbling experience. After a

Figure 9.9 The upper contact of the Manaslu leucogranite exposed near the Larke-la (4,930 m). The white granite has been intruded beneath the South Tibetan Detachment shear zone along the base of the Tethyan sedimentary rocks.

Figure 9.10 The deep gorge of the Burhi Gandaki River leading up to Manaslu.

special meal of *dhal bhat* cooked over a fire of wood and dung, stories of village life were passed on, of how a tiger had killed one of the buffalo calves, of how the boys used to spend days out in the forest looking after the animals, sleeping in rough 'cow houses', how they used to scare away bears that came and ravaged their corn fields, and how very sick people had to be carried for three days down winding forest tracks to Gorkha to get any help. Despite living in deepest poverty, these people were some of the most hospitable I have ever come across.

Suka Ghale became a good friend and through him I learnt a lot about the life of the rural villagers. Their life was a tough one, revolving around planting and harvesting the rice, lentil, and corn crops. Schooling was minimal and jobs were almost non-existent. Many young Nepalese were forced to sign up as labourers for construction or factory jobs in Middle Eastern countries or Malaysia. They were forced to pay exorbitant bribes to a Nepalese middle-man and more to the foreign middle-man. When they arrived, they were frequently treated like cattle, working fourteen hours a day, sleeping in large dormitories that looked rather like concentration camps, with no fans in the stifling heat. They were not even given a salary until their air-fares and bribes had all been paid off. It was quite shocking to know that some of the richest countries in the world were responsible for this exploitation of the Nepali poor.

For those who stayed in Nepal, portering for trekking or climbing expeditions was often the only work around, and that only during the short spring and autumn trekking seasons. Sometimes trekkers were individually generous, but more often some of the larger trekking companies exploited the porters, paying a meagre daily wage and often not seeing that they had even the most basic equipment like decent shoes or sunglasses, crucially needed for crossing the high snow-bound passes. Up in the remoter hill villages like Kashigaon, there was a money-less economy with people relying on growing their own food season by season, and if there was a little surplus, they could barter it for clothes or whatever they needed. Cooking fuel was usually wood or yak dung; electricity was just arriving along the valleys, but up in the hills kerosene lamps were still used. Old women would go out collecting wood returning bowed down under enormous loads. It was a hard existence.

The Manaslu Leucogranite

Leaving Kashigaon we descended back into the main Burhi Gandaki Valley and met up again with Laurent and Kyle. They had been stopped by another Maoist 'checkpoint' and relieved of a hefty 'donation' for the Maoist cause. Gorkha was one of the centres of Maoist uprising, so the entire countryside was united against the Kathmandu government, and the local hero was a man called Baburam Bhatterai, the second-in-command of the Maoist revolutionaries. We left Laurent and Kyle and trekked on up towards the Larke-la. Staggering up to the dizzy heights of 5,000 metres I tried to carry out my geological mapping in between nauseous headaches. Suka took my rucksack as I climbed higher to try to get a close look at the granite contact. Across the pass a steep track descends between the glaciers and moraines and finally we reached the beautiful green meadows of Bimtang, the campsite across on the western side of the Larke-la. Above Bimtang, 3,000-metre-high cliffs of pale white granite soar up towards the summit of Manaslu, draped in pristine white icefields and hanging glaciers (Figure 9.11). Everywhere around Bimtang were clean outcrops showing granites laden with black needles of tourmaline, big red garnets, and white muscovite micas, Igneous and metamorphic textures of the gneisses showed every stage of the melting process from *in situ* sweats of granite within the gneiss to accumulations of granite melt into dykes and sills, and finally into the massive granite (Figures 9.12 a–d and 9.13a, b, c). Descending down to the Marysandi from Bimtang, the trail wound through magical forests with thundering white water streams. These pristine forests were filled with the sound of bird song, white-faced langur monkeys crashed through the canopy, and

Figure 9.11 The impressive 3,000-metre-high cliffs of granite on the south face of Manaslu from Bimtang meadows.

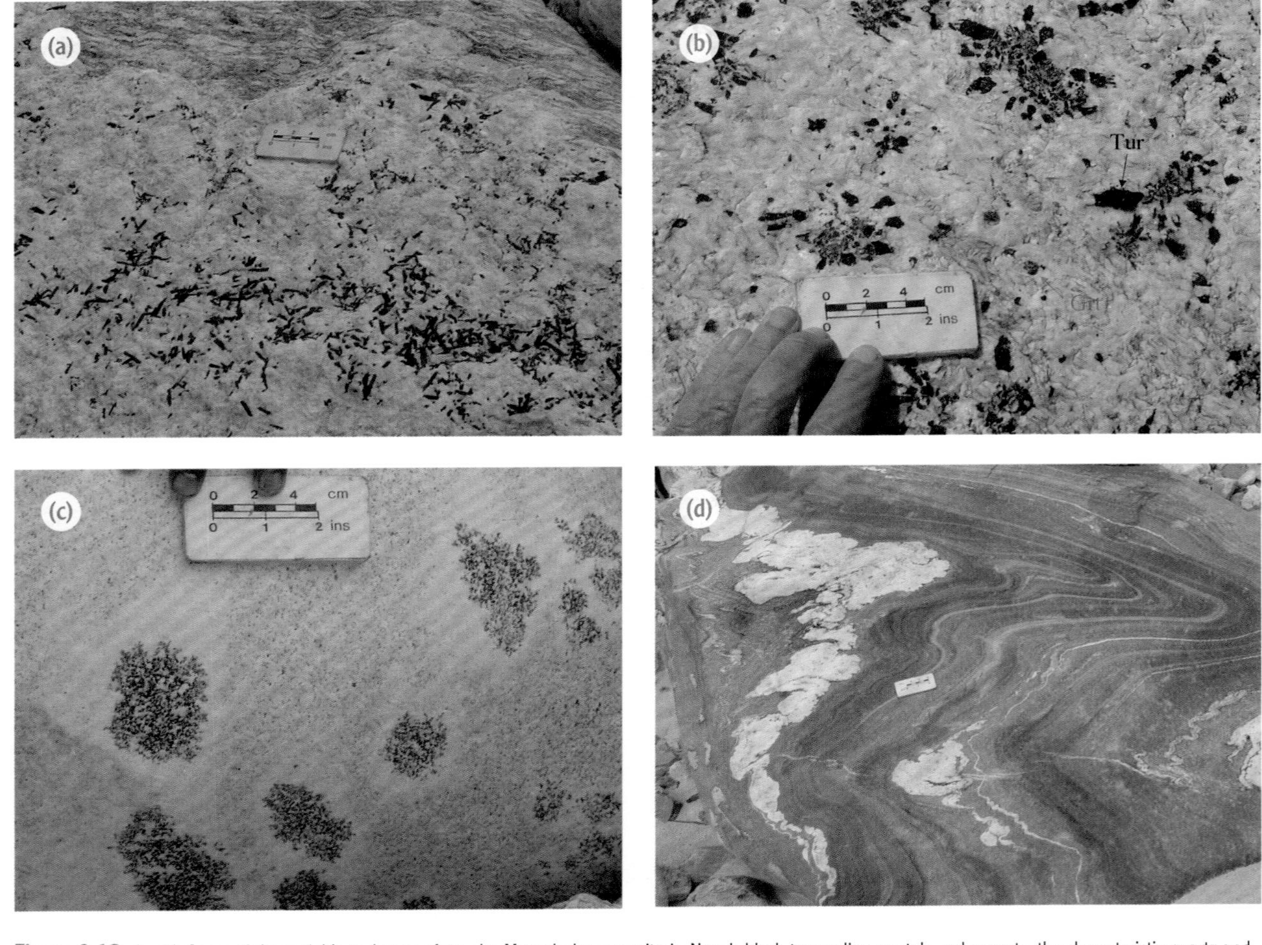

Figure 9.12 (a–d) Some of the variable rock types from the Manaslu leucogranite in Nepal; black tourmaline crystals, red garnets, the characteristic quartz and tourmaline 'stars', and the wavy textured migmatites, results of partial melting to form granite.

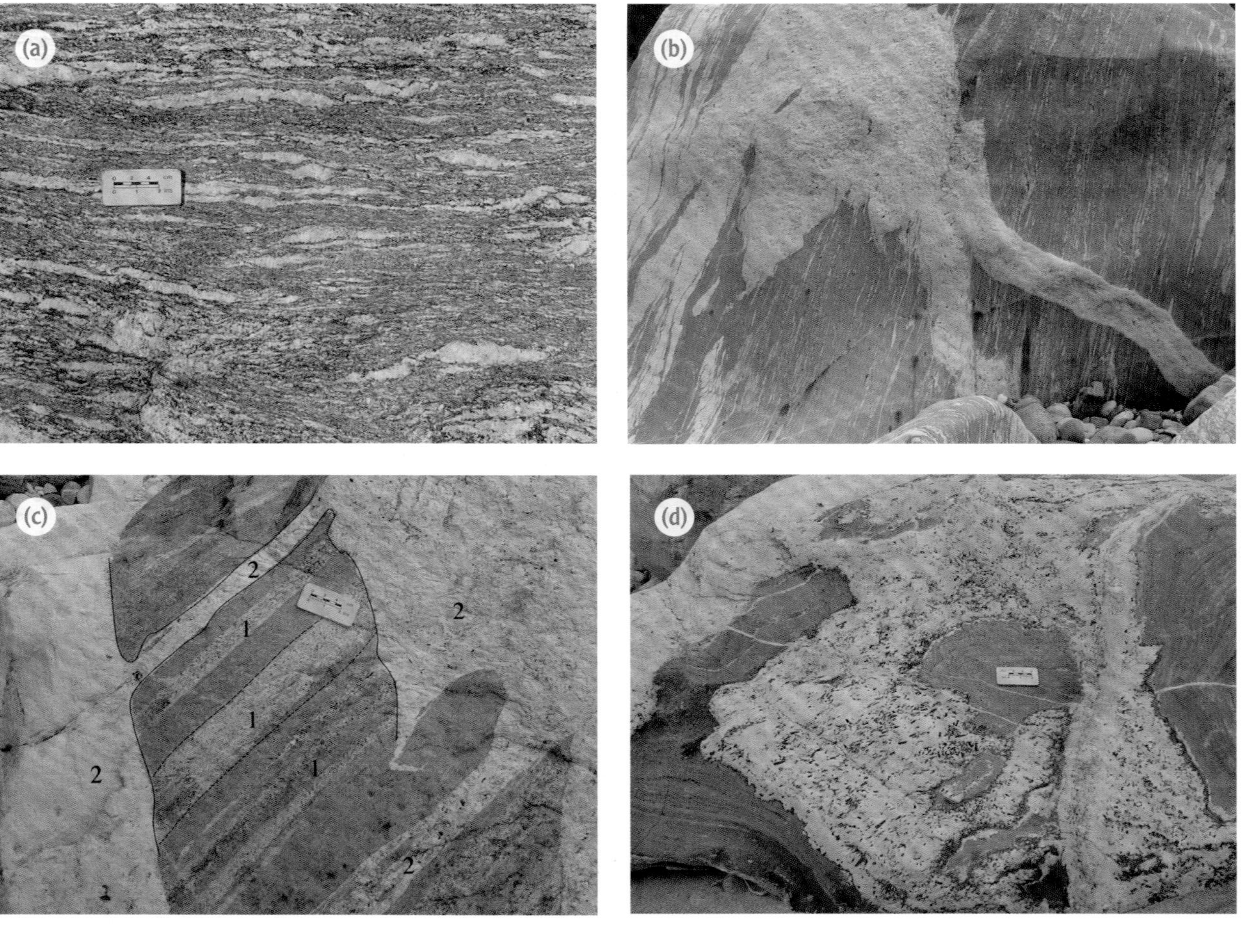

Figure 9.13 Typical migmatite textures from the Manaslu region in Nepal. (a) Partial melting with light-coloured granite melts sweating out of darker gneisses; (b) late granite dykes cutting across the layered migmatites; (c) two phases of cross-cutting granite dykes emanating from the Manaslu granite; and (d) tourmaline granites enclosing dark-coloured xenoliths of gneiss.

the meadows abounded in colourful flowers. Lower still, we entered the bamboo forests and then the higher farms and first village settlements.

Geologically the circumnavigation of Manaslu had proved extremely useful. It had reinforced our earlier hypothesis that the Manaslu granite is built up of major kilometre-thick sheets of granite dipping to the north beneath the Tibetan Plateau and wholly within the metamorphic carapace of the Greater Himalayan metamorphic sequence. The granite was capped and truncated by the stretched and heated mylonites and fault rocks along the low-angle normal fault that extended eastwards to the Annapurna Detachment. Above this fault lay the unmetamorphosed, highly folded sedimentary rocks of the northern Tethyan Himalaya. The structural geometry was indeed very similar to that I had already seen and mapped in the Ladakh-Zanskar and Garhwal Himalaya of India to the far west and the Everest massif to the east. The whole structure of the Himalayan Range was beginning to fall very nicely into place in my mind.

Descending through the forest trails from Bimtang we finally rejoined the main Marsyandi Valley on the Annapurna circuit at the little village of Dharapani. Once again we were stopped at a Maoist 'checkpoint' and asked for a 'donation', which Suka and I spent thirty minutes negotiating down to a reasonable amount. We found that we were suddenly in another country; the Maoists had declared this whole area the new 'Tamuwan Autonomous Republic'. We trekked on down to Besisahar, the village at the start of the Annapurna circuit, and then caught a bus to Pokhara for a few days of well-earned rest and recuperation, eating, drinking, and paddling around the Phewa Lake, with its unsurpassed panorama of the entire Dhaulagiri, Annapurna, and Manaslu Ranges to the north.

The Annapurna Detachments

Undoubtedly the Kali Gandaki Gorge between the great 8,000-metre-high peaks of Dhaulagiri and Annapurna was one of the best transects across the Himalaya to study the amazing geological structures (Figure 9.14). The valley and its tributaries enable the geologist to get a unique three-dimensional view through these structures. I spent two further field seasons working in this area, mapping in greater detail and exploring up several side valleys.[7] I was particularly interested in the low-angle fault structures—the South Tibetan Detachment System—along the upper contact of the Greater Himalaya. Earthquakes very rarely occur on faults aligned at angles less than 30°, and it was generally assumed that these faults were passively rotated from higher angles after they ceased to move. Although the Annapurna low-angle faults were not presently active, it seemed very unlikely that

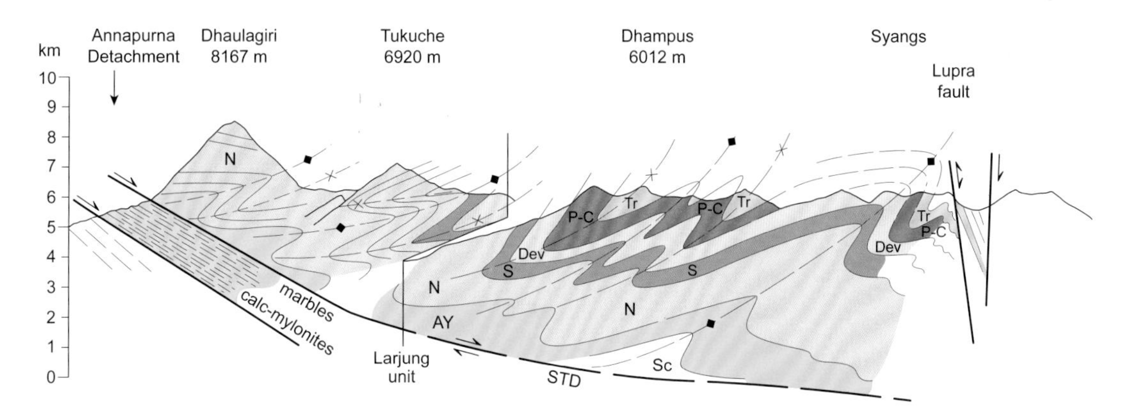

Figure 9.14 Geological cross-section across the Dhaulagiri–Dhampus Range, west of the Kali Gandaki.

they initially formed at higher angles and were later rotated simply because the crust above would have to have been unrealistically thick. There must have been some other mechanism to allow these faults to move at low-angles, sometimes even almost horizontally. The answer soon became apparent in the rocks immediately beneath the faults (Figure 9.15).

The Kali Gandaki Valley is geologically unusual in that the highest peaks, Dhaulagiri and the entire Annapurna Range, are actually composed of limestones and dolomites of the unmetamorphosed Tethyan Himalaya like the very top of Everest and not the crystalline rocks seen in Manaslu and along the Greater Himalaya elsewhere in Nepal and India. The South Tibetan Detachment low-angle normal fault actually cuts across the base of Dhaulagiri and Annapurna. Beneath the fault the rocks rapidly became extremely highly metamorphosed, showing first the index minerals sillimanite and cordierite, then kyanite in the rock assemblage. Each one of these is also associated with the beginnings of partial melting. At temperatures above about 750°C the rocks were actually starting to melt, forming at first a partially melted rock called a migmatite, and then on complete melting, a granite. I realized that it must have been the granitic melts that were lubricating the channel and allowed the entire mid-crustal layer to flow southwards, exactly as we had discovered in the Zanskar Himalaya to the west. Dating the metamorphic and granitic rocks had revealed that the peak of high-temperature and high-pressure metamorphism had occurred between about 30 and 20 million years ago, again very similar to the ages we had acquired from Zanskar and Garhwal in the western Himalaya.

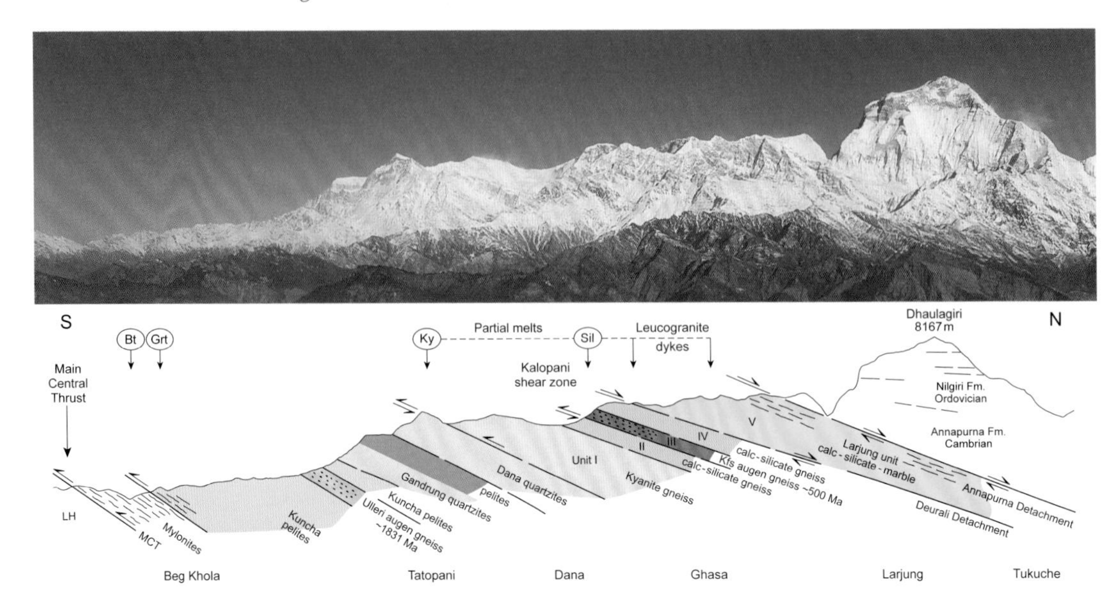

Figure 9.15 Panorama of the Dhaulagiri Himalaya viwed from Poon Hill across the Kali Gandaki. Coloured rocks are the metamorphic rocks of the Greater Himalaya. Metamorphic isograds are marked: biotite (Bt), garnet (Grt), kyanite (Ky), and sillimanite (Sil).

As India and Asia continued to push into each other after the collision, folding and thrusting had led to crustal thickening. The crustal thickening had resulted in increased temperature and pressure, producing metamorphism and finally even partial melting. The melts had allowed the entire 10–20-kilometre-thick middle crust to flow and had in places lubricated the low-angle ductile shear zones above. As the once deeply buried metamorphic rocks were extruded southwards and upwards they started to cool (Figure 9.16a, b). When the partially molten rocks cooled, ductile flow gave way to brittle faulting and the rocks were carried piggy-back along giant thrust faults such as the Main Central Thrust. Geochronological results showed that the entire metamorphic core of the Himalaya had cooled dramatically some 14 million years ago when the Greater Himalaya almost locked solid as thrust faulting propagated southwards to the active Lesser Himalayan Range and the Main Boundary Thrust. This is the huge thrust fault that runs along the southern margin of the Himalaya, placing Himalayan rocks over the young sediments of the Siwalik basin. All the historic earthquakes known from the southern Himalaya have occurred along this major thrust fault that accommodates the underthrusting of the Indian plate northwards beneath the Himalaya and Tibet.

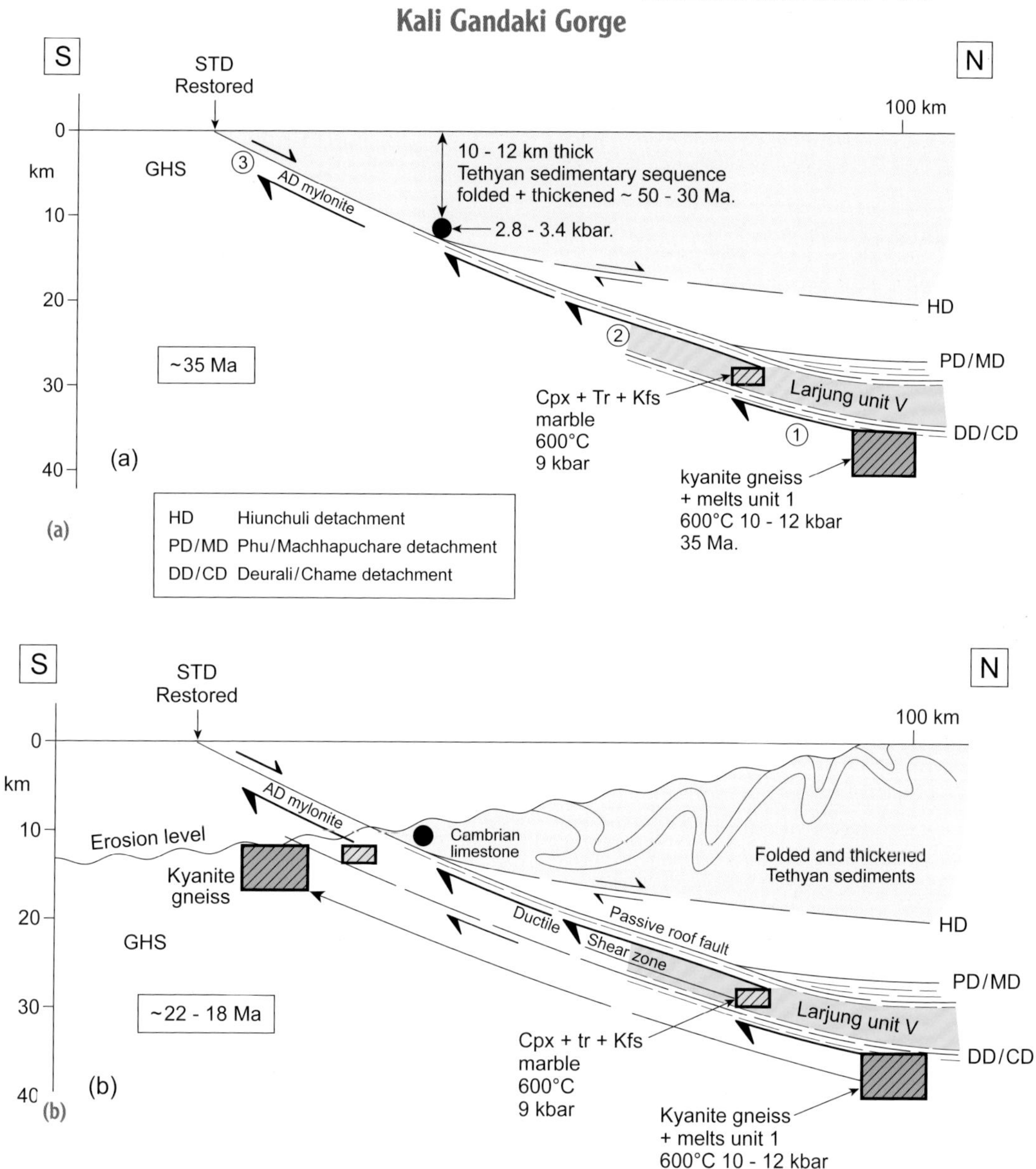

Figure 9.16 (a,b) Restoration of the South Tibetan Detachment in the Annapurna Himalaya. Blue colours are the unmetamorphosed sedimentary rocks of Dhaulagiri, yellow are the metamorphosed marbles along the ductile shear zone of the South Tibetan Detachment, and the green rocks are the deeper kyanite-bearing gneisses. (b) Arrows show the movement of these metamorphic rocks up beneath the low-angle normal fault from their original depth to their present juxtaposition.

Figure 9.17 A small plane flying from Pokhara to Jomoson passing across the south face of Machapuchare, the 'Fish Tail' peak.

In the spring of 2011 I revisited the Kali Gandaki Gorge for the third time, with the aim of trying to sample in detail across the South Tibetan Detachment. Suka and I flew into Jomoson from Pokhara, the wings of the Buddha Airways Fokker Friendship skimming frighteningly close to the hanging glaciers of Machapuchare and then beneath the massive south face of Dhaulagiri, surely one of the worlds' most dramatic flights (Figure 9.17). It was a shock to see jeeps and trucks in Jomoson. The Nepalese had built a dirt track road along the Kali Gandaki from Beni and now black smoker trucks belching pollution and noisy honking jeeps rubbed shoulders with trekkers along the Kali Gandaki. Apparently the Chinese had already built a road down from Tibet to Lo Mantang, the capital town of Mustang, and were about to link it south to Jomoson.

What used to be the most beautiful trek in Nepal through flowing fields of green barley, quiet villages, and whispering apple orchards was now destroyed forever. Suddenly there were no more trekkers, the lodges were empty; tourists did not want to trek along roads with trucks and jeeps roaring past, belching fumes, with all the noise and pollution that they bring. Other road projects were in the pipeline too, mostly with Chinese finance, along the Trisuli River from Tibet to Kathmandu, and even along the Arun, the deep gorge that cuts through the Himalaya east of Makalu. It seemed incomprehensible to me why the Nepalese authorities were destroying the most unique and wonderful asset they have—the pristine Himalaya.

Suka and I trekked north along the windswept Kali Gandaki to Kagbeni then turned south, collecting samples all the way down. We were based in the villages of Larjung and Tukuche whilst we climbed high on the eastern flank of Dhaulagiri, sampling the limestones downwards towards the ductile shear zone where they gradually became mylonites and then, abruptly below, metamorphosed marbles and gneisses. I was hoping that we could obtain precise pressure and temperature conditions of formation so that we could demonstrate the big jump in metamorphic conditions across the shear zone. For two days we scrambled along steep and dangerous scree slopes above Marpha and Larjung with spectacular views to the east where giant folds were magnificently exposed along the flanks of the Nilgiri peaks. Then on the flat valley trail below Larjung I slipped on a shortcut route and fell over twisting my ankle and grazing my leg. Both swelled up like balloons and I hobbled down to the nearest lodge at Koketani, by which time I could barely walk. Suka wanted to cut out all the dead blood in village fashion, but I thought better of this drastic remedy and left it to blow up. For two days I sat with my leg up, sipping Marpha apple brandy, watching the skyscapes above Dhaulagiri, and then decided I needed to get it seen by a doctor. As I could not walk on my leg, Suka went all the way back to Pokhara and came back a day later with his motorbike. I climbed on the back and for two extremely uncomfortable days bounced down the trail back to Beni and then Pokhara. I have to admit that I was very glad that the road was there! Therein lies the dilemma: the local people generally want the road for trade, and easier access to the markets of Beni and Pokhara. The winners are the Chinese businessmen who truck their cheap goods down to flood the markets of Nepal, the apple orchard owners, and the jeep drivers; the losers are the porters who are no longer needed, and the lodge owners whose tourists have deserted them. Above all, once the road had been built, the environmental degradation was irreversible. With a new road comes all the dirt and squalor and pollution of the plains. It is actually so easy for a few men and a bulldozer to destroy some of the most unique and beautiful places on Earth. For trekkers visiting Nepal, what used to be number one on the agenda, the Annapurna circuit, is now well down the list.

CHAPTER 10

Around the Bend

Nanga Parbat, Namche Barwa

As far as the western world was concerned we were exploring country of which nothing was known, but much was speculated; one of the last remaining secret places of the Earth, which might conceal a fall rivalling the Niagara or Victoria Falls in grandeur.

Frederick M. Bailey, *No Passport to Tibet*, 1913

From the geological mapping, structural, and metamorphic investigations along the main Himalayan Range from Zanskar in the west through the Himachal Pradesh and Kumaon regions of India and along the whole of Nepal to Sikkim, a similar story was emerging. The overall structure and distribution of metamorphic rocks and granites was remarkably similar from one geological profile to the next. The Lesser Himalaya, above the Main Boundary Thrust was composed of generally older sedimentary and igneous rocks, unaffected by the young Tertiary metamorphism. Travelling north towards the high peaks, the inverted metamorphism along the Main Central Thrust marked the lower boundary of the Tertiary metamorphic rocks formed as a result of the India–Asia collision. The large Himalayan granites, many forming the highest peaks, lay towards the upper boundary of the 'Greater Himalayan sequence'. North of this, the sedimentary rocks of the Tethyan Himalaya crop out above the low-angle normal fault, the South Tibetan Detachment. The northern ranges of the Himalaya comprise the sedimentary rocks of the northern margin of India. The two corner regions of the Himalaya, however, appeared to be somewhat different. The Indian plate has two major syntaxes, where the structural grain of the mountains swings around through ninety degrees: the western syntaxis, centred on the mountain of Nanga Parbat in Pakistan, and the eastern syntaxis, centred on the mountain of Namche Barwa in south-east Tibet.

Nanga Parbat—The Naked Mountain

Nanga Parbat (8,125 m) is a huge mountain massif at the north-western end of the great Himalayan chain (Figure 10.1). It is most prominent seen from the Indus Valley and the hills of Kohistan to the west, where it seems to stand in glorious isolation, ringed by the deep gorges carved by the Indus and Astor Rivers (Figure 10.2), before the great wall of snowy peaks forming the Karakoram to the north. The northern Rakhiot face is a massive wall of rock and ice laced with hanging glaciers that falls

Figure 10.1 Landsat photograph of Nanga Parbat (8,125 m), Pakistan. The Indus River is in the top left with Raikhot face in shadow to the north, the Diamir face to the east and the Rupal face to the south.

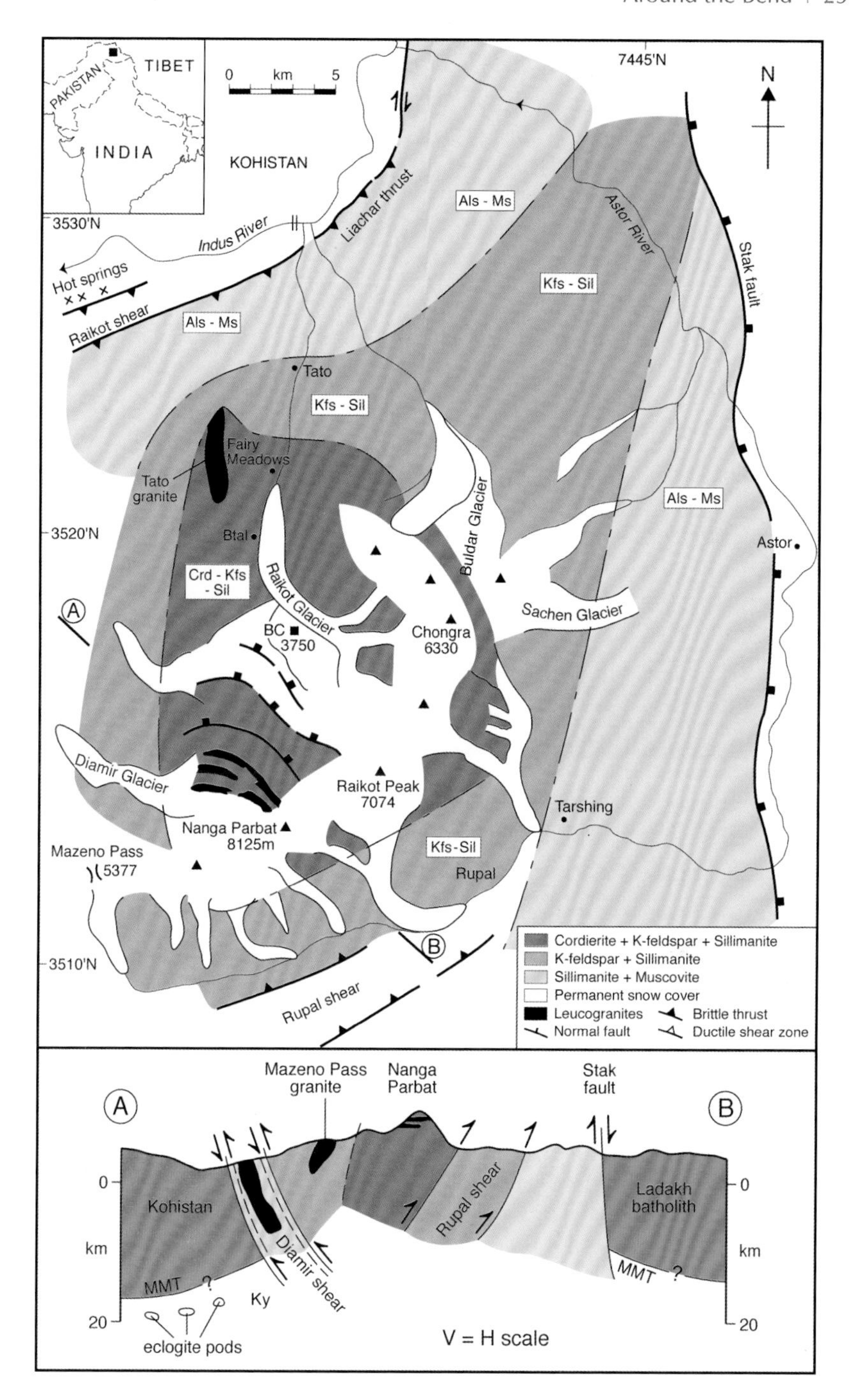

Figure 10.2 Detail of the geological map of Nanga Parbat.The massif is bounded by two large-scale faults: the Raikhot fault system to the west and the Diamir fault to the east. Rocks to the west are units of the Kohistan island arc, and to the east are older granites of the Ladakh batholith; rocks to the north are the young metamorphic rocks of the Karakoram.

Figure 10.3 Hanging glaciers on the Rakhiot face of Nanga Parbat falling over vertical normal faults flanking the rapidly uplifting core region of the mountain.

4,000 metres down to the beautiful alpine pastures of Fairy Meadows, the site of the base camp (Figure 10.3). Both the Rakhiot face and the western Diamir face fall directly down from the 8,125-metre-high summit to the Indus River at less than 2,000 metres' altitude, making these slopes the highest known anywhere (Figure 10.4). The awe-inspiring southern face of Nanga Parbat, the Rupal face, falls 4,700 metres down to the Astor Valley in a complicated series of ice cliffs and buttresses (Figure 10.5). The Indus River, rising around Mount Kailas on the Tibetan Plateau, and running for hundreds of kilometres through the Ladakh Himalaya, has carved probably the most spectacular canyon of all, right through the middle of the Nanga Parbat and Haramosh massifs. The water of the Indus thunders through the gorges and over waterfalls dropping two or three kilometres down to the barren brown hills of Gilgit and Chilas.

The first attempt to climb Nanga Parbat in 1895 was by the great British mountaineer Alfred Mummery, who disappeared on the attempt on the Diamir flank of the mountain. In the early 1900s a series of tragedies on the Rakhiot side of the mountain resulted in the deaths of several leading mountaineers, including Willo Welzenbach and Willy Merkl, and in one terrible accident,

Figure 10.4 Migmatites and granites collected from 6,000-metres elevation on the Raikhot face of Nanga Parbat have some of the youngest ages known anywhere on Earth, as young as 600,000 years old, and show the fastest exhumation rates, up to 1 cm/year.

an avalanche buried sixteen men. The 1934 expedition ended in disaster when a ferocious storm hit the mountain and six Sherpas and three Germans died during a harrowing retreat. In 1939 attention turned to the Diamir face. War broke out while four German climbers were on the mountain. All four were interned by the British after they returned, but Peter Aufschnaiter and Heinrich Harrer managed to escape to Tibet, where they very sensibly remained for seven years whilst Europe was busy destroying itself.[1] Nearly twenty years later, the first ascent of Nanga Parbat was made by an Austrian climber, Herman Buhl, who reached the summit alone on 3 July 1953 via the Rakhiot face east ridge route in an astonishing feat of mountaineering.[2] Later epic climbs by Reinhold Messner who, in 1978, was the first to climb the mountain solo, have also gone down in mountaineering legend.[3]

In the Nanga Parbat and Haramosh regions, a great tongue-like projection of the Indian plate's lower crust protrudes north of the main Himalaya. The first geologist to identify the great 'western syntaxis', or the 'Nanga Parbat bend', was

Figure 10.5 (*opposite*) Nanga Parbat (8,125 m) more than six kilometres vertical elevation above the Indus Valley to the west and Rupal Valley to the south, the highest face on Earth.

Darashaw Wadia, a field geologist from India who spent many summers during the 1920s roaming around the mountains of Kohistan and the Deosai Plateau around Nanga Parbat. Wadia's 'Geological Map of part of the Gilgit District' at a scale of one inch to four miles included all the territory around the peak of Nanga Parbat.[4] This work included the first cross-section across the Nanga Parbat massif from the Indus Valley in the west to the Rupal Glacier in the east. Wadia recognized that the massif was composed of gneiss with interbedded 'Salkalla Series' surrounding a central summit region composed of 'granitoid gneiss'. He assumed, quite reasonably for the time, that these crystalline rocks were part of the very old Archaean basement of India. In those days, without the benefit of modern dating techniques, almost all metamorphic rocks were simply assumed to be Precambrian basement. Wadia would have been amused to know that now the Nanga Parbat gneisses are understood to show the youngest metamorphic event dated by U-Th-Pb anywhere in the world!

The large German–Austrian expedition of 1934 led by Willy Merkl included the geologist Peter Misch. Misch was the first person to make a detailed petrological study of the rocks, and described the increasing grade of metamorphism towards the central summit region of the mountain.[5] He noted the first indications of partial melting of the rocks with wispy veins and pods of granite seeming to sweat out of the host gneisses. Misch rightly proposed that the rocks were flushed through with, and altered by, high-temperature fluids and that this process, called metasomatism, might explain the origin of these special types of granites. At the time this went against all the established ideas that granites were formed from very hot magmas deep in the crust.

Melting of the Crust: The Origin of Granite

In the 18th century granites were widely accepted as being solely a result of crystallization from a liquid magma or some sort of aqueous solution. During the 1940s, the 'granite controversy' split geological opinions between the Vulcanists and the Neptunists. The Neptunists or 'magmatists' argued that granites were formed by intrusions of liquid magma, a point forcibly argued by experimental petrologists like N. L. Bowen, of Bowen's reaction series fame.[6] The Vulcanists or 'granitizers' thought that granites formed through the chemical alteration or metasomatism of water-rich fluids recrystallizing metamorphic rocks, a point of view ably championed by H. H. Read, a field geologist.[7] Read had noted that in the

Scottish Highlands, granite melts started to appear at the highest metamorphic grade. He proposed that it was possible to produce a granite melt from the end product of regional metamorphism. Several Finnish geologists, notably Jakob Sederholm and Pentti Eskola, made similar observations working on the Precambrian rocks of Scandinavia. There appeared to be a conflict between the petrologists looking down their microscopes and the field geologists. Nanga Parbat played a major role in the development of arguments concerning the origin of granites through the work of Peter Misch. He outlined the crux of the problem in his classic paper of 1949: 'This is the problem confronting us: do the granitic gneisses which form the bulk of Nanga Parbat massif owe their origin to intrusion of granitic magma or to metasomatic granitization?' Misch noted in Nanga Parbat that the 'most striking feature is the intimate relationship [between granite and gneiss] and simultaneous increase of synkinematic granitization' and that 'the rise of temperature and increasing infiltration by granitizing solutions go in parallel'.

Pleistocene Metamorphism and Melting

During the 1980s it was becoming clear that the Himalayan granites were a very special and almost unique type of granite. They are very white (hence the term leucogranite), and composed mainly of quartz and feldspar, but they also contain very distinctive minerals, notably black tourmaline, red garnet, abundant white mica, muscovite, and occasional black mica, biotite. The geochemistry of these granites was also proving to be very unusual. The ratios of strontium isotopes ($^{87}Sr/^{86}Sr$) suggested that they were pure crustal melts. Although many Himalayan granites were initially described as intrusive plutons it was obvious that they did not form buoyantly rising domes such as those in the Andes. Rather, they were formed as a series of layer-parallel sills emanating from a migmatite zone, a zone of partial *in situ* melting from the surrounding gneisses. As we have seen, most Himalayan granites were formed during the Early Miocene period, between about 24 and 15 million years ago.[8]

The youngest thermal events are recorded in the two syntaxes—the corners of the indenting Indian plate, the north-west Nanga Parbat syntaxis in Pakistan, and the north-east Namche Barwa syntaxis in south-east Tibet. In both these areas, Indian basement gneisses went through the whole Cenozoic Himalayan thermal history of metamorphism resulting from the collision, and then continued with even younger sillimanite-, garnet-, tourmaline-, and cordierite-bearing migmatites—the result of very high-temperature partial melting of the crust that happened during Pliocene and Quaternary times.[9] Maybe these rocks were actually recording the thermal conditions of the deep levels of the Himalaya today, and the only reason they were exposed here was because of the special structural controls of the syntaxes.

I seemed to have known Nanga Parbat for a long time. Every time I drove our jeep along the Karakoram Highway north to Gilgit and Skardu on the way to the Karakoram, we passed the mountain. At the Rakhiot Bridge we always stopped to look at the incredible outcrops along the road where deep crustal, highly metamorphosed Nanga Parbat gneisses were thrust over almost unconsolidated gravels of the Indus River. This one outcrop seemed to summarize the incredible geology of the Himalaya, where rocks that formed deep down in the crust were sitting above Pleistocene river gravels that formed so recently in the past. Even more incredible was the history of the Nanga Parbat gneisses. Isotopic and dating studies had revealed that these rocks did initially belong to the Indian crust basement. They had undergone high-temperature metamorphism during earlier mountain-building episodes during the Precambrian and then happily settled at the base of the crust for more than 500 million years until India and Asia collided. The crust was then thickened again by repeated thrust stacking, until once again the rocks were subjected to high-grade metamorphism during the Himalayan orogeny. Finally during the recent and active uplift of Nanga Parbat, the once deeply buried and hot gneisses of the lower crust were thrust up and out onto the recently deposited gravels of the Indus River. This was all part of the great crustal recycling scheme.

After our Karakoram expeditions of the late 1980s and early 1990s I had driven along the Astor Road and across to the Deosai Plateau, and trekked over to the Diamir face and up to base camp along the Rakhiot Valley. In 1995 I decided to take a closer look at this amazing mountain. By then a new single-track road had been blasted up almost to base camp (Figure 10.6). I trekked further on from Fairy Meadows along the glacial moraine until I was forced onto the Rakhiot Glacier (Figure 10.7). The glacier drains the entire north face of Nanga Parbat and amongst the boulders I found abundant migmatites with a bright green mineral in the melt zones as well as the ubiquitous black tourmaline, flaky white muscovite micas, and red garnet. This new mineral was cordierite, a characteristic mineral of high-temperature but low-pressure metamorphism. Both sillimanite and cordierite were part of the magmatic assemblage making up the granites (Figure 10.8). By following the moraines upwards I could get a rough idea of exactly which part of the mountain was providing the abundant boulders. The cordierite-bearing migmatites seemed to be coming from the central part of the massif around the summit cliffs. I managed to sample the sidewalls of the glacier, but the only way I could sample the upper parts of Nanga Parbat was to collect boulders in the

Figure 10.6 The Tato road from the Indus Valley up to Nanga Parbat Base Camp.

glaciers that came from the highest cliffs. Altogether I collected about twenty-five samples from the Rakhiot Glacier basin and brought them back to Oxford for detailed studies on the mineral assemblages. I sent five samples off to my friends from MIT, Jim Crowley and Sam Bowring, for U-Pb dating. The dating results were quite incredible.

These Nanga Parbat rocks, the youngest migmatites known anywhere on Earth, reached peak metamorphic conditions of 720°C and pressures of 5–3.5 kilobars, equivalent to depths of burial of between 20 and 10 kilometres' depth in the crust, between 1.7 and less than one million years ago. They are now at elevations of

Figure 10.7 The Rakhiot face of Nanga Parbat. Cliffs above show Pliocene-Pleistocene-age migmatites and granites, the youngest crustal melt rocks known anywhere on Earth.

Figure 10.8 Cordierite (green mineral) and garnet-bearing granite dykes formed only seven or eight hundred thousand years ago intruding along ductile shear zones cutting young partially melted migmatites, from the core region of Nanga Parbat.

7,000 to 8,000 metres on Nanga Parbat. The youngest U-Pb ages ever measured are tiny grains of monazite from a garnet-biotite-sillimanite melt vein that records ages as young as 790,000 years, and four xenotime crystals that have ages between 720,000 and 690,000 years old.[10] These rocks record the highest exhumation and erosion rates known anywhere on Earth—more than 11–13 mm per year (Figure 10.9).

Why does Nanga Parbat show these extremely young metamorphic ages and extremely high uplift rates? Whereas most of the Himalaya records a head-on collision, with the Indian plate subducting and underthrusting the Asian margin, in the north-west Pakistan region, the corner of the indenting Indian plate is being compressed from every direction. Nanga Parbat and the Haramosh massif to the north are being squeezed in both a north–south direction and an east–west direction, so the only place to go is up. It is evident from the abundance of hot springs around Nanga Parbat that the crust is completely infiltrated with boiling fluids. The origin for the youngest Nanga Parbat granites was indeed high-temperature metasomatism, exactly as was first proposed by Peter Misch. Whereas in normal crust the upper 10–20 kilometres shows brittle deformation and earthquakes, at Nanga Parbat the uplift rates are so high that the hot, deep, and therefore ductile, deforming lower crust is being pushed rapidly upwards.

Figure 10.9 Aerial photo of Nanga Parbat.

Landslides and Floods

The geomorphology of Nanga Parbat is determined by its geology. It forms a huge isolated mountain massif that is rising faster than anywhere else on Earth. The young uplift has created extremely steep slopes, and glaciers draining the central core of the massif have steep icefalls that cascade down active normal faults. Rock-falls and avalanches occasionally block the Indus River. In 1841 a series of earthquakes resulted in a massive landslide along the north-western flank of Nanga Parbat between the Raikot and Liachar Valleys. The landslide blocked the Indus and over the course of several weeks a huge lake, 50 kilometres long, built up behind the dam. The dam finally failed and the lake emptied in twenty-four hours, resulting in an incredible flash flood. A massive wall of water swept down the Indus Valley and out onto the plains in the Peshawar Valley. Thousands of people died, including most of a Sikh army that happened to be camping on the banks of the Indus. Frederick Drew traced the source of the Great Indus Flood during his travels between 1862 and 1872 and discovered the remnants of the enormous landslide above Raikot.[11] Over a hundred years later, Rob Butler and Dave Prior from the University of Leeds mapped the region in detail and discovered an active thrust fault, the Liachar thrust, that places the high-grade metamorphic gneisses of Nanga Parbat westwards over the remnant landslide debris and the practically unconsolidated Quaternary sediments of the banks of the Indus River.[12] The story behind this amazing outcrop, in which some of the oldest rocks in the Himalaya are thrust over some of youngest, can now be understood. The Precambrian gneisses of Nanga Parbat, formed through metamorphism in ancient mountain-building events, were buried and metamorphosed again during the Cenozoic Himalayan orogeny, subsequently subjected to sillimanite-grade metamorphism and partial melting during the Pliocene-Pleistocene, then exhumed from depths of about 20 kilometres, and finally thrust out onto the recently deposited, unconsolidated river gravels along the banks of the Indus River.

Namche Barwa Eastern Syntaxis

The Eastern Himalayan syntaxis is centred on the mountain of Namche Barwa (7,756 m) in south-east Tibet (Figures 10.10, 10.11, 10.12). This is probably the most inaccessible part of the entire Himalaya, with thick jungle-clad mountains rising from tropical bamboo forests right up to serrated icy mountain peaks over 7,000 metres high. The valleys are extremely steep with vertical canyon walls, and rivers tumble down impassable cascades and waterfalls. The earliest accounts of the mist-wreathed valleys of Pemako came by word of mouth from the Mishmi and Monba hunters who penetrated the land of aboriginal Lopas to hunt for takin,

Figure 10.10 Landsat photo of SE Tibet and the eastern Himalayan syntaxis centered around the peak of Namche Barwa (7,756 m).

musk deer, tigers, and medicinal plants. The earliest explorers were the pundits like the famous Kintup, plant-hunters like Frank Kingdon-Ward, or spies and map-makers, players of the Great Game like Frederick Bailey. The first attempt at mapping the region was made by Frank Kingdon-Ward, a plant hunter who made more expeditions to the area than any other Westerner.[13] Although not a geologist he was a careful observer and wrote a detailed journal of his travels: 'Every day the scene grew more savage; the mountains higher and steeper . . . the great river plunging down, down, boring ever more deeply into the bowels of the earth. The snow peaks enclosed us in a ring of ice. Dense jungle surged over the cliff.' He made

Figure 10.11 Aerial view of the incredible array of mountains in SE Tibet.

some astute observations of how the uplift of the Himalaya and Tibet affected the monsoon and climate, and also carried out some secretive spy missions to map out the imperial borders of British India.

Like the western syntaxis at Nanga Parbat, Namche Barwa shows several features unique to these two regions. The rocks belong to the northern margin of the Indian plate and, in common with the rest of the Himalaya, have been through the entire Himalayan metamorphic cycle during the period 50–20 million years ago. Namche Barwa and Nanga Parbat also show very young Pliocene-Pleistocene high-grade metamorphism and melting, resulting in the formation of partially molten migmatites and granites. There are high levels of seismicity and active faulting around the syntaxis region, indicating compression in almost every direction. There are extremely high geothermal gradients, with many hot springs indicating a vigorous fluid circulation system. Like Nanga Parbat, Namche Barwa shows extremely high erosion and exhumation rates (possibly up to 1 cm per year in places) and extremely high river incision rates.

Conventional geological mapping in the Namche Barwa region is almost impossible. The entire region shows extreme topography with precipitous mountains and deep gorges with thick jungle vegetation. More than two-thirds

Figure 10.12 Major rivers have cut deep gorges through the eastern part of the Tibetan plateau around the east Himalayan syntaxis.

of the world's plant species occur in the region that spans over 7 kilometres of elevation contrast. Unlike my annual failures to get permission to work in the Namche Barwa region, Peter Zeitler from Lehigh University and Page Chamberlain from Stanford University in the US did both manage to get permission from the Chinese authorities and raise the large amount of grant money needed to work there. They and their colleagues mapped and collected samples along the Yarlung Tsangpo Gorge. Peter and Page had carried out all the early dating work in the Nanga Parbat syntaxis in Pakistan so their work in both syntaxes provided most of the raw data for interpreting the tectonics of the syntaxes regions. Their dating results proved that high-grade metamorphism and partial melting had occurred in the Namche Barwa syntaxis since at least the mid-Miocene 12–15 million years ago. U-Pb ages on monazite and titanite crystals indicated that active metamorphism and granite melting occurred between 3 and 10 million years ago.[14] These ages were similar to the Nanga Parbat ages, but not quite so young.

Like Nanga Parbat, the eastern Himalayan syntaxis is closely associated with the course of a major Himalayan river and a deeply eroded canyon (Figures 10.13 and 10.14). In the western syntaxis the Indus River plunges through a deep canyon cutting directly between two extremely high peaks, Nanga Parbat and Haramosh. In the eastern syntaxis, the Yarlung Tsangpo River flows through a similar deep gorge in between the two giant peaks of Namche Barwa and Gyala Peri. Peter Zeitler noted the temporal and spatial association between young metamorphism

Figure 10.13 Aerial view of the eastern Himalayan syntaxis.

and fluvial erosion and proposed a model termed the 'tectonic aneurysm' model.[15] In this model, metamorphism, melting, and rapid exhumation are the result of erosion and down-cutting of the major rivers, all driven by climate. The 'tectonic aneurysm' model was similar to the theoretical-numerical model of Chris Beaumont to explain Himalayan channel flow, where in a similar fashion, focused erosion along the Himalaya caused the mid-crustal layer to flow southwards from under southern Tibet (Chapter 8). There is no doubt that there must be an erosional, thermal, and mechanical coupling in both syntaxes regions, but how could climate be the cause of these geological phenomena?

A number of factors seem to argue against the tectonic aneurysm model. For starters the name: an aneurysm is a blockage or enlargement (of a blood vessel, for example). In Nanga Parbat and Namche Barwa there is no blockage; if anything there is a vertical 'fountain' of rapidly exhuming rock material caused by compressional squeezing from all sides. The course of the Indus River does not correspond to the area of highest exhumation and youngest metamorphism. Both the Indus and Yarlung Tsangpo Rivers are antecedent to the rise of the Himalaya and were initiated at least 40 million years prior to the youngest metamorphism in both syntaxes. As the Himalayan mountains uplifted, the river incised deeper, creating

Figure 10.14 Huge glaciers flow from the high mountains of the eastern Nyenchen Tangla range SE Tibet.

the gorge. The Indus and Yarlung Tsangpo Rivers cut across the metamorphic core of both massifs, just as they cut across the Greater Himalaya along the main part of the range. How can climate, rainfall, or erosion cause uplift of the mountains? Rainfall cannot suck rocks out of the ground, and besides Nanga Parbat, the western syntaxis region that shows the youngest metamorphic-melting ages, sits in a desert environment where rainfall is very scarce. Clearly all these processes are linked and probably concomitant in time. There cannot be high mountains without accompanying erosion, and high mountains create their own climate. It seems much more likely that tectonic forces cause uplift of mountains, and that uplift results in the initiation of rivers, and impacts on climate.

Yarlung Tsangpo Gorge

The Yarlung Tsangpo River rises in south-west Tibet from Lake Manasarova near Mount Kailas and flows 1,700 kilometres eastwards across southern Tibet before plunging through one of the deepest and most spectacular gorges on Earth, between the mountains of Namche Barwa (7,756 m) and Gyala Peri (7,150 m). Only 21 kilometres separates these two Himalayan peaks as the lammergeier flies, and the river has cut a chasm over 4 kilometres deep between them. In 45 kilometres' horizontal distance the river cascades from the Tibetan Plateau through the chasm to disgorge out of the Abor Hills onto the plains of Assam at only 700 metres' elevation. Along this deep and largely inaccessible gorge are several magnificent waterfalls, the 25-metre-high Rainbow Falls and the Hidden Falls. The great bend of the Tsangpo–Brahmaputra is certainly one of the great wonders of the Himalaya.[15]

Early travellers were puzzled by the geography of the Tibetan rivers. Near the village of Pei in south-east Tibet the great Yarlung Tsangpo River flowed straight into the heart of the Namche Barwa Range through an impenetrable-looking gorge. No-one knew where the river actually went, with speculation that it linked north-east towards the Yangtse, south-east towards Yunnan and the Red River, or south into Burma. One of the most remarkable pundit–explorers in the service of the Indian Survey was a man called Kintup, alias K.P., an illiterate tailor from upper Sikkim. In 1878 Kintup reached the village of Gyala at the top of the Tsangpo Gorge, and finding no way down, retraced his route back to India. The Survey of India had mapped the great Brahmaputra River in Assam, but no one knew where this huge river had come from. Some speculated that it must have joined the Irrawaddy in Burma, others that maybe the Salween or the Yarlung Tsangpo was the continuation. Kintup was sent to find the riddle of the Tsangpo Gorges and set off from Darjeeling in 1880. Four years later he came back with a remarkable story. He was sold into slavery, escaped, and almost died on several occasions. Having once more failed to descend the gorges of the Tsangpo, he sent a letter to Captain Harman of the Survey of India with a wonderful plan to test the source of the Brahmaputra. Kintup would throw fifty logs a day for ten days into the Tsangpo, and someone was to stand on the banks of the Brahmaputra in Assam and look to see if they came down. Unfortunately Captain Harman had left for England before the letter arrived and his successors in the Survey of India believed Kintup was either arrested or dead.

It was left to Frederick Bailey's clandestine expedition of 1911–12 to finally prove that the Tsangpo and the Brahmaputra Rivers were one and the same.[16] Frank Kingdon-Ward made an epic expedition into the heart of the Tsangpo Gorge with Lord Cawdor during 1924–25, but failed to find the mysterious 'tallest waterfall' that was rumoured to exist by local Monpa hunters.[17] In fact there are a series of waterfalls along the gorge. Ian Baker, Hamid Sardar, and Ken Storm finally discovered the 25-metre-high Rainbow Falls as recently as 1998.[18] The Yarlung Tsangpo is the Everest of kayakers' dreams and a few expeditions have attempted the descent. A Japanese kayak expedition attempted the gorge in 1993 but lost one member, and a National Geographic Society sponsored expedition in 1998 also abandoned the attempt after their most experienced kayaker was killed.[19]

The river drainage patterns around southern Tibet are complex, with long-lived river systems like the upper reaches of the Yarlung Tsangpo antecedent to the Himalayan mountain uplift, and more recent rivers related to the uplift of the Himalaya. Rivers have been depositing conglomerates along the course of the Yarlung Tsangpo River since at least Oligocene times (38 million years ago) as recorded in the Kailas conglomerates and Indus molasse. As the Himalayan mountains were uplifting to the south the rivers in Tibet aligned west–east, collecting drainage from the northern slopes of the Himalaya. Many other rivers initiated along the southern slopes of the Himalaya during the earliest rise of the mountains around 45 million years ago, flowing south from the watershed towards the Ganges basin. These rivers steepened as the Himalaya rose and cut back, incising northwards into Tibet. Many of these great Himalayan rivers in Nepal, Bhutan, and Arunachal Pradesh have incised so far back that they have gone right through the highest axis of elevation and are almost on the point of cutting right through to the Yarlung Tsangpo. When this does happen, catastrophic changes in river systems will occur. Predicting the geological future is always tricky, but at some stage the Arun River in eastern Nepal and the Manas River in Bhutan will cut northwards and capture the drainage of the Yarlung Tsangpo River in Tibet. In the next chapter we shall examine the geological evolution and uplift of Tibet, the greatest high-altitude plateau on Earth.

CHAPTER 11

Roof of the World

Tibet, Pamirs

'Today the map has no more secrets.' Idle minds repeat that parrot phrase but who knows all Tibet or its far away frontiers in western China? Even its own prayer-muttering tribes know only their own bleak wind-swept valleys.

Joseph Rock, *Seeking the Mountains of Mystery*, 1930

The Tibetan Plateau is by far the largest region of high elevation, averaging just above 5,000 metres above sea level, and the thickest crust, between 70 and 90 kilometres thick, anywhere in the world (Figure 11.1a, b). This huge plateau region is very flat—lying in the internally drained parts of the Chang Tang in north and central Tibet, but in parts of the externally drained eastern Tibet, three or four mountain ranges larger and higher than the Alps rise above the frozen plateau. Some of the world's largest and longest mountain ranges border the plateau, the 'flaming mountains' of the Tien Shan along the north-west, the Kun Lun along the north, the Longmen Shan in the east, and of course the mighty Himalaya forming the southern border of the plateau. The great trans-Himalayan mountain ranges of the Pamir and Karakoram are geologically part of the Asian plate and western Tibet but, as we have noted before, unlike Tibet, these ranges have incredibly high relief with 7- and 8-kilometre-high mountains and deeply eroded rivers and glacial valleys.

The western part of the Tibetan Plateau is the highest, driest, and wildest area of Tibet. Here there is almost no rainfall and rivers that carry run-off from the bordering mountain ranges simply evaporate into saltpans or disappear underground. Rivers draining the Kun Lun flow north into the Takla Makan Desert,

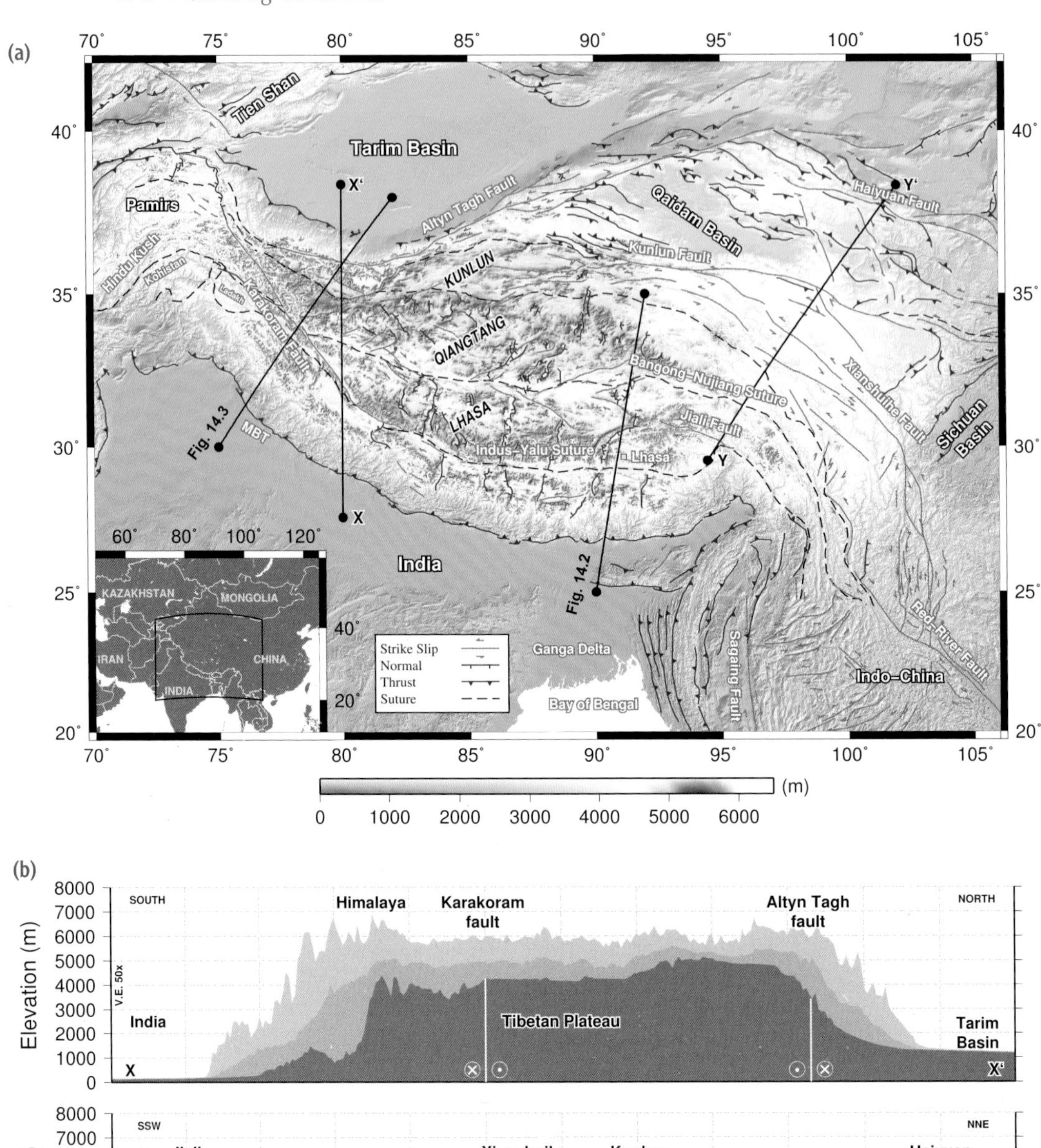

Figure 11.1 (a) Digital Elevation model of the Tibetan model and surrounding mountain ranges. (b) Topographic profiles along western and eastern Tibet.

forming seasonal marshlands in the wet season and a dusty desert when the rivers run dry. The discovery of fossil tropical leaves, palm tree trunks, and even bones from miniature Miocene horses suggest that the climate may have been wetter in the past, but this is also dependent on the rise of the plateau. Exactly when Tibet rose to its present elevation is a matter of great debate. Nowadays the Indian Ocean monsoon winds sweep moisture-laden air over the Indian sub-continent during the summer months (late June–September). All the moisture is dumped as the summer monsoon, the torrential rains that sweep across India from south-east to north-west. The Himalaya acts as a barrier to the monsoon so the southern slopes are heavily forested and highly eroded, whereas the northern slopes are dry and arid. The Himalaya is one of the most sudden and spectacular climatic-geomorphic divides known anywhere on Earth. Occasionally strong summer monsoons cause the heavy rains to spill over into Tibet where they can cause extensive damage and turn dust and sand into Land Cruiser-swallowing lakes of mud.

Three great rivers rise in south-western Tibet at the holy Mount Kailas, revered by Buddhists and Hindus alike as the spiritual centre of the Earth[1] (Figure 11.2a, b). The Indus flows west cutting through the Karakoram and Ladakh Ranges, right through the Nanga Parbat-Haramosh massif and then swings south, cutting through the Pakistan Himalayan Ranges to drain eventually into the Arabian Sea. The river course cuts across the Ayilari and Ladakh Ranges, and has been dragged and offset by subsequent horizontal motion along the Karakoram strike-slip fault for a distance of 120 kilometres. The Indus River was initiated as Tibet began to rise, at least from the timing of the India–Asia collision if not earlier, and the Karakoram fault was formed after 13 million years ago, as our earlier dating of the granites in northern Ladakh showed (Chapter 4). The 120-kilometre offset therefore gives a reliable estimate of the maximum horizontal motion that must have occurred along the Karakoram fault.

The second great river that rises on Mount Kailas is the Sutlej, which rises from Lake Raksas south of Kailas and trickles gently through the barren, dry badlands of Tsaparang, or the ancient Tibetan Kingdom of Guge (Figure 11.3a, b), before cutting abruptly through the Kumaon Himalaya in India, picking up much of the drainage of the western Himalaya and eventually converging with the Indus. As the Himalaya began to rise following the collision with Asia 50 million years ago, the mountains dammed the river. Waters backed up into a vast lake nearly 300 kilometres long and 40 kilometres wide. At some stage the river burst through the Sutlej Gorge, draining the lake and flooding the plains of the Punjab. Historic records show that this great flood of 1841 wiped out the remnants of the army of Alexander the Great that were unfortunately camped on the plains of the Punjab at

(a)

(b)

Figure 11.2 (a) Satellite image of Mount Kailas, source of the three great trans-Himalayan antecedent rivers, the Indus, Sutlej, and Yarlung Tsangpo-Bhramaputra Rivers. Kailas lies north of the twin lakes of Raksas (left, altitude 4,572 m) and Manasarovar (right, altitude 4,586 m) in south-western Tibet. (b) Mount Kailas (6,638 m) composed of layers of sedimentary conglomerates, with boulders derived mainly from the high Gangdese Range granites to the north.

(a)

(b)

Figure 11.3 (a) Satellite image of Gurla Mandata (7,728 m) composed of Indian plate metamorphic rocks rising along the northern Himalaya, south of the twin lakes of Raksas and Manasarovar in south-west Tibet. Dark rocks south of Lake Raksas are the ophiolites of the Indus suture zone. (b) Gurla Mandata (7,728 m), an uplifted dome of Indian plate Himalayan metamorphic rocks, rise above the blue waters of Lake Manasarovar.

Figure 11.4 Late Tertiary lake deposits of the Tsaparang basin lapping onto the northern Himalaya. The juvenile antecedent Sutlej River winds through the Tsaparang basin and drains the lake cutting through the rising Himalaya to the south.

the time. The scattered remnants of this huge lake can be seen in the flat-lying lake sediments that rim the Tsaparang basin in the far south-west of Tibet (Figure 11.4).

The Yarlung Tsangpo is the third great river to rise from the slopes of Kailas. It flows east across southern Tibet, following the line of the India–Asia suture or collision zone before abruptly falling off the edge of the plateau. The river flows along the Indus–Tsangpo suture zone separating the northern Himalayan Ranges comprising Indian plate sedimentary rocks to the south from the Gangdese Range granites that mark the southern boundary of the Asian plate to the north (Figure 11.5). As we have seen in Chapter 10, in far south-east Tibet, the Yarlung Tsangpo River winds through an incredible series of deep gorges as it cuts in between the mountains of Namche Barwa (7,756 m) and Gyala Peri (7,150 m).[2]

Exploration and Espionage

Tibet has always been a difficult, remote, and frequently dangerous place to access. During the 1800s the Russian empire of the tzars was spreading eastwards with the occupation of the vast steppes of Kazakhstan and the mountains of Khirghizia and Tadjikistan. The northern extent of the British empire in India reached up to the Himalaya, although in the west it spread across to the high peaks of the Karakoram and the Hindu Kush. Both empires kept a wary eye on each other and both were anxious to keep Tibet, in particular, free as a buffer zone. The 'Great Game' was the secretive game of cat and mouse espionage played out on the roof of the world by strange and shadowy figures that were to define the international boundaries,

Figure 11.5 Landsat satellite photograph of central south Tibet. The line of the India–Asia collision, the Indus suture zone, is marked, and major faults are shown along the Tibetan rift valleys.

many of which still remain to this day.[3] Beyond the Himalaya, the vast Tibetan Plateau was probably the largest blank on the map. Both Nepal and Tibet were closed, but the temptation of discovering what lay beyond the great ranges of the Himalaya was too much for the inquisitive minds of the British map-makers sitting in Calcutta and Dehra Dun.

The Survey of India began recruiting local Indians who were trained to become secret surveyors. Some were disguised as Muslim fakirs, others as Hindu holy men, and others as travelling Buddhist lamas. They were called the pundits or 'native explorers' and they played a vital role in the earliest explorations of the Himalaya and Tibet.[4] They were trained in surveying techniques and carried a sextant hidden in their robes. They learned to take bearings by the stars and to measure distances by counting paces with the help of Tibetan rosary beads, modified from the Tibetan sacred number of 108 to a round 100. They carried thermometers to determine altitudes by noting the boiling point of water and made copious notes, hidden in prayer wheels or walking sticks. These pundit explorers set out to survey and map the border regions and often disappeared for months, even years. They were known only by their initials, and beyond the borders of the British India were entirely left to their own fate. The pundits were at the cutting edge of the 'Great Game' and frequently acted as spies as well as surveyors. The characters in Rudyard Kipling's classic tale *Kim* clearly come straight out of the shadowy world of the pundits.

Perhaps the most famous of the pundit explorers was Nain Singh, a Bhotia (migrant from Tibet) who lived in the village of Milam in the Kumaon Himalaya. Nain Singh was recruited and trained by T. G. Montgomery of the Survey of India, and made his first journey into Tibet in 1865, travelling with a Ladakhi caravan. Nain Singh traced the course of the Yarlung Tsangpo River from its source near Lake Manasarovar and succeeded in reaching Lhasa, where he had an audience with the Dalai Lama in the Potala Palace. He later explored the headwaters of the Indus River near Mount Kailas, and he discovered the gold mines of Thok Jalung near Gartok in western Tibet. His final journey in 1874 traced a route from Ladakh across the remote and desolate Chang Tang Plateau from Pangong Lake to Lhasa across central Tibet. As we have seen in Chapter 10, in eastern Tibet, Kintup was the most profilic pundit who first explored the great chasm of the Tsangpo Gorges.[5]

Perhaps the greatest Tibetan explorer of all was the Swedish traveller Sven Hedin who, between 1894 and 1908, made three long and eventful expeditions to Chinese Turkestan (now Xinjiang) and Tibet, making topographic maps and collecting rock samples along the way. Hedin travelled all over the Pamir, and explored almost the length and breadth of the trans-Himalayan Gangdese Mountains of southern Tibet. He was a prolific writer and his books chronicle the landscape, geography, and peoples he encountered.[6] During the British Raj, an expeditionary force was sent to Lhasa under Sir Francis Younghusband with the aim of preventing the Russian imperialist advances of the Tsars and bolstering trade. Younghusband entered Lhasa on 7 August 1904 and signed a treaty in the Potala Palace with the Tibetan government.[7,8]

The Russians were more interested in the high Pamir to the west than invading Tibet, but the Chinese, who had long-standing and somewhat dubious claims over Tibet, finally invaded in 1910, and the Dalai Lama fled across the Himalaya to India, where he remained for two years until a popular uprising in Tibet forced the Chinese to retreat. This was a precursor to the latest Chinese invasion of 1950. From

1913 to 1950 Tibet was an independent and autonomous state, ruled by the feudal and lamaist clergy from the Potala Palace in Lhasa. In October 1950 three divisions of the Red Army including 40,000 troops invaded eastern Tibet and marched westwards to cover the entire plateau as far as the Pamir and Tien Shan Mountains. The Cultural Revolution from 1966 to 1976 was a time of unprecedented disaster for Tibet and indeed China. During those ten years almost every single monastery in Tibet was destroyed and looted, valuable Buddhist texts were burned, all religions were banned and many Tibetans went into exile. It was only some time after Mao's death in 1976 that China returned to some sort of normality, but Tibet and Xinjiang remain under complete Chinese control to this day.

Mapping Tibetan Geology

Following the initial explorations of Tibet in the late 19th century Tibet was closed to foreigners. In 1978 the Chinese authorities invited a few Western scientific delegations to visit the plateau and several major collaborative geological programmes were established, some combining surface geology studies with geophysical investigations. During the late 1970s French and Chinese geologists mapped large tracts of southern Tibet around the main collision zone.[9]

Despite the fact that large areas of the plateau remain unvisited by geologists, the major geological events of the past 250 million years are now apparent. Apart from the current collision with the Indian plate, there is evidence for three earlier plate collisions. The earlier plates have left their mark in the form of accreted terranes, on the old southern margin of Asia, that are separated by suture zones similar to the Indus–Tsangpo suture present between India and Asia. From north to south, older to younger, they are known as the Kun Lun, Qiangtang, and Lhasa terranes, and from stratigraphic and palaeontological ages we know they collided with, and accreted to, Asia approximately 250 million years ago, 200 million years ago, and about 160 million years ago respectively. The final collision, that of India with Asia occurred, as already established, around 50 million years ago (Figure 11.6).

In 1984 I was invited to Tibet for a conference in Lhasa and a grand field trip around southern Tibet, following the initial fieldwork carried out by the French and Chinese. It was an exciting time, when the whole plateau suddenly opened up for outside scientific work and we were finally able to see some of the important geological sites. Driving around in ancient open-top Beijing jeeps, led by Paul Tapponnier, Vincent Courtillot, Li Tingdong, and Xhao Xuchang, we were first able to study the ophiolites and deep-sea sedimentary rocks exposed along the Yarlung Tsangpo suture zone, the folds and thrust structures along the northern margin of the Indian plate, and finally see for the first time the low-angle normal fault, the

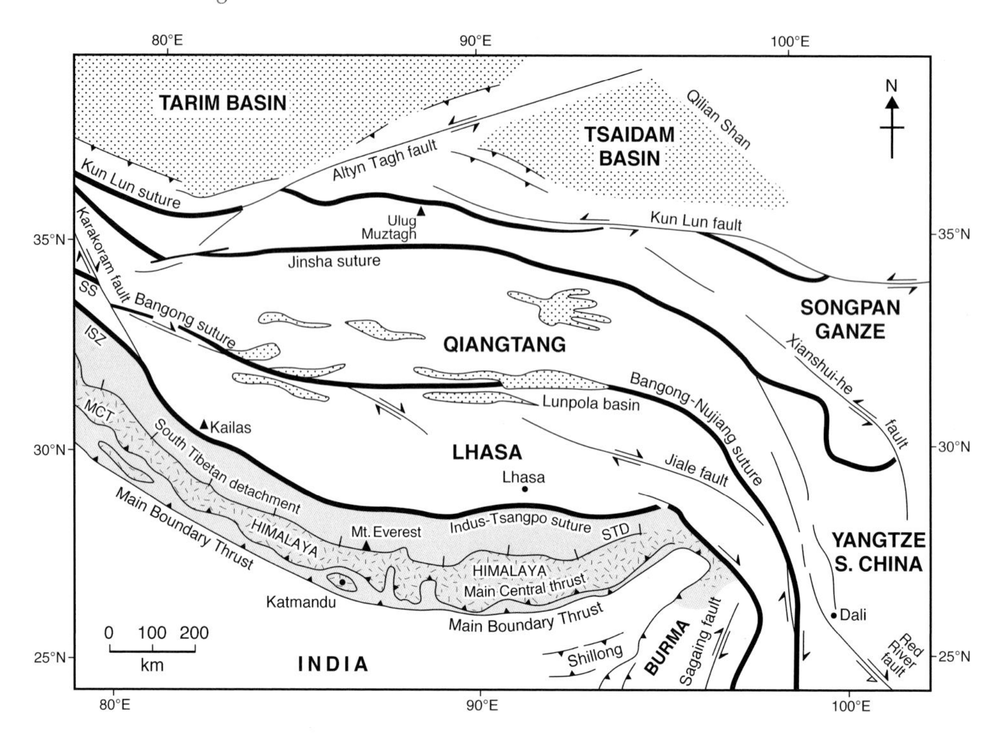

Figure 11.6 Geological map of Tibet showing the terranes and bounding suture zones. The Indian plate Himalaya is shaded and shows the two major shear zones, the South Tibetan Detachment (STD) low-angle normal fault and the Main Central Thrust (MCT). The Main Boundary Thrust (MBT) is the active southern margin of the Himalaya.

South Tibetan Detachment that bounds the northern margin of the Himalaya and the southern edge of the plateau. This low-angle normal fault was first mapped and studied by Jean-Pierre Burg in Tibet[10] and it appeared to be precisely the same structure that I had mapped and seen along the Zanskar Valley in Ladakh.

After our field-trip around south Tibet in 1984, I stayed behind in Lhasa to explore the hills and countryside around that amazing city. I wandered through the narrow alleyways of the Barkhor, the old Tibetan part of the city around the Jhokang Temple, in those days completely devoid of tourists. I walked up to the Potala Palace, and trekked up into the granite mountains around Lhasa (Figure 11.7). It was wonderful to see these places I had read so much about in Heinrich Harrer's classic book *Seven Years in Tibet*.[11] Heinrich Harrer and Peter Aufschnaiter, two Austrian mountaineers were caught in India during the outbreak of World War II, interned, and then escaped, walking across the Himalaya to Lhasa where they befriended the young Dalai Lama. Harrer's book is an evocative account of life in Tibet during the 1940s. It was then I met my Dutch friends Réne Schrama, who had just set up a travel company organizing tourist groups in Tibet and Fred Maddern, who started up the Dunya restaurant, the best watering hole in Lhasa. Together with our local

Figure 11.7 The Potala Palace in Lhasa, home to the Dalai Lamas and spiritual centre of Tibet.

Tibetan guides, Thingley and Sonam Wangdu, they were of invaluable help when we started working in the northern parts of the Tibetan Himalaya in the Everest region, along the Rongbuk and Kangshung Valleys between 2000 and 2004.

In 1985 a Chinese-British geological expedition traversed the plateau from Lhasa north to the Kun Lun Mountains,[12] extending the original French work in the south. Subsequently, several major deep crustal seismic reflection experiments have been undertaken to profile the crustal structure of the plateau. Since most of the plateau exposes mainly sedimentary rocks with only a few exhumed deeper metamorphic rocks (Figure 11.8), Tibet was not a great place to go to study the deeper processes associated with colliding continents. The Karakoram was certainly the place to study these. In Tibet we were reliant on remote geophysical methods, seismic reflection profiling, gravity, magnetics, and magnetotelluric data that enabled us to interpret the structure, density, and fluid content of the lower crust and upper mantle.

Active Tectonics of Tibet

During the 1980s and 1990s satellite technology progressed to such an extent that it was possible to measure with millimetre accuracy precise distances between fixed points on the Earth's surface. Global Positioning System (GPS) satellite data is now

Figure 11.8 Peaks of the Rolwaling Himalaya rising to the south of the desolate and high Tibetan Plateau.

an amazingly powerful tool for geologists. The satellites continually record data that tell us the present-day motions of the Earth's surface. The most recent GPS data shows that the present-day north–south shortening rates across the Tibetan Plateau are 40 millimetres per year with almost 20 millimetres a year east–west motion with respect to India and stable Asia.[13] GPS data only give us information on the motions of the Earth's surface today: it is geology and geochronology that give us information on plate motions back in time. In interpreting the geological evolution of mountain ranges like the Himalaya and Tibet it is important to study both the active and the geological time frames.

GPS stations have been placed all over Tibet and Asia and monitor every detail of the Earths' motions. GPS satellites have now been in operation for about fifteen years so we have a continuous record of the surface motions over that period of time. As GPS data from Tibet became available, the results soon turned out to be remarkable. The pattern that emerged showed that northern Tibet was being extruded eastwards, bounded by a series of strike-slip faults running across south Tibet and the huge Altyn Tagh and Kun Lun strike-slip faults running across the northern margin of the plateau. The GPS motions in south Tibet were more aligned to the north–north-east, although they started to swing eastwards in the north.

These GPS motions tended to support the hypothesis of large-scale eastward extrusion of Tibet originally proposed by Peter Molnar and Paul Tapponnier in the 1970s[14] (Chapter 5), but only really in the northern parts of the plateau. The most astonishing fact about the GPS data was that the motions showed what looked like Tibetan crust 'flowing' eastwards around the old, cold, and stable Sichuan block, and then in Yunnan and Burma the motions appeared to flow right back on themselves swinging around to the south in Yunnan and back west in northern Burma (Figure 11.9). These motions gave rise to the hypothesis of lower crustal flow in eastern Tibet originally proposed by Leigh Royden, Clark Burchfiel, and their colleagues.[15] The eastern margin of Tibet, the Longmen Shan Range, does not expose any lower crustal metamorphic rocks so, unlike the Himalayan boundary to the south, there is no geological evidence for lower crustal flow in eastern Tibet. Much of the deformation and metamorphism in eastern Tibet appeared to be relatively old and it was impossible to know if any lower crust was flowing out of the plateau region. Certainly the geology of Yunnan and Burma gave no indication that this was happening.

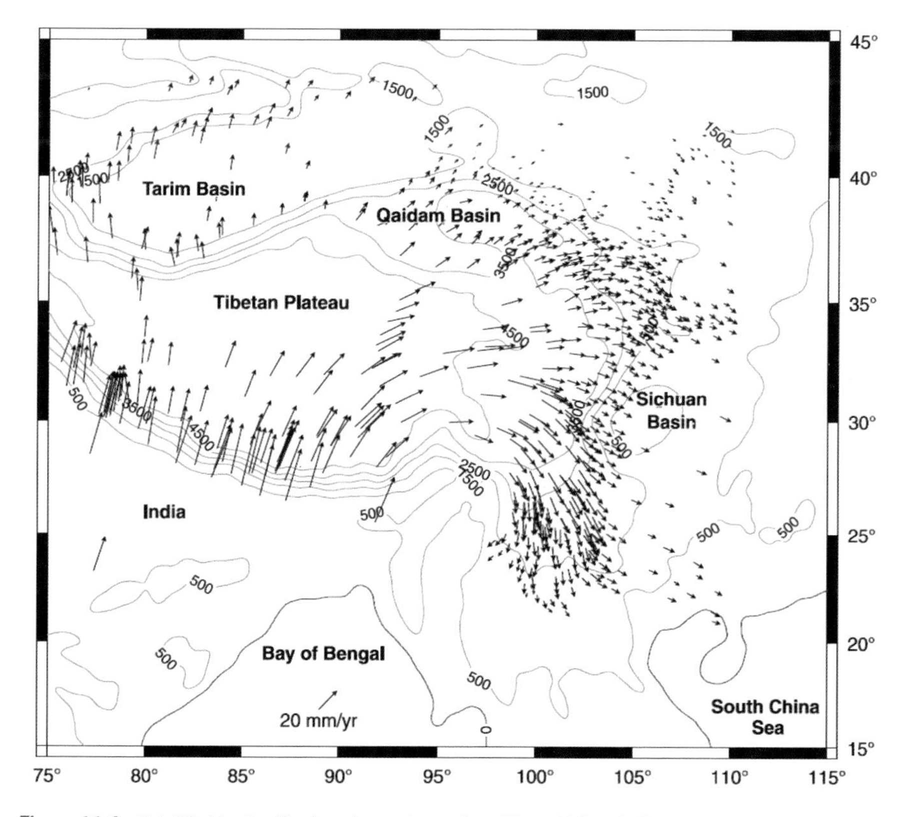

Figure 11.9 Global Position Satellite data show active motion of Tibet with length of arrows proportional to speed.

Earthquakes result from the sudden release of strain that has built up along faults and can give us valuable information on how the crust behaves. Earthquakes are usually common in the brittle, deforming upper 15–20 kilometres of the crust, whereas deformation in the lower crust is more diffuse, with ductile shear zones taking up the strain. With InSAR satellite data it is possible to map the 'shockwaves' around an earthquake, to determine precise amounts of offsets along faults and to predict where along faults would be the most likely place for the next earthquake to rupture. It is now possible to figure out fairly accurately where earthquakes are likely to occur, but not when. The distribution and focal mechanisms of earthquakes show that whereas the margins of Tibet are deforming by north–south compression (thrust faulting) or sideways motion (strike-slip faulting), the entire central part of the plateau above 4,500 metres' elevation is presently undergoing east–west extension. This led many geologists to propose that Tibet had reached its maximum crustal thickness and elevation, and was now starting to 'collapse'. Both the earthquake and the GPS data tell us that the upper crust of the plateau is extending, but what is happening in the lower crust? We know from GPS data that India and Asia are still converging today, and that both the Himalaya and Tibet are still rising. As the Indian plate upper crust shortens by folding and thrusting in the Himalaya, the lower crust that used to underlie those rocks has been slowly and relentlessly underthrusting beneath the southern part of the plateau, jacking it up as it penetrates north. It is quite possible that the extensional earthquakes in Tibet actually indicate that the plateau is still rising and thickening, and not collapsing at all.

Rift Valleys and Hot Springs

One of the most obvious geomorphological features of the Tibetan Plateau is the presence of about seven major north–south aligned rift valley systems, spaced roughly 100 kilometres apart and visible from space. The rift valleys are the surface manifestation of the east–west upper crustal extension. The largest and most prominent of the rift valleys is the Thakhola graben that extends north from the Dhaulagiri and Annapurna massifs of Nepal (Chapter 9). The rift valleys cut across almost the entire width of the plateau but do not cut across the Himalaya to the south or the Kun Lun Range in the north.

The lakes in many of the Tibetan rift valleys are saline due to the very arid climate, high evaporation, and lack of outflow. The rift valleys are also conduits for numerous hot springs and spectacular geysers (Figure 11.10). In the Daggyai tso rift, geysers hundreds of metres high fountain boiling waters out of the rift valley floor. Local Tibetan nomads (*drokpas*) and their yaks rely on these hot springs for keeping warm during the long freezing winter months. It is not surprising that several large

Figure 11.10 An impressive geyser of boiling waters erupting over the Daggyai tso rift valley.

geothermal fields such as Tirtapuri in south-west Tibet are major pilgrim sites. The largest and most extensive geothermal fields at Yangbajing, north of Lhasa, have been exploited for geothermal power and the hundred-metre-high geysers we saw in the 1980s have now been tamed.

Some of the gases emitted from the Tibetan hot springs can be used to establish the involvement of relative degrees of crustal and mantle melting. The main gases ejected are carbon dioxide (CO_2), nitrogen (N_2), methane (CH_4), and helium (He). There are two natural isotopes of helium in the Earth, the common isotope ^4He which resides in the crust, and the far less common one ^3He, which comes from the mantle, and is commonly expelled from volcanic eruptions along mid-ocean ridges. The helium composition of a gas is usually described as ratios of $^3He/^4He$; the higher the ratio, the more the mantle component of the gas. In 1995, during our geological traverse of Tibet, Simon Lamb and Leonora Hoke sampled many of the hot springs across the plateau.[16] They discovered that it was possible to draw a line across Tibet, south of which radiogenic helium derived from the crust, ^4He, was ubiquitous, and north of which mantle helium, ^3He, was common. This line runs roughly along the 30–31°N line of latitude, about 100 kilometres north of the Yarlung Tsangpo suture zone. This fascinating result was interpreted as showing the northern limit of Indian lithosphere underthusting of southern Tibet, whereas the relatively hotter Asian lithosphere was showing mantle-derived helium isotopes underlying much of central and north Tibet.

As India continued to push north following the initial continental collision, the relatively cold lithospheric plate progressively underthrust and pushed north into Asia. The hot mantle that originally underlay the south Tibetan crust was progressively replaced by colder Indian mantle as India pushed northwards. The hot Asian mantle was responsible for the source of many of the small and unusual volcanic eruptions and intrusions of dykes seen across the plateau, so geologists began mapping and dating the volcanics to see whether some correlation could be made between the helium isotopes, and the geochemical composition and age of the volcanic rocks (Figure 11.11).

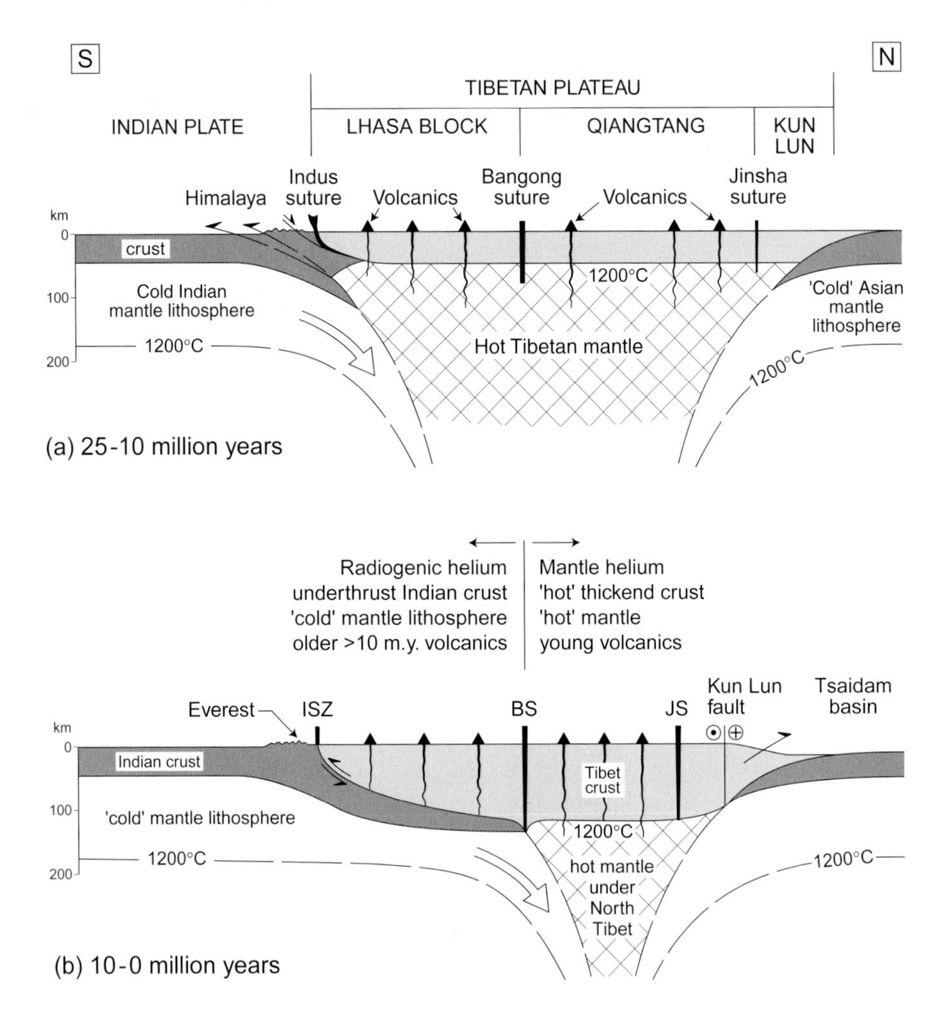

Figure 11.11 Models showing how the progressive underthrusting of India from south to north cuts off the hot mantle source for the Tibetan volcanic rocks.

Tibetan Volcanoes and Intrusions

Tibet has numerous small volcanic centres scattered across the plateau and even more intrusive dykes and plugs of igneous rocks. Volcanic rock emanates from the mantle and its geochemical composition and isotopes can tell us about the source area and temperatures of intrusion. In southern Tibet, a swathe of highly explosive andesite lavas and ignimbrite ash flows are exposed along the Gangdese Range. These very distinctive rocks, called the Linzizong volcanics, are the extrusive equivalents of the deeper level Andean-type granites that range in age from Late Cretaceous to Early Eocene (about 170–50 million years ago), and they are associated with subduction zones. Age dating of the Linzizong volcanics shows that these volcanic rocks were erupted over the period 70–40 million years ago. By comparison to the Andes, it seems that during this time, southern Tibet was an Andean-type plate margin with numerous granite intrusions, violent Mount St-Helens type eruptions, and some continental sedimentation. As the subduction of the Tethys Ocean beneath Asia ended with the India and Asia collision, both granite intrusions and volcanic eruptions ended.

Two very unusual types of volcanic rocks have been found on the plateau, both erupting and intruding into the thickening crust after the continental collision.[17] One type, loosely called shoshonites, is related to a hot and deep mantle source; the other, loosely termed adakites, is related to partial melting of a thickened crust with an eclogite root. Both have ripped off small fragments of the upper mantle and lower crust as the molten magma was forced upwards in volcanic feeder pipes. These small fragments are called xenoliths and their mineral compositions are extremely useful for determining the temperature, pressure, and depth of formation. In southern Tibet some dykes we sampled were dated at 13 million years old and contained small xenoliths of rocks from deep in the continental crust showing high-temperature metamorphism (granulite facies). One of my D.Phil students at Oxford, Gavin Chan from Hong Kong, worked on the pressure–temperature conditions under which these rocks formed and U-Pb dating, and as a result we were able to prove that 13 million years ago, the crust beneath southern Tibet was at least 75 kilometres thick and at extremely high temperatures and pressures.[18] Similar studies had also been done on these xenoliths from central Tibet by Brad Hacker from the University of California and from the Pamirs by Mihai Ducea and colleagues from the University of Arizona, so the interpretation of a thick crust seemed to hold up for the entire plateau.

In southern Tibet these strange volcanic rocks started erupting about 30 million years ago, but ended abruptly 12 million years ago. By then the colder

Indian lithospheric plate had underthrust southern Tibet and shut off the hot mantle source, or rather pushed the hot Asian mantle further to the north. In central Tibet the ages of the volcanic rocks are all younger than 50 million years old, the timing of the India–Asia collision. The youngest volcanic rocks in Tibet were found to lie along the far northern part of the plateau, in the Kun Lun, ranging from about 14 million years ago to less than one million years ago.

Probing the Deep Structure of Tibet

Continental deep seismic surveys are the juggernauts of geophysical experimentation. They require relatively flat, uninhabited terrain suitable for setting off large explosions, and large trucks full of recording equipment. Seismic surveys involve laying out a network of receivers to pick up seismic waves from an energy source, either an explosion of some sort or an earthquake. In the summers of 1994 and 1995 a major seismic survey was initiated, termed the INDEPTH project (International Deep Profiling of Tibet and the Himalaya).[19] The geophysicists found a suitable profile across southern Tibet along one of the large rift valleys, the Yadong-Gulu rift, north of Bhutan. After two years of work, the most prominent feature found on the profile was a large reflector horizon corresponding to the major thrust fault called the Main Himalayan Thrust, along which the Indian plate is presently underthrusting the Himalaya and south Tibet. The fault dips gently to the north and hits the surface along the active southern margin of the Himalaya. Most of the recent earthquakes in north India, Nepal, Sikkim, and Bhutan have nucleated along this fault zone.

Another fascinating discovery on the INDEPTH profile was the presence of a mid-crustal zone of high electrical conductivity determined from magnetotelluric studies (Figures 11.12 and 11.13). These data indicated the presence of pockets of hot fluids at relatively shallow depths beneath southern Tibet. These were colloquially termed 'bright spots' by geophysicists working on the seismic data. Both the seismic and magnetotelluric data, together with high heat flow measurements from lakes, suggested that very high temperatures were present at shallow levels under south Tibet. At these temperatures, the pockets of hot fluids were most likely to be leucogranites. The seismic data showed a mid-crustal layer between about 18 and 30 kilometres' depth that was extremely hot, even partially melted, with pockets of inter-connected fluids. Beneath this, the lower crust appeared to be very coherent and strong. This crustal structure fitted well with the earthquake data from Tibet that showed almost all earthquakes occurred in the upper 20 kilometres of the crust; there were none at all in the mid-crustal partially molten layer, and only a few very deep earthquakes occurred in the lower crust at depths of around 70–90

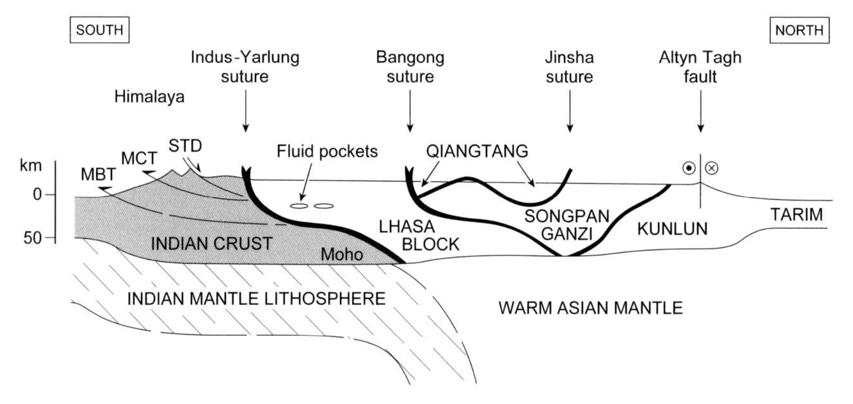

Based on INDEPTH seismic reflection profile (Nelson *et al.* 1996)

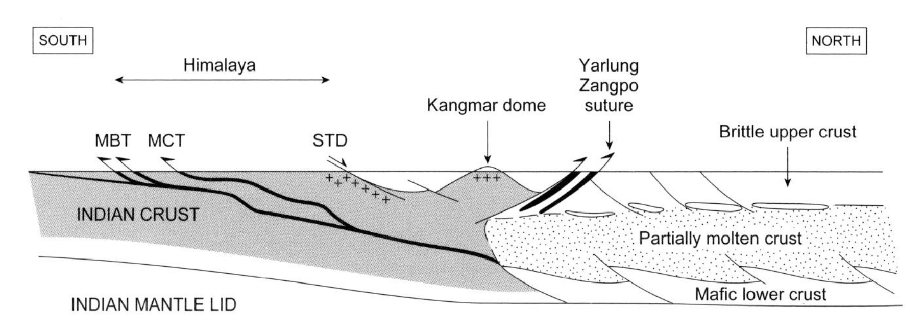

Based on INDEPTH seismic reflection profile (Nelson *et al.* 1996)

Figure 11.12 Cross-section models of Tibet showing the present-day structure with terranes (above) and an interpretation of the INDEPTH seismic profile across the Himalaya (below).

kilometres. Geologists debated furiously over whether these deep earthquakes were located in the lowermost crust or whether they occurred in the mantle.

At an HKT workshop meeting in Switzerland in 1995 at which the INDEPTH profile was first presented, Doug Nelson from Syracuse University in the US, the leader of the INDEPTH experiment, talked about the puzzling seismic and magnetotelluric data from south Tibet that seemed to suggest partial melting at very shallow depths beneath the southern part of the plateau, and the fact that this melting must be occurring right now. How could this vast plateau of Tibet, the highest elevated landmass on Earth, possibly be supported above a layer of partially molten crust? This structure was unknown anywhere else on Earth. In the following session at the Swiss conference, I gave a talk on the Himalayan leucogranites along the Nepal and Indian Himalaya, presenting the structural evidence for their intrusion as long, layer-parallel sills intruding from under Tibet towards the south to the Himalaya, rather than vertical intrusions as they had previously been thought to

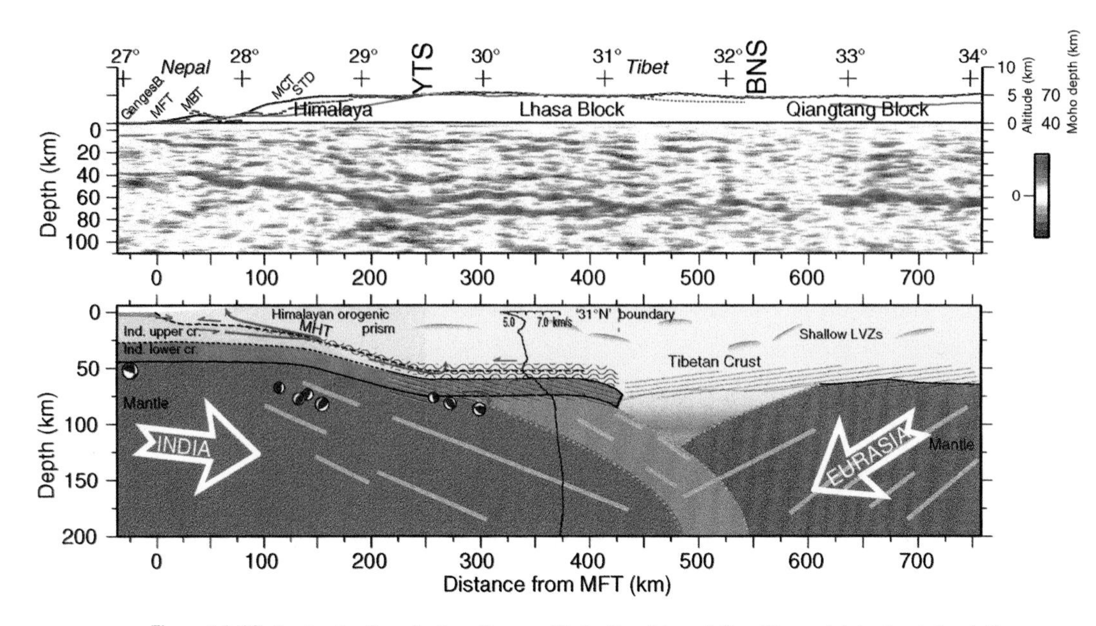

Figure 11.13 Receiver function seismic profile across Tibet with an interpretation of the crustal structure below. Indian lithospheric mantle (in blue) underplates the Himalaya and south Tibet. The deeper, lower crust of India has been transformed to high-pressure eclogite (green). Asia mantle lithosphere (brown) underthrusts northern Tibet. Continued underthrusting of Indian and Asian mantle lithosphere from both sides effectively jacks up the Tibetan Plateau and maintains its high topography and thick crust. Courtesy of György Hetényi and John Nábelek.

be. The Himalayan granites were all emplaced during the Miocene, between about 21 and 17 million years ago and at very shallow depths (18–25 kilometres from our thermobarometric data). I surmised that maybe these Himalayan granites formed in a similar structural setting to the 'bright spots' at the top of the INDEPTH partial melted middle crust beneath southern Tibet, and were then intruded to the south as sill complexes for large distances, maybe even 100 kilometres.

The seismic evidence showed that the crust beneath Tibet appeared to be layered, with a rigid, brittle, seismogenic upper crust where the earthquakes occurred, a very weak, partially molten middle crust, and a strongly layered lower crust. The INDEPTH profile clearly showed the major flat-lying detachments that could be traced southwards to link in with the positions of the major faults in Nepal, particularly the Main Boundary Thrust along the southern margin of the Himalaya. In a wonderful 'eureka' moment, Doug and I traced the boundaries of his partially molten mid-crust layer southwards off the end of the seismic profile, and they linked almost perfectly with the positions of the Main Central Thrust and the South Tibetan Detachment as we had mapped them along the Himalaya in Nepal and Sikkim. Most of the rocks in this region were the Miocene partially molten migmatites and leucogranites of the Kangchenjunga and Jannu massifs.[20] It was a moment when the permanently squabbling brothers of geology and geophysics met in total harmony, certainly a rare moment to savour!

Modelling Tibet

It is almost impossible to find a geological or geophysical paper on Tibet or the Himalaya that does not have a model; indeed it is almost impossible to get a paper published unless one discusses each and every possible model proposed for the evolution of Tibet. Many readers and editors are only interested in the model. Models by definition are simplified representations of the real world. The trouble is that most of the time the real world is far too complex to simplify and define into a meaningful model. Two great quotes come to mind here: 'All models are wrong; some may be useful' is attributed to George Box, a professor of statistics from the University of Wisconsin. The practical question is: how wrong do they have to be not to be useful? The second is unsourced: 'Models are like sausages; you really do not want to know what goes into them'.

The first real model for the crustal structure of Tibet was an inspiration. In 1924 Emile Argand, a Swiss geologist and an early proponent of Alfred Wegener's theory of continental drift, came up with an astonishing insight based on nothing more than a map (Figure 11.14). He proposed a model in which Tibet attained its high elevation because the old and cold Indian lithosphere had underthrust all the way beneath the plateau, a model that pre-dated the plate tectonic revolution by forty years and has retained a large amount of credence to this day.[21] Nearly fifty years later another piece of inspiration came to Kevin Burke and John Dewey, both then working at the University of Albany in the US. They were both real field geologists and had widespread experience of continental geology in North America, Europe, and Africa. They proposed a model that was quite different from that of Argand: that Tibet had attained its double crustal thickness and high elevation by pure shear crustal shortening and thickening. Tibet before the collision with India was originally 2000 kilometres wide and 35 kilometres thick; after the collision it shortened internally and homogeneously to half the width, 1,000 kilometres north–south, and double the thickness, 70 kilometres.[22]

The next main model was the result of the first interpretation of the black-and-white satellite photos of Asia taken from the Landsat satellite, by Peter Molnar and Paul Tapponnier. Theirs has been one of the great defining models for the deformation of Tibet, that of continental extrusion.[14] Several large strike-slip faults appeared to bound the great plateau region of Tibet. The Altyn Tagh and Kun Lun faults along the northern margins of the plateau were moving sinistrally (Tibet moving east relative to the Tarim crust to the north), whereas the Karakoram and Jiale fault systems along the south-west and south-east borders of Tibet were

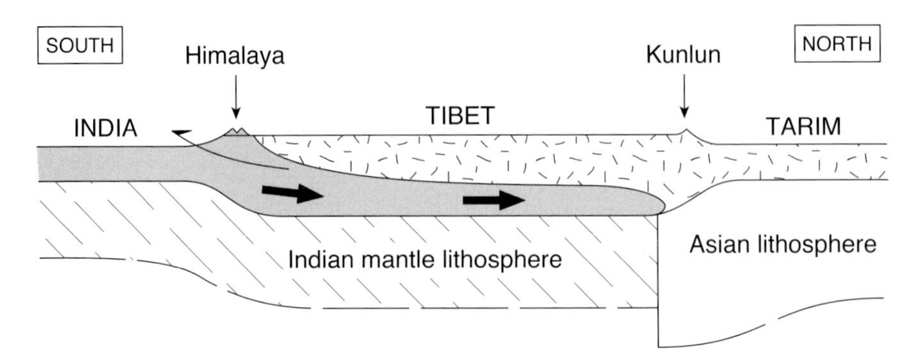

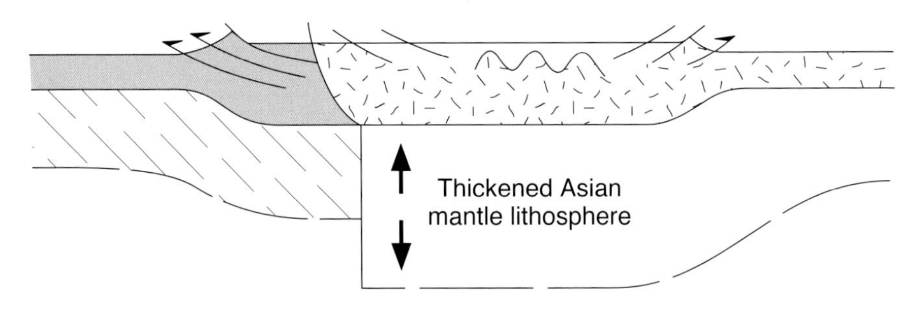

Figure 11.14 Two opposing models to explain the double crustal thickness of the Tibetan Plateau, (a) the Argand underthrusting model and (b) the homogeneous crustal shortening and thickening model.

moving dextrally (Tibet moving east relative to the Karakoram and Himalaya to the south). This showed that the whole Tibetan Plateau was moving east in a pincer movement, bounded by these great strike-slip faults. Peter Molnar and Paul Tapponnier developed this model further, linking all the faults of Indochina and South East Asia to the same system. They proposed that the whole of Indochina was being shunted south-east out of the way of the Indian indentor, bounded by two further huge strike-slip faults, the Sagaing fault in Burma and the Red River fault in Yunnan and Vietnam (Figure 11.15).

For twenty years the continental extrusion model was widely regarded as a pretty good representation of the geological structure of Tibet and Asia. In 1975 when Peter Molnar and Paul Tapponnier came up with the continental extrusion model, Tibet was still a closed country and there was almost no field data on any of the Tibetan strike-slip faults. We did not know the timing of the faults, the amounts of geological offsets, the slip rates, or how deep the faults penetrated.

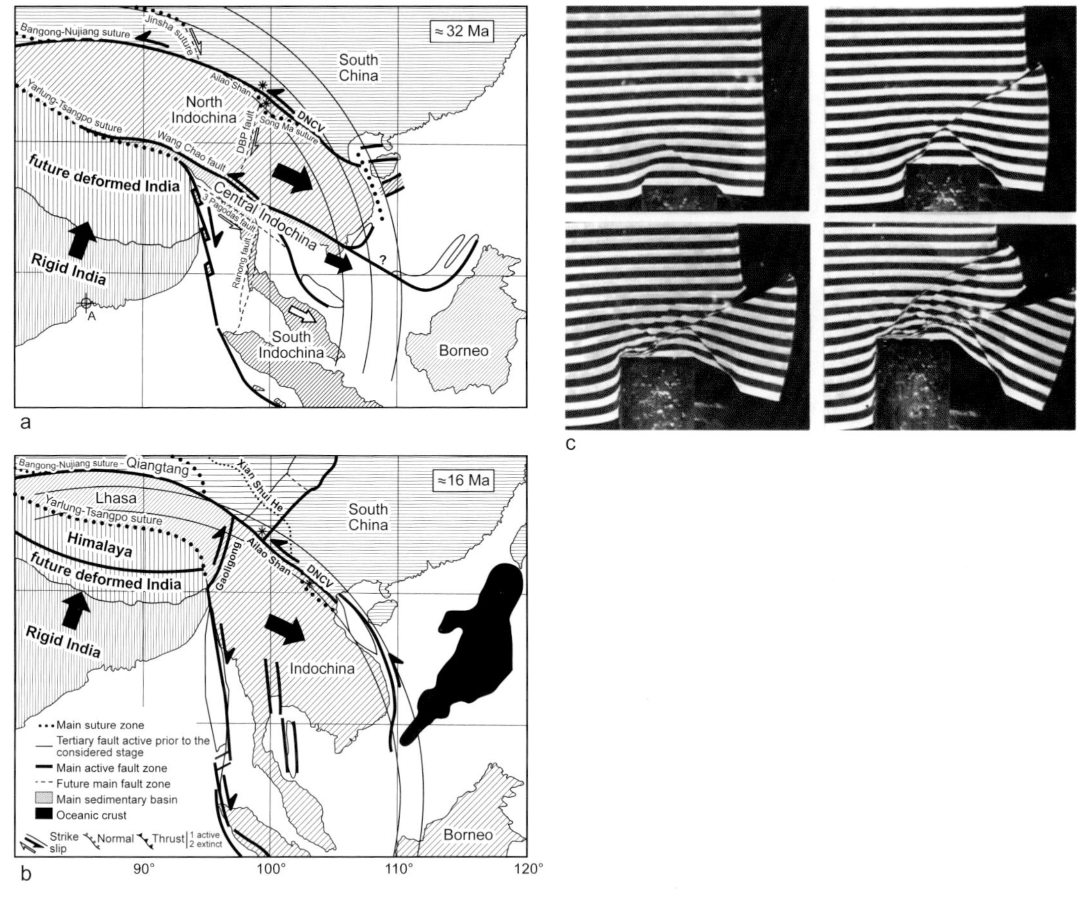

Figure 11.15 The continental extrusion model for the Tibetan Plateau and Indochina.

It was a great model, but was there any geological reality in it? For the next twenty-five years Paul Tapponnier and many of his students from Paris, Gilles Peltzer, Rolando Armijo, Robin Laccasin, and others mapped and sampled many of these faults all across Tibet. Their results were very impressive and seemed to support the model of continental extrusion.[23] Most of the big strike-slip faults in Tibet cut through only upper crustal sedimentary rocks and very young alluvial cover. It was incredibly difficult to find precise geological offset markers on these faults, and even more tricky to get reliable age constraints on the offset features. Consequently, most of their work involved mapping Quaternary and recent sediments and active geomorphological features such as river terraces and alluvial fans. Obtaining precise ages on these young sediments is extremely difficult and the method used—cosmogenic isotope dating of boulders—is not very accurate. The slip rates they obtained were highly variable but their results always seemed to show large amounts of recent offset and large slip rates, thereby supporting the large-scale extrusion model.

Continuum Modelling

Whereas the continental extrusion model assumes that the lithospheric plates behave in a rigid fashion and are internally coherent, with strike-slip faults penetrating the entire crust and upper mantle, another approach assumes the opposite: that the entire large-scale and long-term deformation history can be described as a continuum with a Newtonian or power-law rheology. Rheology is the study of flow in matter. Newtonian fluids (named after Isaac Newton) show a linear relation between shear stress and shear rate (e.g. water), whereas non-Newtonian fluids are more viscous (e.g. custard, honey, or toothpaste). In the extreme, this model describes the Tibetan crust as behaving like a viscous fluid such as treacle or honey, flowing out of the way of the rigid indentor, India. Proponents of this approach, notably Philip England and Greg Houseman, considered the interaction of vertical and horizontal stresses on the strength and flow laws of rocks.[24] The strength of the lithosphere is measured in terms of the Argand number (named in honour of the great Swiss geologist). England and Houseman concluded that most of the convergence between India and Asia was taken up by crustal thickening, and that the strike-slip faults were a relatively recent and minor factor.

A quick look at the earthquake distribution map shows that earthquakes occur all across the Tibetan Plateau region and are not restricted to block or plate boundaries. Clearly the simple view of rigid plate tectonics does not hold up too well in wide areas of active deformation across continental plateau regions like Tibet. It might seem strange to relate the crust of Tibet to a fluid-like medium, but to any geologist working on lower crust rocks who has seen the wildly flowing folds and ductile shear zones in metamorphic gneisses, it comes as no surprise at all. The fact that lower crustal rocks deform in a ductile flowing, non-rigid manner has been known for more than a hundred years. This ductile flow behaviour requires high temperatures, so there is always going to be a magic line in the crust above which rocks deform by brittle fracturing, and below which rocks deform by ductile flow. This line is generally called the 'brittle-ductile transition'.

In 1993 Peter Molnar proposed another exciting new model for Tibet.[25] By then he had jumped over the fence following the evidence and no longer believed in the plate-like behaviour of the Tibetan crust, with faults penetrating the entire lithosphere and large amounts of lateral extrusion. Instead, he joined forces with Philip England and thought that Tibet behaved like a thin viscous sheet. Together they made another leap in theory, with a model linking Tibetan tectonics to climate change. They proposed that removal by convective flow of the lower lithosphere from beneath Tibet around 8 million years ago resulted in its replacement by hotter asthenospheric mantle that explained the intrusions of the strange volcanic rocks of deep origin in Tibet and the rise of the plateau. They also proposed that Tibet may have reached its maximum elevation and thickness at that time, which coincided with the initiation of east–west extension seen across the plateau today, and with initiation of the Indian monsoon system. Later geological data has since disproved the link between extension and volcanism; age dating has shown that the asthenospheric-mantle-derived volcanic rocks were intruded throughout the last 50 million years. Subsequent geophysical data has also disproved the convective removal of lithosphere beneath the plateau. Keith Preistley and Dan McKenzie from Cambridge University have shown quite the opposite: that almost the entire Tibetan Plateau is underlain by cold Indian mantle.[26] The discussion has now gone full circle, back to Emile Argand, with the latest geophysical profiles supporting his original hypothesis of wholesale underthrusting of Tibet by India.

Raising Tibet

The timing of the uplift of the Tibetan Plateau is of fundamental importance, not only for interpreting the tectonics of Asia but also for interpreting the atmospheric circulation systems, the onset of the monsoon, and the chemical and isotopic budgets of the world's oceans. We can be certain that 120 million years ago most of Tibet was at or just below sea level. Marine limestones containing fossils of this age outcrop widely across western and central Tibet. Above this stratigraphic horizon, all the sediments are continental, having been deposited by rivers or lakes on continental crust. We can conclude that the Asian side of the collision zone in Tibet has been above sea level since 120 million years ago, but at what time did it attain its present 5,000-metre-high elevation? As we saw in Chapter 2, the final marine sedimentary rocks along the Indus suture, or Yarlung Tsangpo suture zone as it is called in Tibet, the zone of India–Asia collision that runs across southern Tibet, are earliest Eocene (50 million years old); after the two continents had collided, the Tethys Ocean had closed, and sedimentation was purely continental (fluvial or lacustrine sediments deposited in rivers and lakes) (Figure 11.16).

Geological mapping in the Lhasa region showed that a very prominent unconformity separates folded sedimentary rocks below from flat-lying andesite lavas of the Gangdese Range above (Figure 11.17). Dating of the andesites revealed that they were erupted 55–60 million years ago.[27] This was a puzzle. Tibet's upper crust did indeed seem to be folded by north–south shortening, but this folding must have occurred before the India–Asia continental collision. The southern

Figure 11.16 The transition from pale marine limestones to dark continental sedimentary rocks exposed in the hills near Tingri, in the Indus-Tsangpo suture zone south Tibet. The youngest marine sediments contain small marine foraminifera dated at 50 million years old and are the youngest marine sediments between India and Asia.

margin of Asia prior to the Indian plate collision was an Andean-type margin built of granite magmas and andesitic lavas, showing very similar geology to much of the Andes. In the Andes most of the crustal thickening is thought be due to granite intrusions, so maybe southern Tibet 55 million years ago could have looked very similar to parts of Peru or Chile today.

It is also possible to use metamorphic rocks and geochronology to infer the timing of active uplift. Metamorphism requires crustal thickening by folding and thrusting in order to increase the temperatures and pressures recorded by the rocks. Crustal thickening implies surface uplift (recall the theory of isostasy and the iceberg analogy). The dating of high-grade metamorphic rocks gives us the timing of peak metamorphism and hence timing of burial. Geochronology measures the uplift of the rocks relative to the Earth's surface (exhumation) not the topographic surface, but it is difficult to have thick crust, metamorphism, and high exhumation rates without having mountains. As we have noted earlier in Chapter 4, surface uplift is very difficult to quantify back in time, and there are no accurate methods to date this. There are several indirect methods we can use to infer surface height and give us a clue.

One such indirect method is the subject of palaeobotany, the study of fossil leaves. Leaves are sensitive to two parameters, temperature and humidity (enthalpy). The presence of tropical leaves in lake sediments at altitudes of 4,300 metres in the Lunpola basin, and fossil palm trees in western Tibet suggests that either the elevation was much lower than today or the climate was much wetter and more humid than today, or both. Leaves tend to have different shapes and properties according to the temperature conditions and therefore also altitude. Hot, wet, steamy

Figure 11.17 The unconformity, showing flat-lying andesite volcanic rocks of southern Tibet dated at 55–60 million years old overlying folded red-bed sedimentary rocks beneath. The folding and crustal shortening must have pre-dated the eruption of the andesite lavas and therefore must have occurred prior to the India-Asia collision.

tropical forests under 1,000 metres' elevation have trees with large leaves, smooth margins, and pointed tips, where the water can flow off easily. Trees in cooler, wetter environments tend to have moderate-sized leaves with serrated margins, whereas high-altitude trees growing in dry, cold conditions are stunted and have small, simple leaves. The fossil leaves at Namling were found in lake deposits interbedded with volcanic ash horizons. The ash could be dated and gave ages of 15 million years ago. Fossils recovered from Namling included willows, alders, and maples, together with possible rhododendrons and conifers.[28] Namling today is a high-altitude desert with practically no vegetation growing at all. Clearly the Miocene climate was much warmer and wetter than today, but was this also telling us that the altitude was lower, or was the Earth's climate significantly different from today such that warm conditions could occur at these altitudes? The results were ambiguous.

Exploring South-Eastern Tibet

Geological studies in the uplifted and exhumed metamorphic rocks of the Karakoram Range of north Pakistan had resulted in astounding results, with the suggestion that the crust here had been actively thickening, and sporadically melting to form granites throughout the last 65 million years, from age data we had obtained from the Karakoram Range in particular (Chapter 3).[29] These data suggested that there were probably high mountains along the Karakoram during all of that time. The Karakoram is geologically equivalent to the Qiangtang terrain of central Tibet. In western Tibet, however, the plateau was very flat and showed no exposures of metamorphic rocks. I had surmised that the geology of the Karakoram was telling us what the composition, structure, and age of the crust under the Qiangtang terrain was, but we had no direct evidence from the main part of the Tibetan Plateau. I looked at all the Chinese maps I could find for possible exposures of metamorphic rocks and young granites. North of Lhasa there is one mountain range, the Nyenchen Tangla, that rises to over 7,000 metres. Here the rocks are mainly older granites but a few outcrops of metamorphic rocks and some young granites had recently been dated.[30] There was a suggestion that this range could be exposing the deep metamorphic 'basement' rocks of Tibet (Figure 11.18).

I had been to the western Nyenchen Tangla Range north of Lhasa twice before on quick reconnaissance trips, but what really interested me was the eastern continuation of this mountain belt. In eastern Tibet the plateau changes dramatically from the flat, barren deserts of the west to the gently rolling forested hills of south-east Tibet, and then the enormous snow-capped mountain ranges that fan out across the eastern part of Tibet. This was spectacular country and almost completely unexplored by geologists. Only a few intrepid explorers had

Figure 11.18 The ancient walled city of Gyangtse in Tibet with the mountains of the Gangdese and Nyenchen Tangla Ranges beyond.

managed to penetrate the tangled web of Chinese permissions and logistics to explore far eastern Tibet. Most notable was the Japanese climber Tom Nakamura, who made more than thirty expeditions to almost every corner of the eastern 'Tibetan Alps'.[31] There was, however, practically no information on the geology at all. Here were three mountain ranges, the length of the Alps and much higher, that were still almost totally unknown. We knew more about the geology of the Moon than we did about eastern Tibet. Tom Nakamura sent me several mouth-watering photos of amazing peaks and accounts of his travels and my appetite was well and truly wetted to go to the area. Chris Bonnington, Britain's most famous mountaineer, had just returned from an attempt to climb Sepu Kangri (6,956 m), the highest peak of the eastern Nyenchen Tangla Range,[32] and he sent me some photos which clearly showed that some of these peaks were made of granites. The major problem was getting permission from the Chinese authorities to work in the region. I had applied three times for permission to trek along the Grand Canyon of the Yarlung Tsangpo River and been refused each time by the Chinese authorities.

My chance came in 2009 when, following an HKT meeting in Beijing, there was a field excursion through the eastern Himalayan syntaxis region from the Bayi-Linzhi region and back to Lhasa led by Jingsui Yang, a professor from the Chinese Academy of Sciences. Marc St-Onge, my Canadian colleague and friend from my early Baltoro expedition, was becoming more and more interested in the Himalaya-Tibetan region since we had written a paper comparing the Precambrian trans-Hudson orogeny in northern Canada to the Himalaya-Tibet.[33] Marc and I signed up for the trip and made a two-week excursion with Jingsui and his colleagues. At the very first outcrop at which we stopped along the banks of the Yarlung Tsangpo, we found sillimanite- and cordierite-bearing gneisses, and then beautiful kyanite-bearing migmatites. Here at last were some high-grade metamorphic rocks on the plateau. Maybe finally we could find out the pressure, depth, and temperature conditions and the time that these rocks had formed, and thereby gain some information on the age of Tibetan Plateau formation and uplift.

Back in Oxford Dave Waters and I recruited two D.Phil students, Richard Palin and Owen Weller, to work on these problems and we all returned in 2010 for a longer field-trip with Songyong Chen, Jingsui's colleague from Beijing, during which we hoped to map and sample the area in more comprehensive detail. We flew to Lhasa and worked our way eastwards along the main highway, stopping off at several localities to study the belt of old, Permian age eclogites (high-pressure metamorphic rocks) discovered by Jingsui and his team. Our first base was at the lake of Basum *tso*, a beautiful, tranquil, blue lake surrounded by high peaks of the eastern Nyenchen Tangla. We drove and trekked north along a magnificent valley that cut right into the heart of the range. Small hamlets of traditional wooden houses surrounded by gardens of multicoloured flowers adorned the valley. We found superb exposures of metamorphic rocks and granites all the way up the valley. It was a perfect place to study the deeper geological levels of Tibet.

Our enthusiasm was abruptly curtailed when after a few days we were stopped by the local police south of Bayi and taken to the police station for several hours of questioning. The police network of spies had seen us trekking north of Basum *tso* and the policemen could not understand what we were doing up there. Songyong bore the brunt of this aggressive questioning. We had permission from the Chinese authorities and the army, but apparently not the right permission from the local police. We also had geological maps of the border area with India which, although freely available, were regarded as highly suspicious. After five hours of questioning we were told to return to Lhasa. It seems that the open-door policy of the Beijing government did not extend to this remote part of Tibet. Our expulsion from Linzhi county was a real shame as we had grand plans to explore much more of these fascinating mountain ranges of the eastern plateau, but we did thankfully manage to extricate our rock samples.

Deep Levels of Tibetan Faults

Only a very few of the Tibetan strike-slip faults expose deep-level metamorphic rocks of the middle or lower crust. The Karakoram fault in the far north of Ladakh is the best of all, where the fault slices right through high-grade gneisses and granites in the Tangtse Valley and Pangong Range. It was here that I set up a project in 1996 to map the entire length of the fault in the Nubra and Tangste valleys and my D.Phil student Richard Phillips was able to obtain accurate U-Pb ages from granites along the fault. Our work along the Karakoram fault has been described in Chapter 5. We were able to use deformed granite sills intruded parallel to the sheared mylonite fabrics before the fault slipped, and later granite dykes that cross-cut the mylonite fabrics to constrain the age of slip along the fault. Our work, it will be recalled, did not support the Molnar-Tapponnier large-scale extrusion hypothesis for Tibet. We found that the total geological offsets were much smaller, up to a maximum of 120 kilometres only, than required by the hypothesis, and that the timing of motion also was considerably younger, at less than 13 million years ago, than the timing of the India–Asia continental collision.[34]

The Jiale fault in the far south-east of Tibet is another strike-slip fault that shows exhumed deep crustal rocks. This fault runs from the central plateau region south-east, to swing around the eastern Himalayan syntaxis and then run southwards into northern Burma. In the 1990s this part of Tibet was off-limits to foreigners and even then the access problems and roads were extremely primitive. In 2009

Marc and I had driven down to the Jiale fault, deep in the bamboo forests along the Sichuan Highway, where it was more of a muddy track. We had seen spectacular mylonites along the almost inaccessible cliffs above the river. We had planned to carry out a thorough study of the Jiale fault in 2010 but the local policemen at Bayi had put an abrupt end to our hopes. It seemed that this whole region around the eastern Himalayan syntaxis was politically impossible to work in.

The next major part of the puzzle was the extrusion of South East Asia. The original Molnar–Tapponnier model had proposed that the entire crust of Indochina had been extruded south-east out of the way of the indenting Indian plate. The two longest strike-slip faults bounding Indochina are the Sagaing fault in Burma and the Red River fault that runs from the south-east corner of Tibet through Yunnan and Vietnam to the South China Sea. Burma was a difficult country in which to carry out any sort of geological work and there was very little information on the geological offsets, slip rates, or timing along the Sagaing fault. A large tract of metamorphic rocks runs along the eastern margin of the fault and this is where many of the country's gemstones, rubies, sapphires, and jade, come from. The Red River fault, the largest strike-slip fault in South East Asia, also shows exhumed metamorphic rocks of the lower crust along its trace. One of Paul Tapponnier's students, Herve Leloup, had spent four years mapping and studying the Red River fault in Yunnan and North Vietnam during the late 1990s in great detail.[35] As with Tibet, their interpretations suggested very large-scale movement along the Red River fault (up to 1,000 kilometres) since 35 million years ago. If true, this would surely clinch the large-scale continental extrusion hypothesis. I felt I needed to go both to Burma and to Vietnam to check out these important faults.

CHAPTER 12

Extruding Indochina

Burma, Vietnam, Yunnan, Thailand

The explorers fought their way from the equatorial forests of Cambodia and Laos to climb from the badlands of remotest Burma onto blizzard-swept tundra along the China–Tibet border. They took to the jungle, riding on elephants, bullock carts and horses; mostly they just slashed through the monsoons, knee-deep in mud and festooned with leeches.

John Keay, *Mad about the Mekong*

Geographically, Indochina consists of the South East Asian countries Thailand, Laos, Cambodia, and Vietnam. Geologically, Indochina includes all the land bounded by two very large-scale strike-slip faults—the Sagaing fault, which runs down the length of Burma, and the Red River fault, which extends more than 1,100 kilometres from the south-eastern corner of Tibet south-east through Yunnan and North Vietnam to Hanoi and the Gulf of Tonkin (Figure 12.1a, b). Both faults are active, and show that Indochina is moving south-east relative to both the Burma micro-plate to the west and the South China block north of the Red River fault (Figure 12.2). The unresolved questions were how far Indochina was extruding away from the India–Asia collision zone and when these faults became active.

The eastern margin of the Indian plate lies along the Burma–Andaman–Sumatra–Java trench, where the Indian oceanic plate is subducting beneath the great island arc chain of Indonesia. Behind the island arc, a new oceanic basin has formed in the past 5 million years, with basaltic ocean crust forming along a small active spreading centre in the Andaman Sea.[1] The northern extension of the

(a)

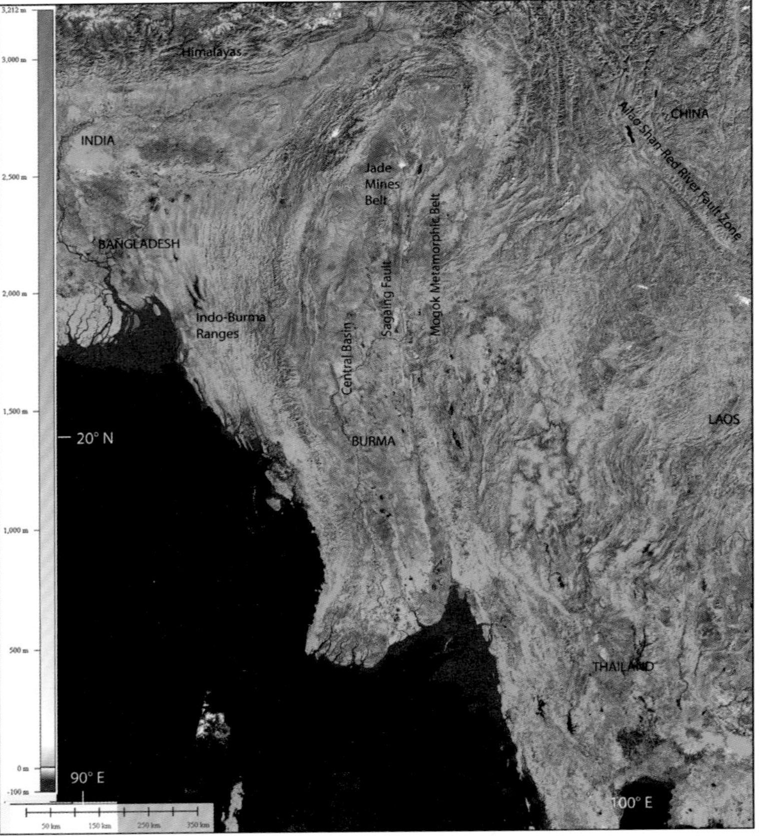

(b)

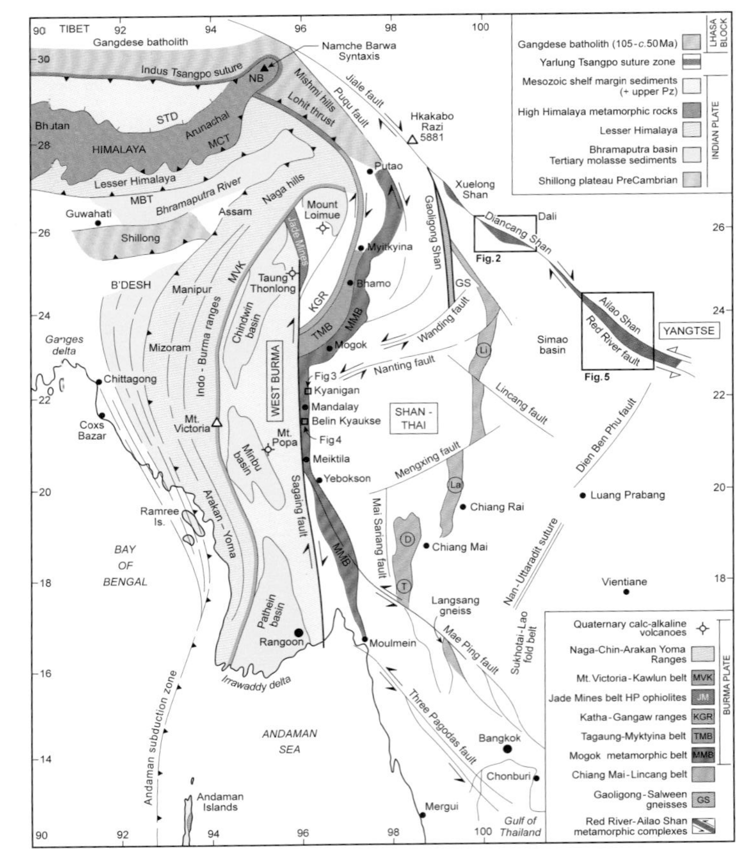

Figure 12.1 (a) Landsat photograph of Burma, Thailand, and Cambodia showing the major fault lines. (b) Geological map of Indochina showing the major geological structures and terranes.

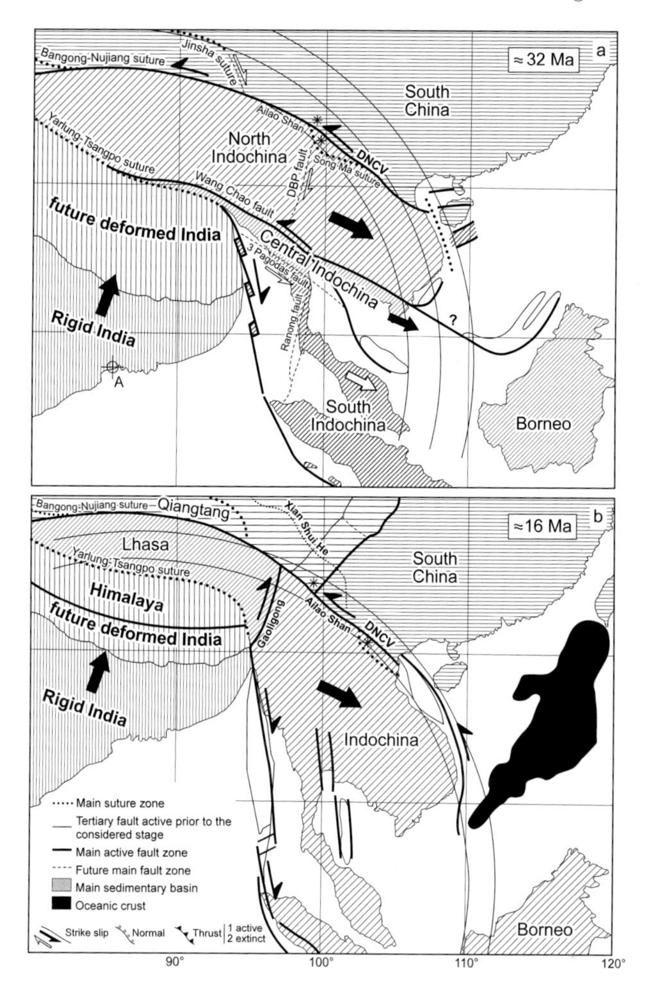

Figure 12.2 The continental extrusion model for Indochina, showing the early eastward extrusion of North Indochina around 32 million years ago and the later south-eastward extrusion of Indochina around 16 million years ago.

Andaman trench extends into the Arakan-Yoma Hills of western Burma, but the nature and location of the transition from oceanic lithosphere beneath the Bay of Bengal to continental lithosphere in Burma is poorly known. In the south of Burma, where the Irrawaddy River drains into the Andaman Sea, a vast delta has built up with over 10 kilometres' thickness of sediments eroded off the mountains of Burma. The Sagaing fault continues offshore and is connected to the young oceanic spreading centre in the Andaman Sea. In northern Burma the fault passes close to the cities of Meiktyla and Mandalay and then splays into several branches that terminate in the Jade belt and other mountain ranges that ripple northwards towards the eastern Himalayan syntaxis.

Burma: Land of Jade, Rubies, and Sapphires

Burma is a hauntingly beautiful country of serene landscapes, golden pagodas, green rice fields, range upon range of distant hills, teak forests, and wide muddy rivers. It is also a land of great mineral riches. In the hills and rivers of Mogok are some of the most beautiful gems known anywhere on Earth—rubies, sapphires, jade, spinel, tourmaline, peridote, smoky quartz (Figure 12.3a, b). In the Jade mine belt of northern Burma, conglomerates along the river banks are loaded with rounded boulders of the most wonderful green jade from pebbles to house-sized. Pure jadeite is a sodium-rich pyroxene, the product of high-pressure metamorphism on original ophiolite mantle rocks that have been subducted to great depths and returned to the Earth's surface. Rubies come from marbles, the result of high-temperature metamorphism of limestones. Some Burmese rubies are the size of an adult fist; the largest stone from Mogok is the Nawata ruby, a flawless 504-carat stone found in the Dattaw Mine in 1990, and beyond value.[2] Spinel, tourmaline, and topaz are all derived from high-grade metamorphism and melting. Blue sapphires are the spectacular product of assimilation of lower crust material in deeply derived syenite intrusions or alkaline volcanic rocks. Northern Burma has been blessed with these magnificent natural gemstones in abundance. Barren red earth, laterite, covers the bedrock, and the rubies and sapphires, both corundum minerals with a hardness second only to diamond, have weathered out and are contained within the soil. Hundreds of miners comb the earth and dig down underground adits to find these precious gemstones which are then sold in the markets of Mogok and Kyatpyin.

For years Burma had been a closed country. Following the assassination of Burma's independence leader, Aung San, and most of the government officials in 1947, the country had been ruled by a ruthless military dictatorship. General Ne Win took control in 1962 and nationalized the ruby mines. In 1989 the State Law and Order Council (SLORC) changed the name of the country to the Union of Myanmar. The National League for Democracy party led by Aung San Suu Kyi, the widely respected and loved daughter of Aung San, won 80 per cent of the vote in the elections of May 1990, but the military government under General Than Shwe refused to accept the results and placed Aung San Suu Kyi under house arrest, where she remained for nearly twenty years. Unfortunately, despite international sanctions against the military regime, ruthless businessmen were busy plundering the forests and mineral wealth of Burma. Entire teak forests in the eastern Shan and Kachin states that had taken a thousand years to grow were logged out in a few years, leaving a trail of ecological devastation where there were once magnificent rain forests. Convoys of trucks laden with huge timbers headed north to China.

Figure 12.3 Burmese gemstones: (a) red rubies, (b) green jade.

Practically all the mining rights were sold to Chinese companies. It seemed like the generals were selling off everything that Burma had to offer. Whilst every other South East Asian country was experiencing rapid growth and rise in living standards, Burma languished at the bottom of the league, the poorest of all Asian countries. In 2012 Aung San Suu Kyi was released and the government relaxed restrictions and even declared that elections would take place later that year.

In 1999 I had been on a diving expedition from Phuket in Thailand through the Similan and Surin islands to the Mergui archipelago off the southernmost province of Burma. These beautiful, largely uninhabited and heavily forested islands have some of the most pristine coral reefs, and the sea around Mergui is renowned for its oceanic sharks, whales, and all sorts of other tropical fish. The Similan islands and some of the Mergui islands are composed of granites—continental rocks that confirm that the entire western coast of Thailand and Burma extends more than 200 kilometres offshore as a submerged continental shelf. Burma had been high on my list of priorities to visit for many years. My father had been in Burma with the Indian army at the end of World War II and had regaled me with stories of the country. An Oxford-based friend, Andrew Mitchell, was working in Burma for a mining company that was developing a newly discovered gold deposit in the western Shan state and also had a large copper mine out in western Burma. Andrew was a leading expert on mineralization throughout South East Asia and had been working in Burma for more than twelve years. I asked him whether I could visit and look at any outcrops of the Mogok metamorphic belt that ran along the eastern margin of the Sagaing fault. Andrew managed to get permission for me to cross at least the first few barriers of army checkpoints so I flew into Rangoon in spring 2000. Rangoon was a delightful city, with quiet, tree-lined streets, old colonial buildings, very few cars, little or no pollution, and a slow pace of life, in complete contrast to the fast and furious city of Bangkok with its skyscrapers, multiple-lane motorways, and crazy traffic. The one advantage of having been cut off from the

Figure 12.4 The golden pagodas of the Shwedagon in Rangoon, the most sacred Buddhist pagoda in Burma, with a history that is 2,500 years old.

outside world with almost no development for more than twenty years was that Rangoon remained a very pleasant city. The magnificent Shwedagon Pagoda in the middle of Rangoon was a forest of golden spires, the gold having been mined from the Shan states in eastern Burma (Figure 12.4).

Sagaing Fault and the Mogok Belt

The Sagaing fault is one of the most active strike-slip faults in Asia, responsible for several major earthquakes in the historic record. A huge magnitude 8.0 earthquake occurred just east of Mandalay in 1912, and two large earthquakes of magnitude 7.0 and 7.6 rocked the town of Mogok in 1931 and 1956 when the Sagaing fault ruptured (Figure 12.5). The Mogok belt, the zone of metamorphic rocks and granites from which all the precious gemstones are derived, runs along the eastern flank of the fault for most of the length of Burma. Andrew Mitchell was very familiar with the geology of Burma and we spent two weeks travelling along the Mogok belt from our base at Meiktyla, visiting quarries where the rocks were exposed. I mapped out the field relationships and collected samples for dating from Meiktyla north to Mandalay and Kyanikan. Despite all my efforts to travel north to Mogok and the jade mines, it proved impossible to get permission from the military authorities.

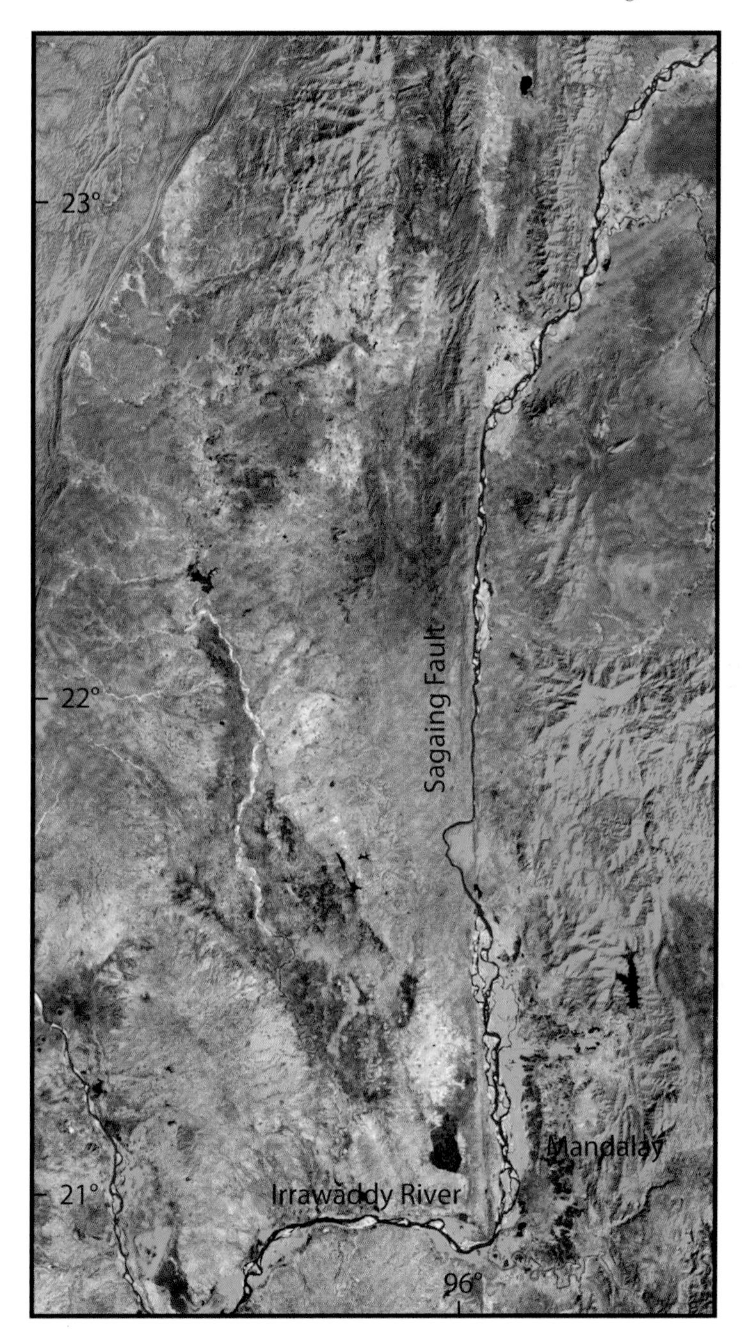

Figure 12.5 Landsat satellite photograph showing the north–south Sagaing fault deflecting the course of the Irrawaddy River in central Burma.

Instead, I travelled east to the Shan Plateau, at an average elevation of 1,000 metres, near the border with Thailand, and west to the old ruined city of Bagan. Back in Rangoon some Burmese friends showed me around the jade market and the fabulous golden pagodas of the Shewadegon, encrusted with gold from the Shan States and gems mined from the Mogok belt.

Back in England we sent the Mogok samples up to the uranium-lead dating labs of the British Geological Survey in Keyworth, and my student John Cottle, together with Matt Horstwood, managed to date the tiny zircons and monazite crystals in the rocks. The results were very surprising. Our new ages showed a series of events related both to oceanic subduction prior to Indian plate collision and to post-collision crustal thickening and metamorphism. Long-lasting Jurassic–Early Cretaceous metamorphism and magmatism was associated with subduction of the Indian oceanic plate beneath Burma. We found that regional metamorphism along the Mogok belt must have occurred prior to 59 Ma—the age of cross-cutting granite dykes. This was puzzling as it preceded the India–Asia collision 50 million years ago, so we surmised that the Burma micro-plate accreted to Asia earlier than the Indian plate in a similar way that the Kohistan arc in Pakistan had accreted to the Karakoram plate before the final India–Asia collision. Another phase of high-grade metamorphism occurred between 37 and 29 million years ago, similar to early Himalayan metamorphism. Localized partial melting from garnet–tourmaline leucogranites at Kyanikan quarry suggested that partial melting of the rocks was concomitant with peak metamorphism 24 million years ago, about the same time as the Himalayan granites were forming.[3]

The Sagaing fault abruptly cuts all the metamorphic rocks of the Mogok belt, so the metamorphism must have been related to earlier collision, crustal thickening, and heating events. Once the Burma plate had accreted and sutured onto South East Asia or Indochina, the Sagaing fault became the active eastern margin. India, with Burma hitching a ride along its eastern margin, continued to travel north, pushed by the oceanic spreading centre in the Indian Ocean and indenting into Asia beneath the Himalaya. Earthquakes rattling along the Sagaing fault showed that the plate to the west was moving north relative to the eastern side. As the fault slipped over thousands or hundreds of thousands of years, it dragged the course of the Chindwin River with it. Before the Sagaing fault initiated, the course of the upper Irrawaddy River flowed further west along the Chindwin. As India and western Burma slid north, the river was captured by the fault, changed course, and flowed south along the trace of the Sagaing fault along what is presently the course of the Irrawaddy (Figure 12.6).

Figure 12.6 The Irrawaddy River around Mingun township, central Burma.

Burma Seismic Zone

One of many remarkable facts about Burmese geology is the occurrence of numerous earthquakes along a narrow east-dipping zone down to depths of 200 kilometres running along the length of central Burma. The configuration of the earthquake zone suggests a subducting slab hanging beneath central Burma. Narrow zones of earthquakes extending deep into the mantle are common along ocean trenches such as the Sumatra–Andaman trench, but very rare along continental plate margins. Originally I thought that this Burma subduction zone was similar to the deep Hindu Kush seismic zone along the Afghan–Pakistan border—a narrow zone of subducting continental crust. Unlike the Hindu Kush, this seismic zone has three large active strato-volcanoes above (Figures 12.7 and 12.8).

One of these volcanoes, Mount Popa, is on the road to Bagan and I spent a day clambering all around it and looking at the rocks. The lavas are andesites, typical of Andean-type continental margins, with a dacite plug intruding right up through the central caldera. On top of the plug is a beautiful golden pagoda and a small Buddhist shrine. The composition of these lavas strongly suggests that the seismic zone is actually old oceanic lithosphere subducting beneath central Burma, not continental crust. A closer look at the earthquakes indicated that the Burma plate was both subducting towards the east and at the same time being dragged northwards by the

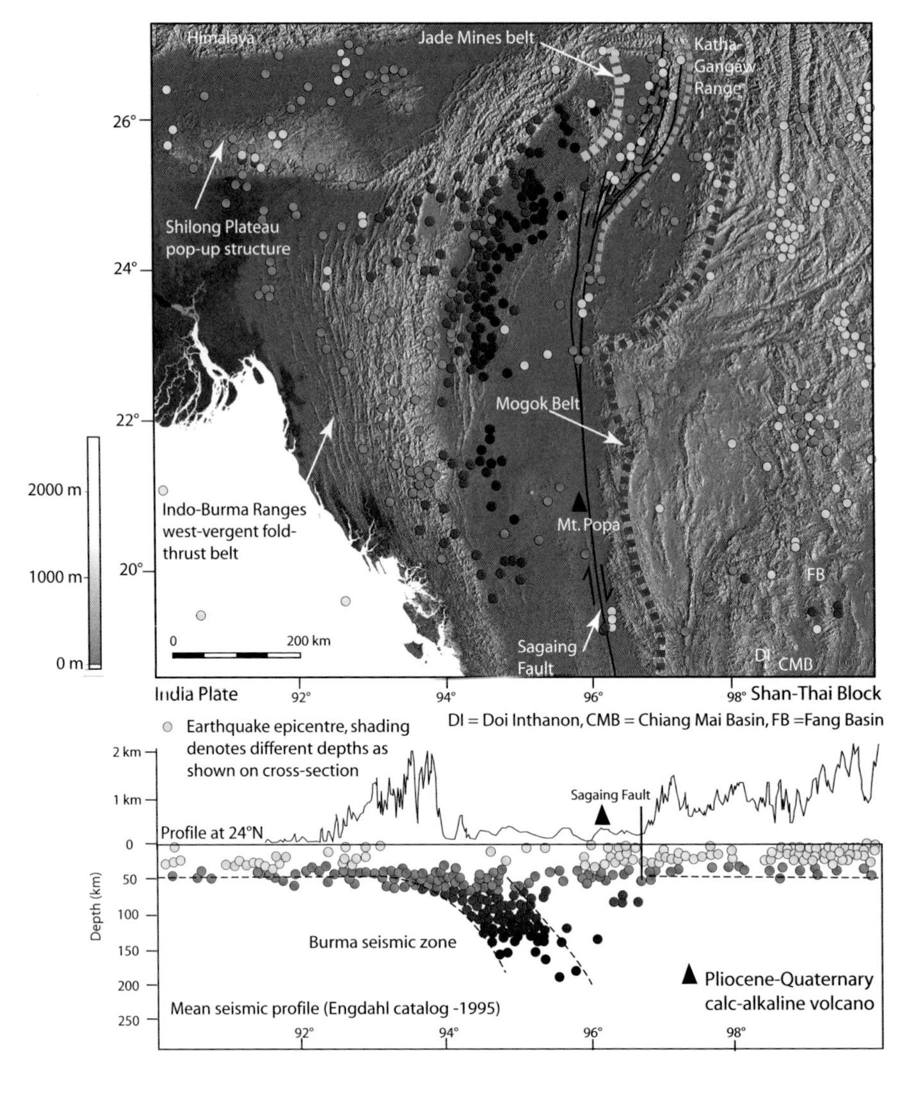
Himalaya
Jade Mines belt
Katha-Gangaw Range
Shilong Plateau pop-up structure
Mogok Belt
Indo-Burma Ranges west-vergent fold-thrust belt
Mt. Popa
Sagaing Fault
FB
DI
CMB
2000 m
1000 m
0 m
26°
24°
22°
20°
0
200 km
India Plate
92°
94°
96°
98°
Shan-Thai Block
Earthquake epicentre, shading denotes different depths as shown on cross-section
DI = Doi Inthanon, CMB = Chiang Mai Basin, FB =Fang Basin
2 km
1 km
Profile at 24°N
Sagaing Fault
Depth (km)
Burma seismic zone
Pliocene-Quaternary calc-alkaline volcano
Mean seismic profile (Engdahl catalog -1995)

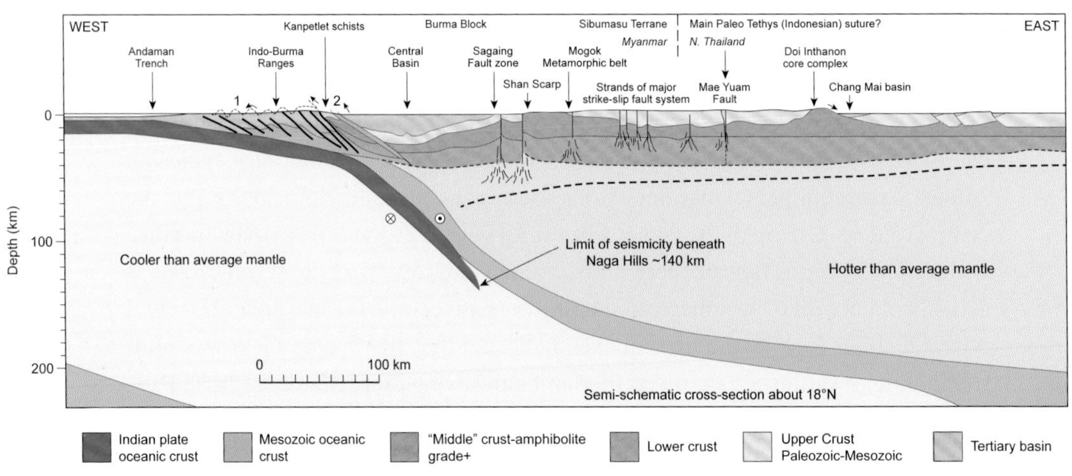
WEST
Kanpetlet schists
Burma Block
Sibumasu Terrane
Main Paleo Tethys (Indonesian) suture?
EAST
Myanmar
N. Thailand
Andaman Trench
Indo-Burma Ranges
Central Basin
Sagaing Fault zone
Shan Scarp
Mogok Metamorphic belt
Strands of major strike-slip fault system
Mae Yuam Fault
Doi Inthanon core complex
Chang Mai basin
Depth (km)
Cooler than average mantle
Limit of seismicity beneath Naga Hills ~140 km
Hotter than average mantle
0
100 km
Semi-schematic cross-section about 18°N
Indian plate oceanic crust
Mesozoic oceanic crust
"Middle" crust-amphibolite grade+
Lower crust
Upper Crust Paleozoic-Mesozoic
Tertiary basin

Figure 12.7 *(opposite top)* Map and cross-section showing the location of all earthquakes along the Burma seismic zone; shading darkens with increasing depth.

Figure 12.8 *(opposite bottom)* Geological section across Burma showing the eastward dipping subduction zone.

motion of India. Earthquake focal mechanisms showed both east–west compression and that north–south aligned strike-slip horizontal motions were occurring at depths extending to 200 kilometres.[4] Towards the north, nearing the eastern Himalayan syntaxis, the deep earthquakes suddenly died out, probably because the temperatures were too high for brittle failure of the rocks. As we have seen in the Namche Barwa region of the syntaxis, extremely young high-temperature and pressure rocks are exposed at the surface meaning that erosion rates are exceptionally high.

Mae Ping and Three Pagodas Faults, Thailand

Along the southern part of the Sagaing fault, two major faults splay off towards the south-east into Thailand, the Mae Ping fault and the Three Pagodas fault. Both these large-scale faults extend eastwards to Cambodia and the Gulf of Thailand. Sedimentary basins forming in between strands of the fault have been wrenched and cut by combinations of normal extensional faults and horizontal strike-slip faults. Many of these basins, such as the Phitsanulok and Ayutthaya basins of central Thailand, contain oil reserves and their internal geometry has been probed by seismic profiles. A friend of mine, Chris Morley, working with the Thai National Oil company PTT, had been interpreting these seismic lines both in Thailand and off the Burmese coast into the Andaman Sea. On several of my trips out to South East Asia I stopped off in Bangkok, and over about three years Chris and I visited most of the major faults and basins extending north and west of Bangkok. The Mae Ping fault was especially interesting as it had exhumed a mountain range of metamorphic rocks in a similar way to the Karakoram fault in Ladakh. In the Lansang National Park we measured structural fabrics in the rocks which clearly showed that the shearing was superimposed on earlier metamorphic and igneous rocks, in exactly the same manner as I had seen along the Karakoram fault in Ladakh and the Sagaing fault in Burma. The Three Pagodas fault did not appear to exhume deep crustal metamorphic rocks but did show some spectacular active faulting features.

In 2008 I drove from Tak north to Mae Saraing and Chiang Mai following the Thai–Burma border. There had been mass demonstrations in Burma which by this time was completely sealed to outsiders as the military attempted to mop up resistance. At the border village of Mae Sot, some desperate Karen refugees even tried swimming across the swollen waters of the Salween River to escape the brutality. It was hard trying to map the geology whilst cracks of gunfire ricocheted across the muddy waters of the Salween. In northern Thailand, large batholithic mountains of granites and migmatites are exposed at Doi Inthanon, the highest mountain in Thailand. Scrambling along the river courses and beautiful thundering waterfalls of Doi Inthanon I noted several different types of granites, but rock exposures were few and far between in the forests, so it was not really possible to determine the intrusive histories.

Chris and I tried to find out exactly what the finite offsets of mapped geological markers were along these faults, in order to work out the movement of the fault. This proved difficult, but along the Mae Ping fault we noted that the north–south-aligned older granite batholith rocks of north-west Thailand were offset up to 150 kilometres by horizontal motions along the fault. This figure was similar to the estimates of offset along several other major Tibetan strike-slip faults, so a pattern was beginning to emerge across Tibet and Indochina. Continental extrusion was occurring, but on an order of magnitude less than was originally proposed 100–150 kilometres, not thousands of kilometres.[5]

Red River Fault, Vietnam

The Red River fault is one of the longest and most prominent faults in all Asia (Figure 12.9). It stretches more than 1,100 kilometres from the south-eastern corner of the Tibetan Plateau across the mountains of Yunnan and North Vietnam to the South China Sea. Four, long, linear mountain ranges composed of deep crustal metamorphic rocks parallel the fault, the Xuelong Shan, Diancang Shan, and Ailao Shan in western Yunnan, and the DayNuiConVoi complex in North Vietnam. The area had been mapped by Chinese geologists, and one of Paul Tapponnier's students, Herve Leloup, had spent several years working on these rocks, including dating the various phases of granite. Paul had always said that the Red River fault showed all the characteristics supporting his model of large-scale horizontal extrusion. Tapponnier and Leloup proposed that the fault extended hundreds of kilometres down into the mantle, was responsible for between 500 and 1,000 kilometres of offset, that shear heating was the cause of the high-temperature metamorphism, and that the fault was responsible for the opening of the South China Sea.[6] Their papers were long and detailed and their arguments appeared to be strong. In 2005 I decided to take a look for myself and flew from Bangkok to Hanoi.

I caught an overnight train from Hanoi up to Lao Cai, a small town on the border with China, and hired a motorbike to go exploring. The position of the Red River fault was obvious, a huge and prominent geomorphological gash that cut diagonally across the whole of North Vietnam. Here, the Red River fault

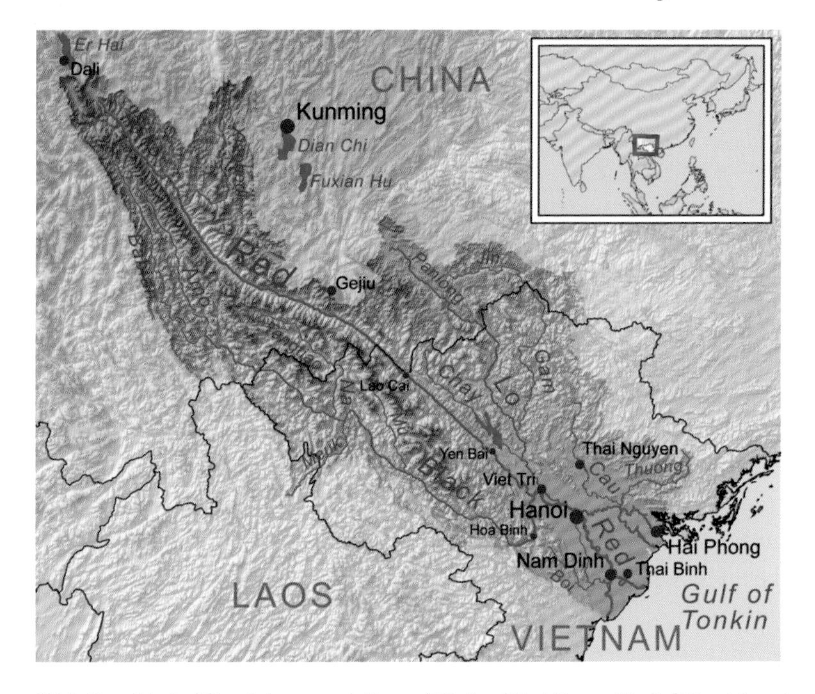

Figure 12.9 Map of the Red River drainage area in Yunnan (China) and North Vietnam. The Red River follows the trace of the Red River fault from its source near Dali to the Gulf of Tonkin, South China Sea.

split into two branches, each showing brittle deformation and in between were metamorphic rocks up to 10 kilometres wide in places, showing extremely high-temperature metamorphism, even partial melting. Along this shear zone were some spectacular sheared gneisses and mylonites, rocks formed by the crushing and grinding along active faults. Earthquakes and GPS data told us that the Red River fault was moving dextrally today (the Indochina block to the south-west was moving north-west relative to the South China block to the north-east). On close inspection, though, all the deeper, more ductile fabrics in the sheared rocks told the opposite story, that the bulk of motion was sinistral, with Indochina moving south-east relative to South China. The fault must have changed slip direction in the recent past. The ages obtained by Leloup and his colleagues suggested that the fault initiated some 35 million years ago, and about 5 million years ago reversed direction.

After two weeks of studying all the rocks I could access along the Lao Cai region I started to have severe doubts about the French model. None of their offset markers appeared to me to be reliable, so the 700 to 1,000 kilometres offset was questionable. These markers included offsets of the Khorat basin in Thailand with the Sichuan basin, two separate basins with different histories that cannot have been conjoined, and dubious correlations of basaltic rocks that never provide a precise pinning point. But there was another troubling point. How could shear heating along the fault produce so much heat to transform the rocks to migmatites and even granite? None of the many other large-scale faults, including the most well-known ones, the San Andreas fault in California, the Dead Sea fault in western Arabia, and the Altyn Tagh and Kun Lun faults of Tibet, showed metamorphic rocks formed by shear heating. I thought it far more likely that the metamorphism was a regional event and that the fault cut through metamorphic fabrics that had formed earlier. I collected key samples for U-Pb zircon- and monazite-dating which could prove useful.

The French model suggested that the age of initiation of the Red River shear zone was given by the age of a major igneous intrusion of syenite granite called the FanSiPan intrusion. Mount FanSiPan is the highest mountain in South East Asia, a tall pyramid cloaked in bamboo forest rising out of the steaming jungles of North Vietnam. On its southern flank is the beautiful hill station of Sapa. I decided to ride the bike up to Sapa and set up base there for a while. Sapa is located on a hilltop with glorious views north to the FanSiPan range of mountains and the lush tropical forests. The town was full of ethnic minority tribes including the colourful Hmong who thronged the busy open-air markets. On the way to Sapa I found large tracts of marbles and gneisses—high-grade metamorphic rocks that were tens of kilometres south-west of the Red River fault and clearly not the result of frictional heating along the fault. Doubts about the accepted model turned rapidly to disbelief. Even the FanSiPan syenite was located more than 15 kilometres away from the fault, and the syenites were completely undeformed except where some discrete shear zones subsequently cut through the rocks. It seemed obvious to me that the syenite was unrelated to fault motion and the age of intrusion could not possibly date the initiation of the fault. I gathered all the structural data I needed and proved to myself beyond doubt that the model proposed twenty years ago and widely accepted throughout the geological community had no basis in the rock record at all.

I decided to attempt to climb Mount FanSiPan and found a friendly local guide in Sapa, Phan Ngoc Hai, to show me the way (Figure 12.10). We set off with one porter, a local Black Hmong tribesman for the three-day hike. It was incredibly hot and humid in the steaming jungle. The first night we set up the tent in a jungle clearing beneath the steep ridge that leads up to FanSiPan. Hai and the Black Hmong porter disappeared into the forest for several hours and came back with about twenty frogs, neatly skewered on a bamboo stake, and a selection of grasshoppers and newts. The frogs were split open and cooked over a fire for our supper, along with wild mushrooms, bamboo shoots, and some sticky rice (Figure 12.11). Next morning Hai and I started early for the climb, the path through the bamboo jungle finally fading out as we resorted to hacking our way up the densely forested ridge.

Figure 12.10 Mount FanSiPan, the highest peak in North Vietnam.

Figure 12.11 Phan Ngoc Hai and a Black Hmong cook up roasted frogs on the FanSiPan climb.

We eventually reached the summit early in the afternoon with clouds swirling all around us. In between the clouds we could peer west down towards the valleys leading to Dien Bien Phu and the Lao border.

South-east of Hanoi the Red River fault splays out into a series of minor faults that are buried beneath the wide delta. After a month's work in North Vietnam I decided to take a short break in the magnificent World Heritage Site of Ha Long Bay. The area off the north-east coast of Vietnam is stunningly beautiful, with thousands of rocky islands, most with huge limestone cliffs rising sheer out of the sea. Large Vietnamese junks converted to carry tourists were plying between the jetty and the islands, cruising around the archipelago. Ha Long Bay was made of Carboniferous and Permian limestones and the islets were dotted with wonderful caves, grottos, arches, and reefs (Figure 12.12). We kayaked around the islets, swam, and explored this enchanting place. Back in Hanoi our friendly guide Tuan took us out for dinner at a local specialty snake restaurant. As the guest of honour I had to select a live snake from the cages. A swift slit from head to tail and the snake was turned inside out and cooked. The heart was cut out and, still pulsating, was given to me to eat whilst all eyes looked on. This was washed down with wine glasses of pure snake blood looking for all the world like a fine claret. Along with snakes, the Vietnamese eat dogs, rats, scorpions, locusts, and just about anything that moves. Vietnamese cuisine certainly takes some getting used to.

Figure 12.12 Ha Long Bay in North Vietnam showing spectacular karst islands composed of Permian-Carboniferous limestones.

DayNuiConVoi: Return to the Red River

After my first foray along the Red River fault in North Vietnam I wrote up my initial findings.[7] At a conference I made the acquaintance of Sun-Lin Chung, a professor of geology at the National Taiwan University in Taipei who, like me, had similar doubts about the extrusion model for Indochina. Sun-Lin was a geochemist and had worked for ten years all across the Tibetan Plateau and Yunnan and Sichuan provinces. He had written one of the most important seminal papers reviewing all the geochemistry and isotope data from Tibetan volcanic and intrusive rocks.[8] These data showed that some of these rocks had very unusual compositions that could only be explained by derivation from a hot mantle source, and others that must have been derived from a garnet-bearing lower crustal source. Sun-Lin had also dated almost a hundred samples from all over Tibet and Indochina and proved that these rocks were formed over a wide range of ages from the time of the India–Asia collision right up to recently active volcanoes along the Kun Lun. His data finally disproved for certain the model of sudden uplift of the Tibetan Plateau only 7 million years ago.

Sun-Lin came over to Oxford for a year's sabbatical visit with his family, and during that we time we planned a long-term collaboration across South East Asia. After his visit to Oxford, Sun-Lin arranged a sabbatical visit to Taipei for me, and during my stay in Taiwan we returned to North Vietnam with another friend, Mary Yeh, a structural geologist also from Taiwan. We wanted to do more geological work along the DayNuiConVoi complex, the high-grade metamorphic rocks along the Red River shear zone, and we also wanted to look at the geology of the country rocks away from the fault.

After a few days in Hanoi we drove north towards Lao Cai, stopping off frequently to look at rocks, but also to haul the jeep out of muddy rivers, and have extended lunches with *beer hoi*, the small glasses of beer served out of a nozzle just like a petrol filling station in the wonderful Vietnamese pavement cafes. We tended to avoid other Vietnamese delicacies—bear paw, snake blood 'wine', sheep's testicles, pickled lizard. We found some excellent examples of mylonite-fault rocks along both strands of the Red River fault and even better metamorphic rocks in between the two faults. These rocks were clearly widespread regional metamorphic rocks, not formed along the fault by shear heating, but simply brought up from the basement by being squeezed upwards during the final stages of exhumation. Indeed, we found extremely high-temperature marbles up to 50 kilometres away from the fault. These spectacular white marbles mainly comprised pure white calcite, but also had small red rubies and bright green tourmalines. The local people were quarrying the stone to make the most beautiful polished ornamental slabs. It became even more obvious to us that the metamorphism was regional in extent, older and unrelated to the fault. The shearing fabrics were superimposed on the already highly metamorphosed rocks during later movement. We now set about trying to obtain precise ages on the gneisses, the few cross-cutting granite dykes, and cooling ages to attempt to piece together the entire history of the rocks.

Ailao Shan and Diancang Shan, Yunnan

Our work in Vietnam had convinced us that the widely accepted model for the Red River fault was not correct. The geology told us that the metamorphic rocks had formed prior to shearing, that the offsets were far less than the 1,000 kilometres previously proposed, and that the fault was a later upper crustal feature, not one extending deep down into the mantle. The connection with the mantle-derived FanSiPan syenite was incorrect, so the age of initiation, 35 million years ago, was also not valid. There were three further metamorphic complexes aligned along the Red River fault in Yunnan, southern China, so the next trip that Mary, Sun-Lin, and I made, in 2007, was to the Ailao Shan, just north of the Vietnamese border and the Diancang Shan, west of the town of Dali. We flew from Taipei to Kunming and met up with a local geologist, Professor Wang from the Yunnan Geological Survey.

Dali is a delightful town on the shores of Urhai Lake. The old quarter with its narrow paved streets had been restored and, in complete contrast with most Chinese cities, had a lot of character. The backstreets of Dali had several small rock-cutting factories where great slabs of Dali marble were cut and polished. Just like the marbles of North Vietnam the metamorphic rocks around Dali were very widespread and had nothing to do with the Red River fault. Diancang Shan was an elongated massif with low-angle normal faults flanking a core of high-grade metamorphic rock. Despite being aligned along the Red River fault, there appeared little trace of the fault in the rocks. Instead it seemed the whole massif was an uplifted slab of basement gneisses. Some trails wound up the mountain but exposures were not great. We were able to map parts of the eastern and southern end of the mountain and examine rocks in detail in a few quarries. South of Diancang Shan the metamorphic rocks disappeared altogether before remerging in the Ailao Shan further south.

The Ailao Shan was even more forested than the Diancang Shan and exposures were very limited (Figure 12.13). Compared to the almost complete exposures I was used to in the Karakoram and parts of the western Himalaya, this area was extremely frustrating to work in. It was impossible to make a decent map and also to link one outcrop with an adjacent one (Figure 12.14). The forest cover was hiding unknown critical exposures and contacts. We did find one excellent outcrop along

Figure 12.13 The forested slopes of the Diancang Shan along the Red River fault, above Dali in Yunnan (China).

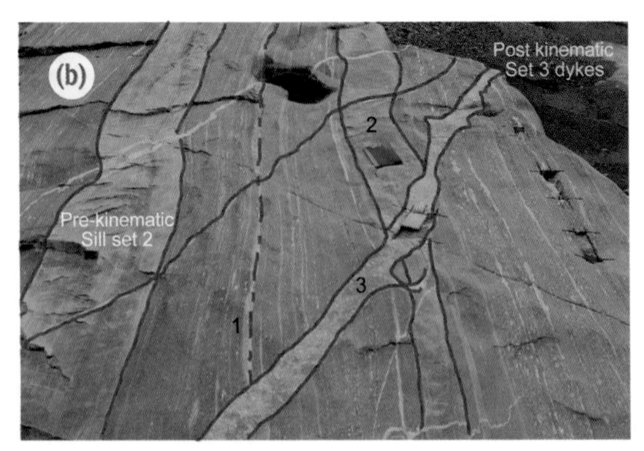

Figure 12.14 Typical outcrop of the Red River gneisses cut by younger granite dykes. The age of the dykes gives a maximum age of ductile motion along the Red River fault.

the floor of a small river where water-worn slabs had polished the rock. Here we could map out three periods of granite dyke intrusion into the gneisses. The earliest set was deformed and intruded parallel to the gneissic fabric, whereas the later two sets cross-cut the fabrics and the earlier dykes. We collected samples from each set of granites to date so that we could bracket the age of the gneiss fabric as being older than the first set of dykes but younger than the final set.

Our results from the Ailao Shan and Diancang Shan backed up our ideas formed earlier in North Vietnam (Figure 12.15). The geological offsets proposed previously were spurious. We still had no idea of the total geological offset along the fault, but it was certainly nowhere near the 1,000 kilometre figure. Our ages confirmed that the metamorphism and early granites were regional events probably formed during the Triassic Indosinian orogeny over 200 million years ago, a mountain-building episode related to the collision of the South China block with Indochina. Only the younger set of granite dykes that cross-cut the fabric could have been related to movement along the Red River shear zone.[9] The shear zone moved rocks south-east relative to the northern margin, but despite being such a large and impressive shear zone, the movement was actually only a minor amount. The continental extrusion theory was a fine theory but on close inspection of the rocks did not stand up to scrutiny.

Eastern Tibet and Lower-Crustal Flow

Whereas the southern (Himalaya) and northern (Kun Lun) margins of the Tibetan Plateau are abrupt and obvious both in topography and crustal thickness, the eastern margin of the plateau is much more irregular. In south-eastern Tibet the plateau has no obvious boundary, with topography decreasing gradually from

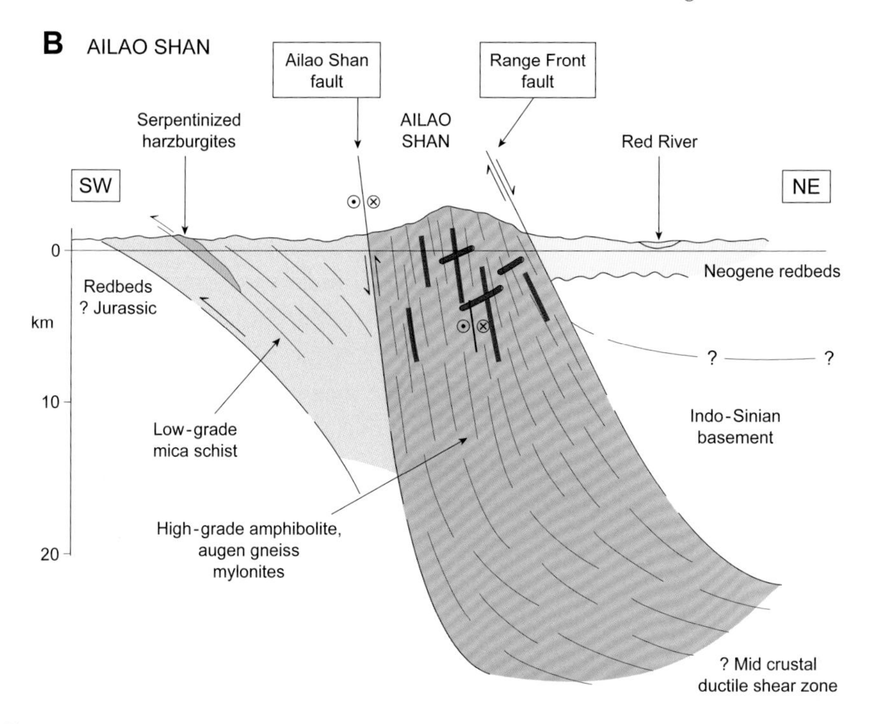

Figure 12.15 Geological section across the Red River gneisses in the Ailao Shan, southern Yunnan. The extent of the gneisses and the outward-dipping normal faults bounding the Ailao Shan show that the gneisses are uplifted basement rocks and were not formed by shear heating along the fault.

5,000 metres to less than 1,000 metres over hundreds of kilometres. In the LongMen Shan Mountains along the western boundary of the Sichuan basin the topography rises very rapidly from less than 600 metres in the Sichuan basin up to peaks more than 6,500 metres high over a distance of only 50 kilometres. The plateau area west of the escarpment is high and flat with only low-relief topography. North of the Sichuan basin the topography becomes slightly more diffuse again from elevations of 5,500 metres decreasing away from Tibet in the mountain ranges of the Qilian Shan bordering the north-eastern plateau.

The LongMen Shan along the eastern margin of the Tibetan Plateau is a very different mountain range from the Himalaya. Whereas the Himalaya shows many hundreds of kilometres of crustal shortening and widespread young (post India–Asia collision) metamorphism and melting, the LongMen Shan shows very little east–west crustal shortening and no young metamorphism at all. The geology of eastern Tibet and the LongMen Shan shows older rocks with almost all the deformation related to the much older Indosinian orogeny. In common with much of the Tibetan Plateau it is amazing how little effect the India–Asia collision seems to have had on the geology. Most of the deformation and metamorphism is much older and it appears that the plateau was uplifted almost passively by the lower part of India underthrusting and jacking it up, as Emile Argand first proposed back in 1924. Whereas the loading of the Himalaya caused the rigid Indian plate to flex down and form a foreland basin that rapidly filled up with material eroded from the Himalaya (the Siwalik basin), there was no such foreland basin in Sichuan, east of the Tibetan Plateau.

Some modellers interpreted this enigmatic boundary in eastern Tibet in terms of a model involving lower crustal flow. Leigh Royden, Clark Burchfiel, and Marin Clark from MIT first proposed the model involving the eastward flow of the lower crust from beneath Tibet (Figure 12.16).[10] Like many geologists working in and around Tibet, they found it hard to envisage that Tibet had thick crust and high elevation for a long time, as it is difficult to maintain thickened lithosphere for long periods of time without it becoming thermally unstable. They presumed that the plateau had started to rise around 15 million years ago, based on low-temperature thermochronological data from eastern Tibet.[11] Localized concentration of deformation along the LongMen Shan was a result of the rheological differences between the weak crust of Tibet and the strong lithosphere of the Sichuan basin. The lower crustal flow model seemed to mimic the GPS 'flow lines' of active surface deformation. A number of geological factors argue against this model. Lower crustal flow away from Tibet implies lowering of crustal thickness and decreasing surface elevation, unless there is addition of lower crust from elsewhere. Addition of lower crust into eastern Tibet could have come from underthrusting Indian lower crust, but this would have been old, cold, dry Precambrian granulites incapable of melting or flowing. If Tibetan lower crust were flowing eastwards away from the plateau one would expect a corresponding amount of upper crustal shortening in the Longmen Shan, which is certainly not present.

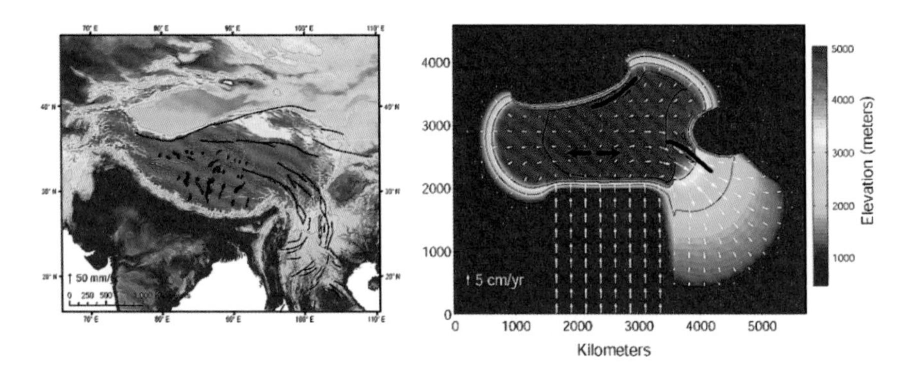

Figure 12.16 Topographic digital map of the Tibetan Plateau and a model showing the northward motion of India and the north-east and south-east 'flow' of Tibetan lower crust along the eastern margin of the plateau.

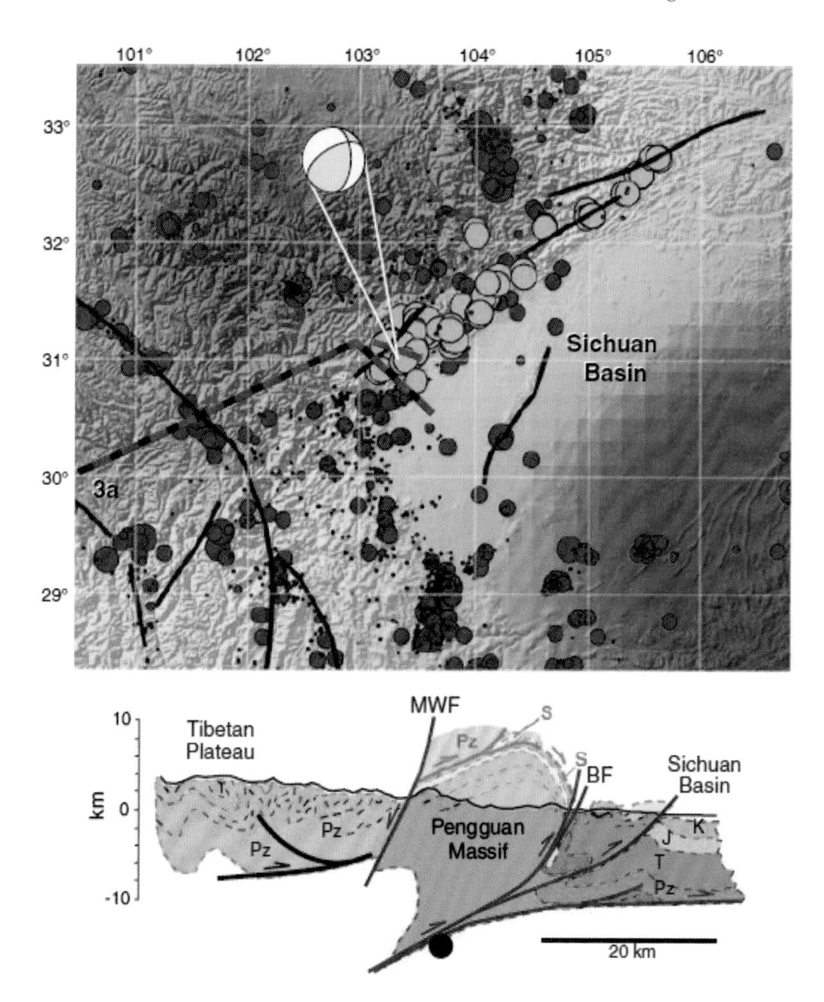

Figure 12.17 The location of the Wenchuan earthquake in the Long Men Shan mountains of eastern Tibet and a cross-section showing the major faults along the western margin of the Sichuan basin.

On 12 May 2008 a massive 7.9-magnitude earthquake occurred beneath the town of Wenchuan in the LongMen Shan Range of eastern Tibet (Figure 12.17).[12] Over 260 kilometres' length of a steep west-dipping thrust fault ruptured with up to 10 metres of displacement showing at the surface. The town of Wenchuan was almost completely destroyed by the shaking and by numerous devastating landslides that followed. The earthquake hypocentre was at a depth of approximately 15 kilometres. Focal mechanisms showed a compressional stress field associated with active crustal thickening and uplift of the plateau. The earthquake indicated that the plateau is still actively rising and thickening. Talk of the 'collapse' of the Tibetan Plateau is clearly premature.

Xianshui-he Fault and Gongga Shan

A glance at the topographic and geological maps of Tibet shows a number of prominent faults curving around the eastern Himalayan syntaxis, suggesting that a clockwise rotation of east Tibet and Asian crust occurred as India indented into Asia and pushed northwards. The Jiale fault cuts across the northern end of the syntaxis, north of Namche Barwa, and swings south into Burma, possibly to connect with the Sagaing fault. Earthquakes along these faults suggest dextral slip, in other words, the block to the north and east of the fault is moving south-east relative to the area to the west of the fault. The Xianshui-he fault to the north-east swings around from central Tibet south-west of the Sichuan basin, and may link via a series of transfer faults to the Red River fault. All these faults show sinistral slip, in other words the Indochina terrane moving south-east relative to the area to the north and north-east. The Red River fault has reversed slip direction in the last few million years, whereas the Xianshui-he fault has remained slipping in a sinistral direction. Three large earthquakes with magnitudes greater than 7 occurred along the Xianshui-he fault in the last century, making this one of the fastest slipping faults in the Tibetan region. One of these, a magnitude 7.5 earthquake, devastated the city of Kangding. During the past ten years or so Kangding has been rebuilt, and now there are lines of twenty-storey buildings strung out along the whole narrow valley, right on top of one of the most active faults in Asia. New Kangding is a disaster waiting to happen.

After one of our annual Himalayan Workshop meetings in Hong Kong I had been on a field trip through eastern Tibet led by Chris Wilson from the University of Melbourne in Australia, who had been working in the area for several years.[13] Chris showed us some spectacular sites along the fault where huge zones of fault gouge—finely crushed rock ground to a powder by slip along the Xianshui-he fault—were exposed. Along the eastern flank of the fault a huge basement complex of Precambrian gneisses was mapped. This was extraordinary. How could the oldest rocks in Tibet be exposed at the surface of Tibet above crust that was double normal thickness? Surely some remarkable structures had to be buried beneath the few kilometres of surface exposures that we could physically see. Maybe the basement was stacked up by a series of unseen thrust faults in order to explain the crustal thickness of 70–80 kilometres. Right at the bend in the fault another extraordinary feature of the fault was apparent. A large granite batholith had been mapped oriented parallel to the fault. The largest mountain in eastern Tibet at 7,756 metres, Gongga Shan (previously called Minya Konka), was composed of this granite.

I was particularly interested in Gongga Shan (Figure 12.18) because a paper by Françoise Roger from the University of Montpellier in France and her colleagues had shown that the granite was very young, only 12.8 million years old, using U-Pb dating of zircons.[14] This was practically the only young granite recorded within the Tibetan Plateau region and could therefore potentially give us important information on the timing of crustal melting hidden at depth. Maybe the clue to finding out the age of Tibetan Plateau uplift was hidden in the granites of Gongga Shan. I thought detailed structural observations around the massif combined with more detailed U-Pb dating might be the way to answer key questions about the timing and offsets along the fault, in much the same way that we had accomplished on the Karakoram fault.

In 2011 we returned to the area with a larger geological expedition, organized by Sun-Lin Chung and Yuan-Hsi Lee from Taiwan together with the China Earthquake Institute in Beijing. We drove from Chengdu to Danba and then up to the high plateau to the old Tibetan town of Barmie, crossing spectacular folded sedimentary rocks of the Triassic fold belt, the Songpan-Ganze terrane. We accessed the Gongga Shan granites along the two main roads to Barmie and Kangding, collecting samples for dating, and noting the various magmatic cross-cutting relationships. Towards the end of the trip we spent several days hiking around the

Figure 12.18 Gongga Shan, the highest mountain of eastern Tibet, rising high above the clouds in the Sichuan basin.

beautiful Yanzigou Valley to the north of Gongga Shan and up in the Hailuogou Valley west of Moxi. Here, large glaciers cascaded directly off the Gongga Shan massif and descended straight down to the bamboo and rhododendron forests of the Hailuogou Valley. The mountain was shrouded in monsoon clouds most of the time but occasionally we got tantalizing glimpses of high peaks through the swirling mists.

It soon became apparent that there were at least four, maybe five, phases of granite intrusions, each with completely different mineral and chemical compositions. From the boulder debris in the glacial moraines draining the flanks of Gongga Shan, most of the mountain appeared to be comprised of hornblende- and biotite-bearing granite, more related to the pre-collision Andean-type intrusions of the Gangdese Range across south Tibet. These granites were completely different from the post-collision garnet and mica-rich granites and later cross-cutting pegmatite dykes. Along the Yanzigou Valley we sampled the later garnet-bearing granites and managed to collect a full suite along most of the length of the batholith for dating.

Extruding Indochina?

Twenty years of continuous GPS measurements have shown that the eastern margin of Tibet is today moving north-east and south-east, apparently around the stable block of the Sichuan basin.[15] The GPS lines of motion in the north-east appear to show the eastward extrusion of crust 'flowing' around the Sichuan basin. South of the Sichuan basin the GPS lines show a clockwise rotation around the eastern Himalayan syntaxis, with motions curving around to the east, then south-east, and finally practically all the way back west in northern Burma. GPS data actually records the present-day motion of the Earth's surface, yet here it appears that the GPS motions are diametrically opposite to many of the active faults in the area, including almost at right angles to motion along the Red River and Sagaing faults. In this region there is a puzzling mismatch between the active GPS data and geological data.

Certainly, the large-scale picture has emerged that some unquantified amount of continental extrusion has occurred in the areas bounded by the dextral Jiale–Sagaing fault system to the west and south, and by the sinistral Xianshui-he–Red River fault system to the east and north. In between these two fault systems in Indochina, some cross-faults curve around parallel to the GPS lines but these are generally quite minor faults compared to the Red River and Sagaing faults. Many of these faults have changed slip sense with time. As India penetrates progressively northwards, east Tibet and Indochina have rotated clockwise to accommodate its motion. Three of these curved faults, the Mengxing, Nam Ma, and Wanding faults, show impressive offsets of river courses. The Mekong River shows an abrupt 24-kilometre offset along the Mengxing fault, and the Salween River shows a minimum offset of 38 kilometres (possibly as much as 60 kilometres) along the Wanding fault.[16] Both the Mekong and the Salween flowed south from their sources on the Tibetan Plateau and the movement along the later faults deflected the river course during horizontal motion along the fault. The rivers actually form very good strain markers or pinning points and can potentially be used, with caution, to derive geological offsets along faults. In the 'Golden Triangle' area where the borders of north Thailand, Burma, Laos, and Yunnan meet, three of the great rivers of East Asia, the Salween, Mekong, and the Yangtze have cut parallel courses close together in the 'Three Gorges' area. Each river has cut deep, narrow gorges only a few kilometres apart, with the intervening mountain ranges of the Gaoligong and Chong Shan rising to over 4,000 metres.

The tectonics of eastern Tibet and Indochina was extremely complicated. A combination of detailed mapping, combined with stratigraphy, structure, and dating of metamorphic rocks and granites by several different groups had begun to unravel the history of the region. It seemed that much of the story involved the older Triassic mountain-building episode (the Indosinian orogeny) when the old Palaeo-Tethys Ocean closed and south China collided with Indochina. Tibet remained an enigma. There was very little evidence of Tertiary crustal thickening or metamorphism after the India–Asia collision. Precambrian rocks were exposed at the surface of double normal thickness crust in parts of eastern Tibet. There was no evidence at all of crustal flow, either mid-crust flow as seen along the Himalaya, or flow of the lower crust which is not exposed. There was strong evidence of clockwise rotations around the eastern Himalayan syntaxis and in Indochina, regions adjacent to the indenting Indian plate. There was good evidence of active faulting along giant strike-slip faults, and the timing of slip, but the precise amounts of geological offsets remained poorly known. We now knew fairly well the consequences of the India–Asia collision along the Himalaya, and across Tibet and Indochina. The final great question was: what happened along the Indian Ocean plate boundary south of Burma?

CHAPTER 13

The Day the Earth Shook

The Sumatra–Andaman Earthquake, December 2004

Civilisation exists by geological consent, subject to change, without notice.

Will Durant

At 00.58 GMT (7.58 local time) on Sunday, 26 December 2004 a massive earthquake occurred off the north-west coast of Sumatra. The earthquake measured between magnitude 9.0 and 9.3 on the Richter scale with its epicentre at 3.32°N, 95.85°E, and occurred at a depth of approximately 30 kilometres. It was the second largest earthquake recorded since instrumental records began and was the deadliest natural disaster in recorded history. The earthquake and the resulting tsunami are estimated to have killed at least 228,000 people across fifteen countries bordering the Indian Ocean. The worst affected countries were Indonesia, Sri Lanka, India, Thailand, Burma, the Maldives, and Somalia.

The earthquake occurred on the subduction zone interface between the down-going Indian Ocean plate and the overriding Burma–Andaman–Sumatra plate. It ruptured approximately 1600 kilometres' length of the plate boundary from Sumatra all the way north to the Burmese coast, travelling at 2–3 kilometres per second. Aftershocks continued unrelentingly for over four months after the earthquake, several reaching magnitude 7.5 as far north as the northern Andaman Islands (Figure 13.1). The seismic waves indicated a thrust fault earthquake that tilted the surface up to the south-west and down to the north-east. The ground surface was elevated as much as 11 metres at the epicentre, with the tilted surface sinking up to one metre further to the north-east, offshore Sumatra. During the

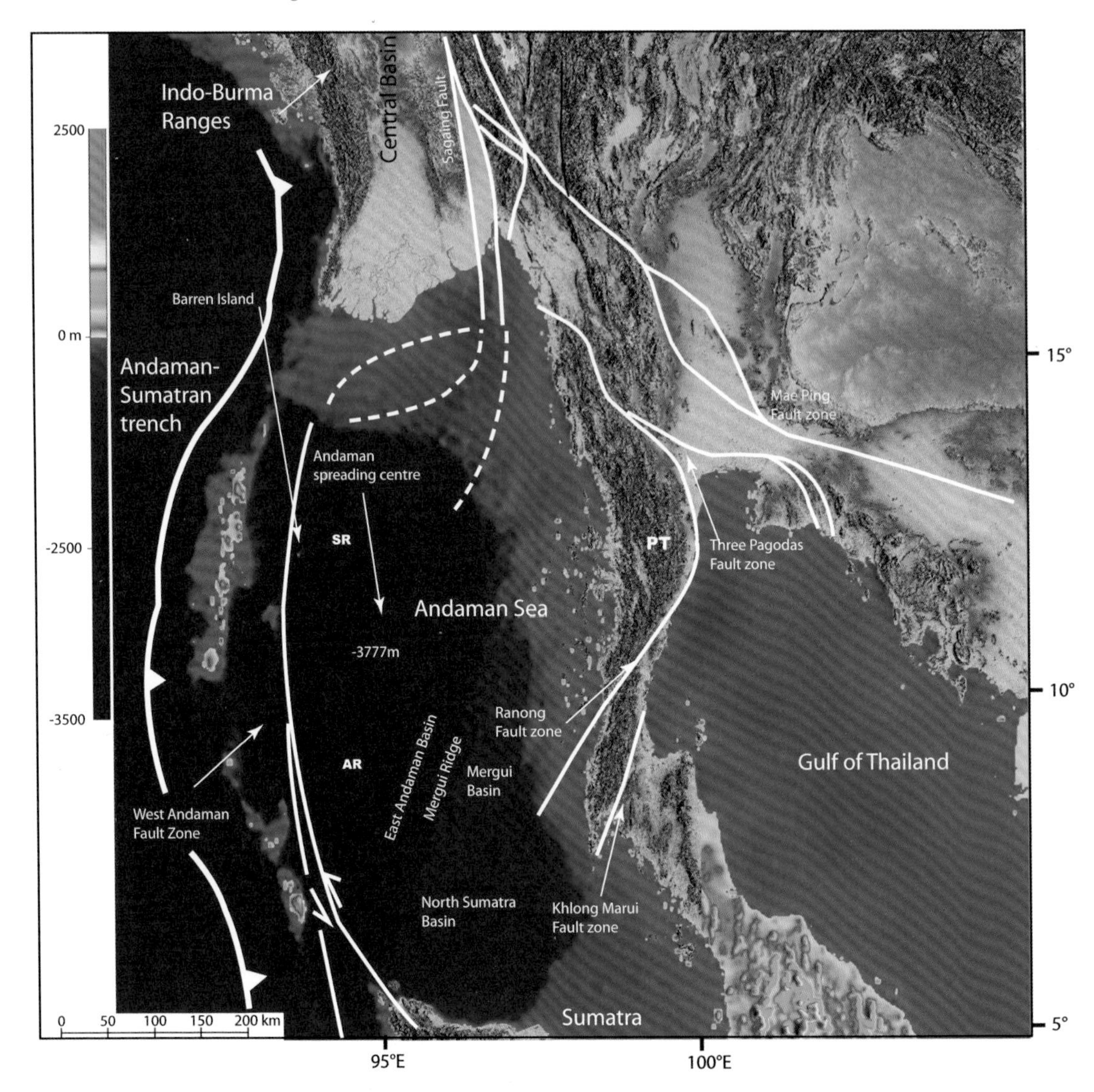

Figure 13.1 Map of the eastern Indian Ocean, Andaman Sea, and Gulf of Thailand region showing the major faults.

rupture, the Burma plate slipped as much as 15 metres horizontally as the Indian Ocean plate slipped beneath.[1]

The force of the quake perceptibly shifted the Earth's axis, raised sea level globally and speeded Earth's rotation. It has been suggested that the earthquake shortened the length of the day by 2.68 microseconds, because of the decrease in oblateness of the Earth.[2] The earthquake caused the Earth to wobble on its axis by up to 2.5 cm in the direction of 145° east longitude. The natural 'Chandler wobble', a small motion in the Earth's axis of rotation (the motion that occurs when the spinning object is not a perfect sphere) can be up to 9 metres over 433 days, so this eventually offsets the comparatively minor wobble produced by the earthquake. The sea floor raised by the earthquake displaced seawater upwards a fraction of a millimetre throughout the world's oceans.

The slip caused by the megathrust earthquake occurred in two phases. The first, lasting 100 seconds, ruptured about 400 kilometres of the plate boundary from Bandar Aceh in Sumatra northwards, the largest rupture ever known in a single earthquake. It travelled at a speed of about 2.8 kilometres per second (10,000 km per hour or 6,200 mph). After a pause of 100 seconds the second rupture travelled northwards to the Nicobar and Andaman islands at a rate of 2.1 km/second (7,500 km per hour or 4,700 mph). The total energy released from the earthquake was estimated to be 65,000 times the energy of the Hiroshima atomic bomb. Earthquakes rattled along the trench, rocking the Nicobar and Andaman islands for months afterwards. Along the rupture itself, deep in the trench south of Sumatra, a wall of rock 11 metres high suddenly appeared thrust up from beneath. A Royal Navy survey ship, HMS Scott, carried out a high-resolution multi-beam sonar survey along the trench after the earthquake and showed dramatic 3D images of the ocean floor. The images show mountainous ridges 1,500 metres high, enormous undersea landslides and an oceanic trench several kilometres wide. When the earthquake ruptured, the column of water above this uplifted region was also uplifted instantaneously, resulting in the largest tsunami recorded in modern history.

The Indian Ocean Tsunami

As the sea floor was raised by up to 11 metres above the epicentre, the entire column of seawater above also rose with it. The elevated ocean fell away in all directions, creating a tsunami that travelled around the entire world. The surface wave swept westwards across the Indian Ocean, slamming head-on with the east coasts of Sri Lanka and Tamil Nadu, sweeping over the low-lying islands of the Maldives, around the Cape of Good Hope and across the Atlantic, around Cape Horn and right across the Pacific Ocean. On the other side of the world, the tsunami wave was even recorded in Hawaii, by then only a few centimetres high. Some twenty-four hours after the earthquake, it swept all the way back to Indonesia, having encircled the entire planet.

The tsunami wave height reached approximately 10–15 metres along the coast of Sumatra and nearby coasts of south-west Thailand and the Nicobar Islands. At Bandar Aceh in northern Sumatra, which took the full force of the earthquake and the tsunami, the wave may have reached as high as 35 metres.[3] It took only minutes to reach Sumatra, where it caused such intense destruction, and less than thirty minutes to reach the south-west coast of Thailand. Approximately forty-five minutes after the earthquake the tsunami reached the Andaman Islands, sweeping up the Bay of Bengal. The Andaman Islands are quite hilly, so people were able to escape the wave comparatively easily. The Nicobar Islands on the other hand are very flat-lying and in places there were few survivors. Less than two hours after the initial earthquake, the wave reached Sri Lanka, wrapping around the east and south of the island. The east coast of India took the full force of the wave along its gently sloping coasts of Tamil Nadu. Fisherman out at sea rode the unusually high waves almost without noticing, but when they returned to their shoreline homes they saw with horror the complete devastation left behind. Tens of thousands were killed in India and Sri Lanka. Four hours after the quake the tsunami wave broke over the low coral atolls of the Maldive Islands and seven hours afterwards, it reached the east coast of Africa and the southern coast of Arabia.

In some low-lying flat islands in the Nicobar group the wave actually went right across the island. At Car Nicobar island an entire Indian airforce base was wiped out, most of the people and aeroplanes having been swept out to sea. The tsunami wreaked its greatest destruction where the coastline forms a gently dipping beach. Where the offshore drop in ocean depth (bathymetry) was steeper, the force of the tsunami was directed more vertically than horizontally. As any surfer knows, the best place to avoid a breaking wave is to dive down beneath it. All the force lies at the surface, which explains why there was remarkably little damage to the coral reefs on the seafloor.

The tsunami travelled across the Indian Ocean at speeds of up to 500 miles per hour. The speed of a tsunami wave is directly proportional to the square root of the depth of seawater through which it is travelling. As the tsunami wave reaches the shoreline, the wave front slows as the water depth shallows, but behind it, out in the ocean the wave is travelling much faster, so the water gathers momentum and elevation, easily lifting huge ships, and with the additional momentum of transported detritus, breaking solid concrete barriers, as seen in Sumatra. Some of the tsunami destruction in Sumatra, Thailand, and Sri Lanka was on an epic scale. Near Hikkaduwa in south-western Sri Lanka, about 1,000 kilometres from the earthquake epicentre, the wave picked up a packed train and hurled the carriages into the nearby trees and swamp. Over a thousand people died in that train alone. Yet, in Koh Phi Phi Island off Phuket, divers only 20 to 30 metres below the surface hardly felt the wave as it passed over their heads. Onshore at Phi Phi Island the unfortunate topography of the island funnelled the tsunami in towards the shore, where it built up, both in height and force, to destroy most buildings on the island.

In Bandar Aceh, the nearest town to the earthquake epicentre, the tsunami wave brought almost total destruction to the coastal settlements. Flimsy wooden houses simply crumpled as the tidal wave swept inland along streets and up river valleys carrying with it an enormous mass of flotsam of wrecked cars, buses, houses, and random debris riding the wave. Pictures from Bandar Aceh show people riding on the flotsam and desperately trying to get off the wave as others ran ahead of the advancing water. The only building left standing on the Bandar Aceh seafront after the tsunami was the concrete mosque, which was well anchored to the bedrock.

Immediately before the tsunami hit the shoreline, the wave was preceded by a drop in sea level by up to twelve metres in places. Early morning swimmers in Thailand looked in astonishment as the sea withdrew hundreds of metres beyond the low-tide mark. Then, ten to twenty minutes later, they saw a wall of water approaching. The wave crashed through sea walls and swept through houses, plantations, or whatever happened to be in the way. After the wave, the sea withdrew once again sucking everything out with it. Altogether four tsunami waves hit the coast of Thailand. Cars, houses, trees, people, animals, and all were simply swept out into the ocean. Some people were found days later clinging to flotsam out in the open ocean. Many thousands of others simply drowned. The force of the tsunami was overwhelming.

Indian Ocean Geology

The great plate boundary between India and Asia traverses the length of the trans-Himalaya in southern Tibet before swinging around the eastern syntaxis in the great gorge country around the peak of Namche Barwa, and then turns south, traversing along the Indo-Burma Ranges into the Bay of Bengal. Here it forms a great trench that extends into the Indian Ocean west of the Andaman–Nicobar Islands and continues to the east, south of Sumatra and Java. Along this abrupt boundary in the southern Indian Ocean the ocean crust is subducting north-eastwards beneath the great island arc chain of the Indonesian archipelago. Above the subduction zone, some of the most explosive volcanoes on Earth are situated in a long line stretching from western Sumatra all the way to the Banda Sea. The volcanic arc continues north of Sumatra with the final two historically active volcanoes, Barren Island and Narcondum, located to the east of the Andaman Islands.

South of Burma and east of the Andaman and Nicobar islands the Andaman Sea is a very young ocean. Magnetic anomalies indicate that an active ocean-spreading centre above the subduction zone and behind the island arc has formed in the past 3 or 4 million years, creating new oceanic crust along a line of recent under-sea volcanism.[4] Such new basins, produced by tectonic forces related to the subduction, are known as back-arc basins. The Andaman Sea is probably the best modern-day analogy for the way in which many ophiolite complexes, such as the Oman ophiolite and the Spontang ophiolite of Ladakh formed, as a small ocean basin spreading behind an active island arc, and above a subduction zone.

Andaman Islands

I had planned to go to the Andaman Islands in 2004, initially to look at the ophiolite rocks exposed along the eastern part of the islands. The basement rocks of the Andaman Islands appeared to be slices of oceanic crust, together with associated sedimentary rocks that had been thrust up onto the land during earlier earthquakes, tens of million years ago. Andy Carter, a colleague from University College, London, who works on dating detrital minerals to determine their source, had booked tickets for us just before Christmas to fly to Calcutta and on to Port Blair, the capital of the Andaman and Nicobar Islands Union Territory, in February. My Norwegian colleague Rolf Pedersen, who specializes in U-Pb dating of zircons from rare ophiolitic granitic rocks, appropriately called trondhjemites, also came with us, as did our colleague from the Geological Survey of India, Pinaki Bandopadhya. After the earthquake and tsunami, it rapidly became apparent that the damage was enormous, not only in the Andaman and Nicobar Islands but across all the Indian Ocean countries. Andy telephoned me to say that he thought we should cancel the trip, but after a few e-mails to Indian friends and colleagues, we discovered that Port Blair had suffered only limited damage and the airport, lying inland, was unscathed, so we decided to go anyway. I was keen to see the after-effects of the earthquake.

Flying into Port Blair it was immediately apparent that severe topographic changes must have occurred during the earthquake. The first thing we noticed was that all the paddy fields and coconut plantations appeared to be under the sea. The land surface around Port Blair must have sunk by several metres, as the trees, fields, and houses were all engulfed every time the tide came in (Figure 13.2). As the Andaman Islands were quite hilly, there appeared to be relatively little damage except around the immediate vicinity of the shoreline. In Port Blair the sea-front road and pavement had been ripped up by the tsunami and it was evident that the entire land surface had sunk. Every high tide now lapped over the sea wall and waves crashed over onto the road. The pier had been destroyed and bizarrely, small boats appeared to have been washed inshore by several hundred metres.

We spent four weeks in the Andaman Islands, travelling throughout the three islands, looking at the ophiolite rocks, and their sedimentary cover, and also at

Figure 13.2 Subsidence of the land as a result of the Boxing Day 2004 earthquake; houses inundated by seawater at high tide, outside Port Blair, Andaman Islands.

the effects of the earthquake. I attempted to figure out how much the land surface had risen or fallen across the islands during the earthquake. At Chiriya Tapu, one unfortunate resident told me how he and his family had felt the earthquake, and ran outside only to see the tsunami approaching. They all escaped by running up the nearby hill and watched helplessly as the wave engulfed their home up to the ceiling of the first floor (Figure 13.3). The tide-line of the tsunami wave could be seen just below the roof. More amazing were the marks he had painted onto his living room wall of high tide in January after the earthquake. These were level with his armchair, nearly two metres above the previous high-tide mark at the end of his road. This could only mean that the land level had sunk by two metres during the earthquake, consistent with the similar amount that Port Blair seemed have suffered. The main road into Port Blair was now under seawater every high tide, almost cutting off the capital from its hinterland.

Figure 13.3 House at Chiriya Tapu, southern tip of Greater Andaman Island, showing high-water marks of the Boxing Day tsunami, and high tides following the 2004 earthquake.

After the earthquake, some reports in the local newspapers quoted local farmers saying that a new volcano had appeared on Baratang Island in the Middle Andaman. On Baratang we were taken to the site of the new 'volcano', which turned out to be a mud-volcano. It was bubbling along nicely with mudflows slowly making their way down its flanks. After the earthquake several mud volcanoes had appeared with methane gas venting from a small crater (Figure 13.4). When an electric storm swept across the islands, lightning struck and ignited the methane gas. I hitched a lift north from Port Blair and drove through the Jarawa Tribal Reserve on the way to Baratang. The Jarawa are one of several aboriginal tribes that have incredibly survived, scattered along the Andaman and Nicobar Islands. They are small, half-naked, dark-skinned aboriginal people with painted faces, and the men carried bows and arrows. The Indian government did not allow any Jarawa outside their 'reserve', partly to protect them from disease. When a

Figure 13.4 A small mud volcano in Middle Andaman Island.

group of five Jarawa leapt out of the forest and stopped our minibus to beg for food it seemed like a clash of history.

The Andaman Islands are home to at least five aboriginal tribes and why they ended up there remains a complete mystery. The Great Andamanese are now mostly integrated with the islanders. The Jarawas, with an estimated population of only about 300, inhabit the west coast of South and Middle Andaman, and now live in a designated 'tribal reserve'. The Onge tribe, numbering only 90, live on Little Andaman Island, and the Shompens, a Mongoloid race, inhabit the Great Nicobar Island. The most remote, isolated, and primitive of all are the Sentinelese, who live entirely isolated on North Sentinel Island.[5] They are very hostile and attack intruders with bows and arrows. They are hunter-gatherers who subsist on fruits, tubers, fish, wild pigs, and turtle eggs and are probably the world's only Palaeolithic people surviving today without any contact with the outside world. Nor do they want any. After the earthquake, the Indian airforce sent a helicopter to check on the status of the island and the Sentinelese rushed to the beach shooting arrows into the air to chase them away.

There was good evidence that a lot of the land along the east and south of the Andamans had sunk during the earthquake, but I wanted to see what differences had occurred along the west coast of the Andamans, the side of the islands closest to the subduction zone. Perhaps the most stunning example of change in land level

I experienced was in Mayabunder, at the north end of Middle Andaman Island. I hired a local fisherman, Titus Bwa, who had lived in Mayabunder all his life and regularly fished around all the islands and inlets in the vicinity. Titus owned a small guest house on the outskirts of Mayabunder and told me that early on the morning of 26 December he had felt the earthquake and immediately ran outside. Cracks appeared in his house and some of the balcony collapsed. He shouted to his family and guests to get out. Some 40 minutes after the earthquake, they suddenly saw the most incredible sight. The entire lagoon of Mayabunder harbour began to empty, leaving acres of raw muddy seabed exposed. Some canoes were swept out to the open sea. Fish were trapped and flapping about. Gulls and herons swooped down for the free meal suddenly presented to them. Everyone came out to look at the incredible sight; no-one had ever seen the ocean retreat like this before. Some people actually ran out to collect the flapping fish. Suddenly, shouts went up as the tsunami wall of water appeared on the horizon, heading rapidly towards the land. As it came nearer, it built up to a great wall of water that rammed against the sea-front. All ran quickly for higher ground. The lagoon filled up as suddenly as it had emptied. Altogether three or four tsunami waves battered the Andaman coast.

Raised Reefs and New Islands

I particularly wanted to take a look at the remote western coastline of the Middle and North Andaman Islands. Titus knew a narrow channel creek connecting the east and west coasts and said he had fished there regularly before. We loaded his dugout canoe with extra fuel, water, and food for the long journey through the mangrove creeks, and set off. Three hours later we broke out of the mangrove creeks to the west coast, and Titus stopped the engine and stared in amazement. He knew this area well, but now it had all changed. We discovered three new islands, coral reefs and sandbanks that had suddenly appeared, and the whole coastline appeared to have risen out of the ocean. With difficulty we found a way through the maze of sandbanks to the passage between Middle Andaman Island and Interview Island.

We motored down south along the coast of Interview Island as Titus took in these stunning changes to the landscape he knew so well (Figure 13.5). We anchored offshore a sandy beach where he used to land only now the beach was a hundred metres inland and acres of stinking, rotting coral reef lay between us and the beach. It was high tide and it soon became clear that the entire reef had been lifted up by several metres, beyond the highest tide mark. I swam ashore and walked over the reefs, astonished at our discovery. Several large coral heads had been upended, probably by the tsunami wave, but most of the reef appeared relatively undamaged. It had been lifted totally clear of the sea during the earthquake. I found hundreds of beautiful small conch shells, cowries, and sea-urchins, dead only for four weeks, starved of the ocean.[6] The entire reef let off an incredible smell of dead corals and rotting fish. This was the making of a new layer of carbonate rock, a slowly solidifying, lithifying coral limestone bed complete with an amazing fossilized reef. This was witnessing the process of fossilization and lithification of coral limestones that formed all the older Permian and Mesozoic limestones that are so wonderfully preserved in the mountains of Oman and the northern Himalayan Ranges of Ladakh, Zanskar, and Spiti.

It was clear to me after a month of travelling the length of the Andaman Islands that there was strong geological evidence of almost instantaneous uplift along the western margin by up to two metres, and subsidence along the south and eastern margin by up to three metres during the 26 December earthquake. At the southern tip of Great Nicobar Island, satellite photos have shown that subsidence of the land may have reached a maximum of 4.25 metres. The base of the Indira Point

Figure 13.5 Stranded coral reefs uplifted above high-tide mark, Interview Island off the west coast of Middle Andaman.

lighthouse built by the British over a hundred years ago at the southern tip of the Nicobar Islands, surrounded by coconut palms, was now several metres beneath seawater. This eastward tilting is similar to that recorded in Sumatra following the earthquake. Could fifty million years of great earthquakes like this eventually have produced the Himalaya?

Sentinel Islands

One of the most stunning examples of instantaneously uplifted islands was North Sentinel Island in the western archipelago of the Andamans. On our second trip to the Andaman Islands, we had flown from Madras (Chennai) to Port Blair directly over the Sentinel Islands. North Sentinel Island is a small, densely forested island surrounded on all sides by a fringing coral reef, on the western margin of the archipelago. From the plane I saw the uplifted coral reefs encircling the whole island, but at the time I did not know how much of the exposure was tidal or due to uplift of the island (Figure 13.6). Google images of North Sentinel Island before the earthquake showed a narrow white sand beach at high tide. Images from the European Space Agency Proba satellite taken in April 2005 after the earthquake show much more extensive uplift reefs completely surrounding the old beach (Figure 13.7). Images taken from the US-Japan ASTER (Advanced Spaceborne

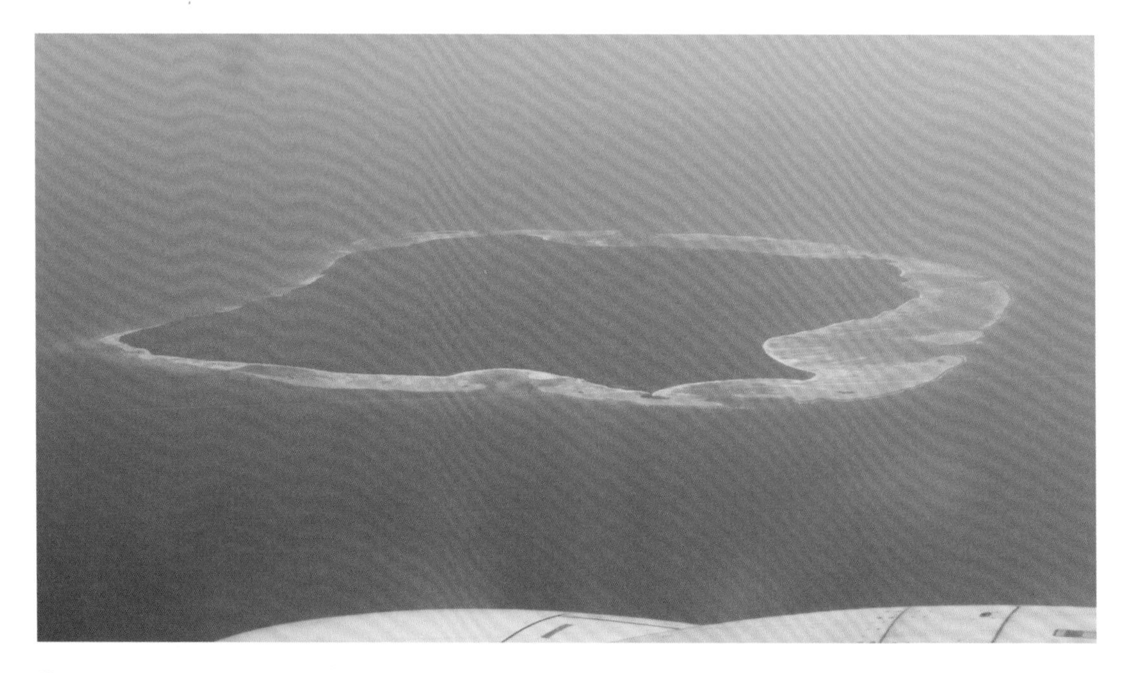

Figure 13.6 North Sentinel Island off the west coast of Andaman Islands showing recently uplifted coral reefs around the island, January 2005.

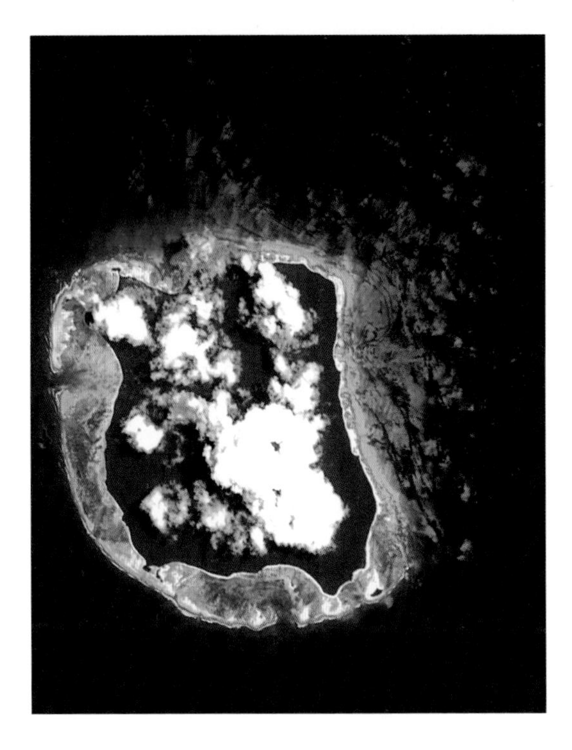

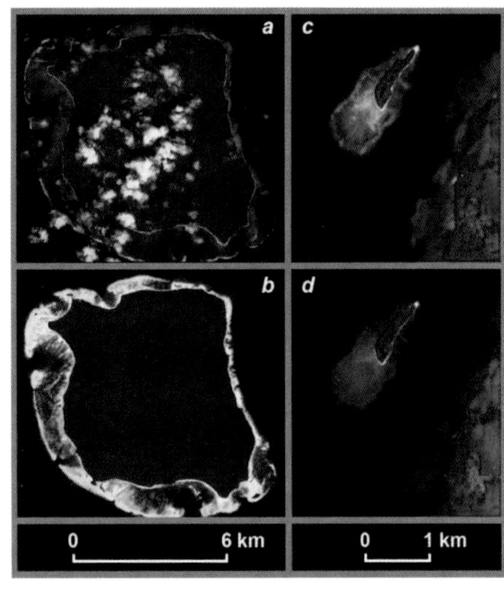

Figure 13.7 (*left*) Landsat satellite image of North Sentinel Island after the 26 December 2004 earthquake showing the newly uplifted land around the fringing reefs.
Figure 13.8 (*right*) Images taken from the US–Japan ASTER (Advanced Spaceborne Thermal Emission and Reflection Radiometer) instrument on NASA's Terra spacecraft show the emergence of the coral reefs surrounding North Sentinel Island and land sinking around a small island off the north-west coast of Rutland Island.

Thermal Emission and Reflection Radiometer) instrument on NASA's Terra spacecraft show the emergence of the coral reefs surrounding North Sentinel Island[7] (Figure 13.8). The tide was 30±14 centimetres lower in the pre-earthquake image (acquired on 21 November 2000) than in the post-earthquake image (acquired on 20 February 2005), requiring a minimum of 30 centimetres of uplift at this locality. Observations from an Indian coastguard helicopter on the north-west coast of the island suggest that the actual uplift could have been more like 1–2 metres. Much of the shallow lagoon may have uplifted in a similar way to that we witnessed on Interview Island. A small island off the north-west coast of Rutland Island, 38 kilometres east of North Sentinel Islands shows submergence of the coral reefs. The tide was higher in the pre-earthquake image (acquired on 1 January 2004) than in the post-earthquake image (acquired on 4 February 2005) requiring subsidence at this site. The pivot line between land emergence and land subsidence must therefore lie between North Sentinel and Rutland Islands (Figure 13.9).[8]

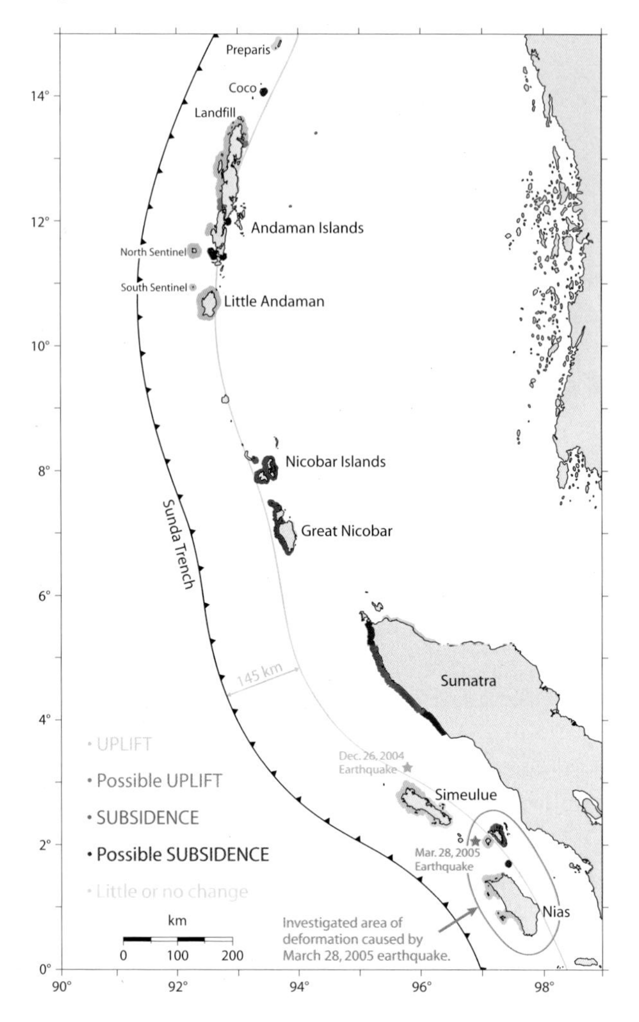

Figure 13.9 Map of the Andaman-Nicobar Islands and northern Sumatra and Nias islands showing the axis of tilt (yellow line). Islands to the west show uplift and emergence of land whereas land to the east shows sinking after the 2004 earthquake.

Nias Island Earthquake, 2005

Three months after the devastating Sumatra earthquake on 28 March 2005, another huge earthquake of magnitude 8.7 rocked the Nias Islands, south of Sumatra, only 50 kilometres along the plate boundary east of the previous earthquake.[8] Seismologists had widely predicted that the Sumatra 2004 earthquake should have taken up most of the pent-up strain along the convergent oceanic trench and that another large earthquake was unlikely to happen for a long time; now, only three months later, their predictions were literally blown out of the water. The Nias Island earthquake also occurred along the same thrust fault, originating at a similar depth to the Boxing Day quake. This time the rupture unzipped 400 kilometres along the trench plate boundary to the east, towards Java, Lombok, and Bali. Nias Island was known as one of the best surfing localities in the world. During the Boxing Day 2004 earthquake ten-metre-high waves pounded the island and more than 120 people were killed. The 28 March 2005 earthquake was initially thought to be a very powerful aftershock, but it is now known to be the second most powerful earthquake ever recorded in Indonesia and amongst the top ten worldwide. More than 800 people were killed on Nias after this second earthquake, one tenth of the entire population of the island.

Measurements from coral atolls and GPS have revealed up to three metres of uplift of the land surface along some of the outer islands and one metre of subsidence to the north-east of Nias. The ENVISAT satellite passed over Nias Island before and after the earthquake and comparison of the satellite synthetic aperture radar (SAR) images has revealed more astonishing details of the ground motions.[9] The entire Nias Island, together with Banyak, a small island north of Nias, rose up to two metres during the earthquake. Ten new islands appeared out of the ocean, the largest up to 1.5 kilometres across. The tsunami resulting from the Nias earthquake was only small compared to the devastating one on Boxing Day 2004. Most of the wave spread southwards across the southern Indian Ocean to dissipate in the wild seas of the Southern Ocean.

Indonesia continued to shake with more devastating earthquakes. On 10 April 2005, a 6.8-magnitude earthquake at 30 kilometres depth occurred beneath Siberut Island (the largest of the Mentawai group of islands) off western Sumatra. This sent another tsunami out into the Indian Ocean, but thankfully, like the Nias tsunami it was directed more to the south-west into the Southern Ocean. On 5 July 2005 Nias Island was once again rocked by an earthquake, this time 'only' a 6.7-magnitude earthquake, but luckily this one produced no casualties and no tsunami. Two earthquakes, a 4.6 event on 26 October 2005 and a 6.5 event on 19 November struck Simeulue Island. Aftershocks continued to occur all the way along the Sumatra, Nicobar, and Andaman plate boundary, some with magnitudes as high as 7.2. The most recent large earthquake was the 7.6 magnitude 30 September 2009 earthquake that occurred offshore southern Sumatra and resulted in over 1,300 deaths, displacement of more than a million people, and a lot of destruction in and around Padang city.

Can Earthquakes Trigger Volcanic Eruptions?

The worrying aspect of both the Sumatra and the Nias Island earthquakes was that above the Sumatra–Java length of the subduction zone lies a string of highly explosive volcanoes, including the infamous Krakatau volcano, west of Java. These

Indonesian volcanoes regularly explode with catastrophic consequences. The Krakatau eruption of 27 August 1883 created several tsunamis of its own which swept the shores of both eastern Sumatra and western Java.[10] It is thought that over 36,000 people lost their lives, most from the tsunami that followed the eruption. Ash that erupted from Krakatau was blown three times around the planet and formed a blanket, lowering the surface temperature by as much as 3°C. Deep red sunsets were observed for several years after the Krakatau eruption and the colours of the sky were so intense that they influenced some of the classic landscape painters of the time, such as Turner. A new island, Anak Krakatau, 'child of Krakatau', first appeared out of the caldera in 1927 and has been growing at a rate of 13 cm a week or more than 6 metres per year sporadically since 1950.

A similar eruption of the volcano Tambora, on the island of Sumbawa in April 1815 was the largest eruption in recorded history (Figure 13.10). Pyroclastic flows spread at least 20 kilometres from the caldera, and so much ash was spewed into the atmosphere that it produced a 'volcanic winter'. Widespread crop failures across the northern hemisphere in 1816, the 'year without a summer', are attributed to the ash cover from Tambora. Other historic earthquakes had occurred along Java, Lombok, Bali, and the islands to the east for at least the past 20 million years.

The most explosive volcanic eruption the world has ever witnessed occurred at Toba volcano in western Sumatra some 73,500 years ago.[11] The crater alone is 100 kilometres long and 35 kilometres wide and is presently infilled with a lake. The eruption lasted fourteen days but its effects lasted decades. It has been estimated that 2,000 km^3 of ignimbrite was erupted and about 800 km^3 which fell as ash was spewed out of the volcano, most of it blown west by the prevailing

Figure 13.10 The volcanic caldera of Mount Tambora volcano on the island of Sumbawa, Indonesia.

winds. Up to 600 metres thickness of ash was deposited near the vent. The ash covered over four million km^2 and the winds carried the ash around the entire globe several times. Ash from the Toba eruption has been found in Greenland ice cores and at the same stratigraphic level all over the world. The bright ash reflected sunlight and, together with volcanic aerosols like sulphur, impeded solar radiation for up to six years. The drop in temperature resulted in an 'instant ice age' or a so-called 'volcanic winter' that lasted six years and may have driven early humans (Neanderthals and the dwarf *Homo floresiensis*) to the brink of extinction.[12] The Toba eruption was two orders of magnitude greater than the largest volcanic eruption in historic times, that of Mount Tambora, also in Indonesia in 1815, which killed at least 71,000 people and also resulted in a volcanic winter.[13] Toba, like Yellowstone in the western US, is a supervolcano capable of global catastrophe. Supervolcanos are those capable of producing eruptions in which the ejecta exceeds 1,000 cubic kilometres and which have a magnitude of 8 on the Volcanic Explosive Index, the highest on the scale. It has been estimated that Yellowstone eruptions on this scale occur roughly at 600,000-year time intervals. The last eruption there was 640,000 years ago.

The Toba eruption blew itself out, but who knows when its magma chamber will fill up again and start to simmer? The timing of a volcanic eruption, like earthquakes, is difficult to predict with certainty. Several prominent volcanoes near well-populated cities are now seismically monitored continually with GPS and geophones recording earth movements. The instruments can tell us when a magma chamber beneath the crater is expanding and could be ready to blow, but the volcano will sometimes blow without any warning. In January 1993 Cerro Galeras, a huge Andean volcano in Colombia, suddenly exploded without any warning, killing nine geologists who were in the crater taking samples of gas and lava.[14] One of those killed was Geoff Brown, a leading British volcanologist who was the head of my old department of Earth Sciences at the Open University where I worked for my PhD.

Earthquakes also cannot yet be predicted with accuracy. We have a pretty good idea of where one is likely to occur, but not when. GPS and satellite-derived InSAR observations tell us the precise amount and direction of movement of the surface of the Earth. Geologists can now measure the build-up of strain and predict where parts of faults are likely to break, but as yet we cannot predict exactly when the earthquake will occur. Could the giant earthquakes ripping the subduction zone 30 kilometres deep beneath the supervolcano Toba and the active and growing Krakatau actually trigger another eruption? Toba's magma chamber must have been shaken by the earthquake but when would it next erupt? . . . Nobody knows.

Barren Islands Volcanic Eruption

In May 2005 I was back in Oxford when I received an e-mail from my colleague in Calcutta: Pinaki Bandopadhya alerted me to the fact that the Barren Island volcano had suddenly started to erupt (Figures 13.11 and 13.12).[15] Barren Island and Narcondum Island are two volcanoes about one hundred kilometres east of Port Blair in the Andaman Sea and are the northern extension of the great Java–Sumatra island arc system of active volcanoes. Along the same chain in Indonesia are hundreds of volcanoes including the giant supervolcano of Toba in Sumatra, and the active volcanoes of Krakatau between Sumatra and Java, Gunung Merapi, Semeru, Arjuna, and Bromo in Java, Gunung Agung in Bali, Gunung Rinjani in Lombok, and Kelimutu in Flores.

Barren Island volcano has erupted sporadically in the past, most recently in 1803, 1991, and 1995. Eye-witnesses reported 'violent spirals of dark smoke and gas' gushing out from the north-east subsidiary vent up to a height of more than a kilometre; fire fountains rose to a height of fifty metres or so on 16 May 1991.[16] Subsidiary craters had all blown out and merged with the main crater by the

Figure 13.11 The eruption of Barren Island volcano east of the Andaman Islands, May 2005. Photo courtesy of Pinaki Bandopadhya.

Figure 13.12 Panorama of Barren Island showing the recently erupted lavas flowing into the Andaman Sea.

following week. The most recent eruption occurred in 1995 when lava from the cone reached the sea along a 5-kilometre-long channel. The lavas erupted from the Barren Island volcano; basaltic andesites, that were remarkably similar to the Cretaceous lavas we had mapped along the northern part of the Kohistan island arc in North Pakistan.

The Sumatra and Nias Island earthquakes were incredible geological events. It was quite amazing to see at first hand the ground motions associated with the earthquake: to witness new islands rising out of the ocean, and land subsiding below the waves. They acted as a powerful reminder that these instances of sudden catastrophes are but a blink of the eye in the normal time scales geologists are used to working with, millions of years. Using Charles Lyell's famous principle of Uniformitarianism, which states that the 'present is the key to the past', we could deduce that the rocks we had seen and mapped in the mountains of Kohistan were once an island arc similar to the arc volcanoes we see today in the Andaman Sea. The principle can, however, also be turned on its head: we can use the upended, tilted, and eroded Cretaceous Kohistan volcanic arc in the western Himalaya to infer what the deeper levels of the Java-Sumatra island arc may look like. Such is the beauty of geology.

CHAPTER 14

The Making of the Himalaya, Karakoram, and Tibetan Plateau

There are known knowns. These are things we know that we know. There are known unknowns. That is to say, there are things that we know we don't know. But there are also unknown unknowns. There are things we don't know we don't know.

Donald Rumsfeld, 2001

My quest to figure out how the great mountain ranges of Asia, the Himalaya, Karakoram, and Tibetan Plateau were formed has thus far lasted over thirty years from my first glimpse of those wonderful snowy mountains of the Kulu Himalaya in India, peering out of that swaying Indian bus on the road to Manali. It has taken me on a journey from the Hindu Kush and Pamir Ranges along the North-West Frontier of Pakistan with Afghanistan through the Karakoram and along the Himalaya across India, Nepal, Sikkim, and Bhutan and, of course, the great high plateau of Tibet. During the latter decade I have extended these studies eastwards throughout South East Asia and followed the Indian plate boundary all the way east to the Andaman Islands, Sumatra, and Java in Indonesia. There were, of course, numerous geologists who had ventured into the great ranges over the previous hundred years or more and whose findings are scattered throughout the archives of the Survey of India. These were largely descriptive and provided invaluable ground-truth for the surge in models that were proposed to explain the Himalaya and Tibet. When I first started working in the Himalaya there were very few field constraints and only a handful of pioneering geologists had actually made any geological maps. The notable few included Rashid Khan Tahirkheli in Kohistan, D. N. Wadia in parts of the Indian Himalaya,

Ardito Desio in the Karakoram, Augusto Gansser in India and Bhutan, Pierre Bordet in Makalu, Michel Colchen, Patrick LeFort, and Arnaud Pêcher in central Nepal. Maps are the starting point for any geological interpretation and mapping should always remain the most important building block for geology.

I was extremely lucky that about the time I started working in the Himalaya enormous advances in almost all aspects of geology were happening at a rapid pace. It was the perfect time to start a large project trying to work out all the various geological processes that were in play in forming the great mountain ranges of Asia. Satellite technology suddenly opened up a whole new picture of the Earth from the early Landsat images to the new Google Earth images. Global Position System satellites (GPS) were able to measure with millimetre precision distances between stations and their relative motions of the Earth's surface. Geophysical methods showed us new seismic images across whole mountain belts deep into the crust, and tomography imaged the temperature structure of the entire lithosphere. Earthquakes were monitored with increasingly greater precision. Geologists can now determine the magnitude, depth, focal mechanism, and rupture speed of an earthquake. It was possible to map out quite easily the plate boundaries in the oceans, but when it came to Tibet and indeed most of Asia the earthquakes showed that the continents were not always behaving in a rigid, plate tectonic fashion. Seismic experiments and magnetotelluric data showed that parts of the Tibetan crust were actually partially molten in certain places. The chemistry and isotopic compositions of volcanic rocks hinted at the nature and composition of the upper mantle and lower crust from which they were derived.

Advances in structural geology, metamorphism, and geochronology have enabled us to extract incredibly detailed pieces of information from rocks. Constructing balanced and restored cross-sections across the brittle-deforming upper crust enable us to place constraints on the amounts of crustal shortening across mountain belts. Advances in microstructural studies using fabric data enable us to interpret strain recorded within the rocks. Advances in technology, particularly in scanning electron microprobes, enable us to get precise chemical compositions from tiny mineral grains. Using these chemical compositions it was possible to work out the precise pressures and temperatures at which many of the metamorphic rocks formed. Thermobarometric data sets enable us to calculate pressure–temperature paths through time so that we can extract most of the burial and exhumation history of an individual rock. We can use recent advances in geochronology and mass spectrometry to date uranium- and thorium-bearing minerals in the rocks so that we can now put time constraints on exact points on the P–T path. Thermal ionization mass spectrometers (TIMS) date individually separated zircon or monazite crystals, whereas Laser-ablation mass spectrometry

(LA-ICPMS) and sensitive high-resolution ion microprobes (SHRIMP) can date crystals *in situ*, thereby linking the age with precise P–T conditions in the mineral assemblage. We can date a variety of minerals that have known closure temperatures using different isotopic systems to obtain precise cooling histories of the rocks. Zircons are known to have the highest isotopic closure temperature in the U-Pb system so that occasionally zircon may be the only mineral that records the 'age' of the original unmetamorphosed rock or protolith in a young metamorphic rock. Zircons and monazites are widely used to date magmatic intrusion events and 'peak' metamorphism. Titanite, rutile, and apatite record lower temperature closure times. The cooling of rocks in the middle temperature range can be ascertained using a dating technique called argon step heating on hornblende, and biotite and muscovite micas. Low-temperature thermochronology involves fission track dating of zircon and apatite crystals and cosmogenic dating on quartz. If all these minerals occurred in one rock sample then potentially the entire history of the rock could be pieced together.

Colliding Continents: The Himalayan View

So, where have these past several decades of work got us? Palaeomagnetic data shows that during Mesozoic times the Indian continental plate was part of the Gondwana supercontinent in the southern hemisphere. India rifted away from southern Africa, Madagascar, and Antarctica about 130 million years ago and travelled at high speeds (about 5 cm per year) northwards, pushed by newly formed oceanic spreading centres in the Indian Ocean, and pulled by a northward-dipping subduction zone beneath Asia, under which the Tethys Ocean was disappearing. The southern margin of Asia during this time was a magmatically thickened Andean-type margin dominated by large granite batholiths and explosive andesitic volcanoes. Between about 90 and 65 million years ago a series of ophiolites and island arc terranes were emplaced onto the northern passive margin of India, simultaneously with the complete Oman ophiolite and the fragments of ophiolites preserved along the western Pakistan borders and the eastern margin in the mountains of Burma and the Andaman Islands. About 50 million years ago the Neo-Tethys Ocean that separated India from Asia closed as the Indian continental crust collided with, and subducted deep beneath, the Asian margin. From the time of collision to this day India has continued to push north, indenting into Asia and creating the high topography and thick crust of both the Himalaya and Tibet (Figure 14.1).

The Himalaya are made from the scraped off, decoupled, strongly folded and thrust upper crust of the Indian plate, with a few uncommon thrust sheets of the remnants of Tethyan oceanic sedimentary rocks and ophiolites. About one

Figure 14.1 Oblique aerial view of the Himalaya viewed from the Space Shuttle, looking along the snow-capped mountains of Bhutan (foreground), Sikkim, Nepal, and India (distance). Photo courtesy of NASA

thousand kilometres of convergence between India and Asia has occurred in the past 50 million years. This includes the internal folding and thrusting seen in the upper crust Tethyan Himalaya, the internal ductile folding and thickening in the Greater Himalaya, and several hundred kilometres of underthrusting along the Miocene Main Central Thrust and the Pliocene–Pleistocene Main Boundary Thrust. No lower crust (Archaean and Palaeoproterozoic rocks of the Indian Shield) rocks are exposed anywhere in the Himalaya, so it must be inferred that the lower crust of northern India has decoupled and underthrust the Asian margin of Tibet. It is this underplating that appears to have been mainly responsible for the uplift of the Tibetan Plateau (Figures 14.2 and 14.3).

The Indian plate rocks of the Himalaya were intensely folded and thrust, processes that resulted in crustal shortening and thickening. Crustal thickening led to increased temperature, pressure, and metamorphism. The initial deep subduction of the leading edge of India led to ultra-high-pressure metamorphism and the formation of eclogite at depths of over 100 kilometres. The garnet- and pyroxene-bearing eclogites were originally basaltic sills emplaced within the Indian crust. As compression continued, the entire upper crust was transformed into metamorphic schists and gneisses leading to high-pressure kyanite-grade metamorphism around 35–30 million years ago and then high-temperature sillimanite-grade metamorphism between 24 and 17 million years ago. This phase was concomitant with widespread partial melting in a layer of the middle crust, as imaged on seismic reflection and refraction profiles across southern Tibet, and observed and mapped along the Great Himalaya. The final granite dykes that formed from melting of the gneisses were intruded up to about 13 million years ago.

The youngest thermal events are recorded in the two syntaxes, the north-west Nanga Parbat syntaxis in Pakistan, and the north-east Namche Barwa syntaxis in south-east Tibet. In both these areas, Indian basement gneisses went through the whole Cenozoic Himalayan thermal history and then continued with very high-temperature partial melting of the crust that happened during Pliocene and Quaternary times, producing the even younger sillimanite-, garnet-, and cordierite-bearing migmatite rocks. These amazing Nanga Parbat rocks, the youngest migmatites known anywhere on Earth, reached peak metamorphic conditions of 720°C and pressures of about 5 kilobars, equivalent to depths of

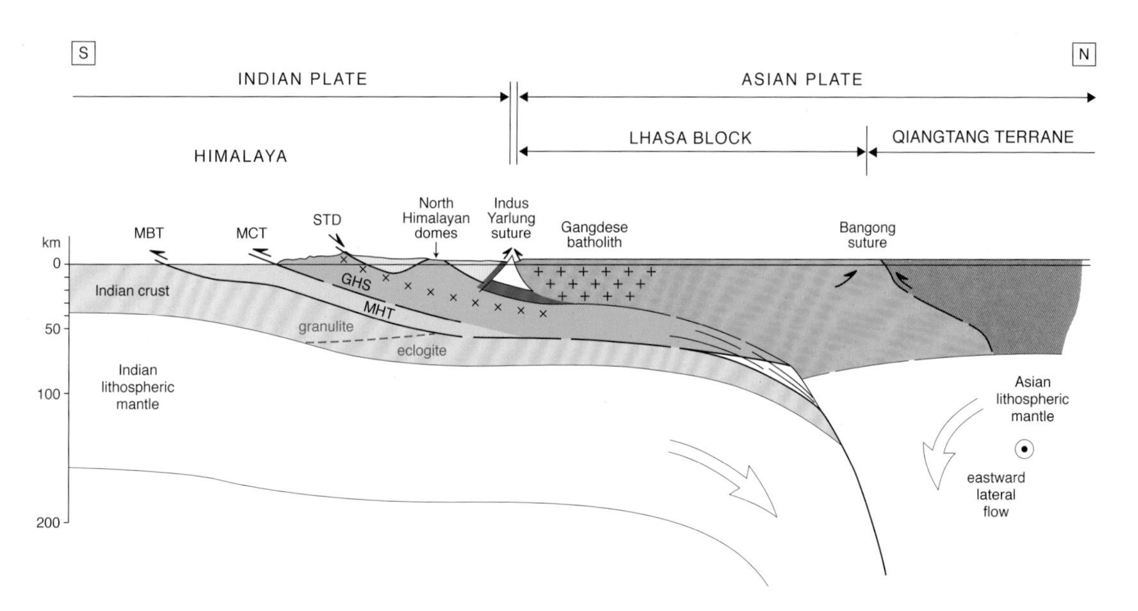

Figure 14.2 Geological section across the central Nepal Himalaya and Tibetan Plateau.

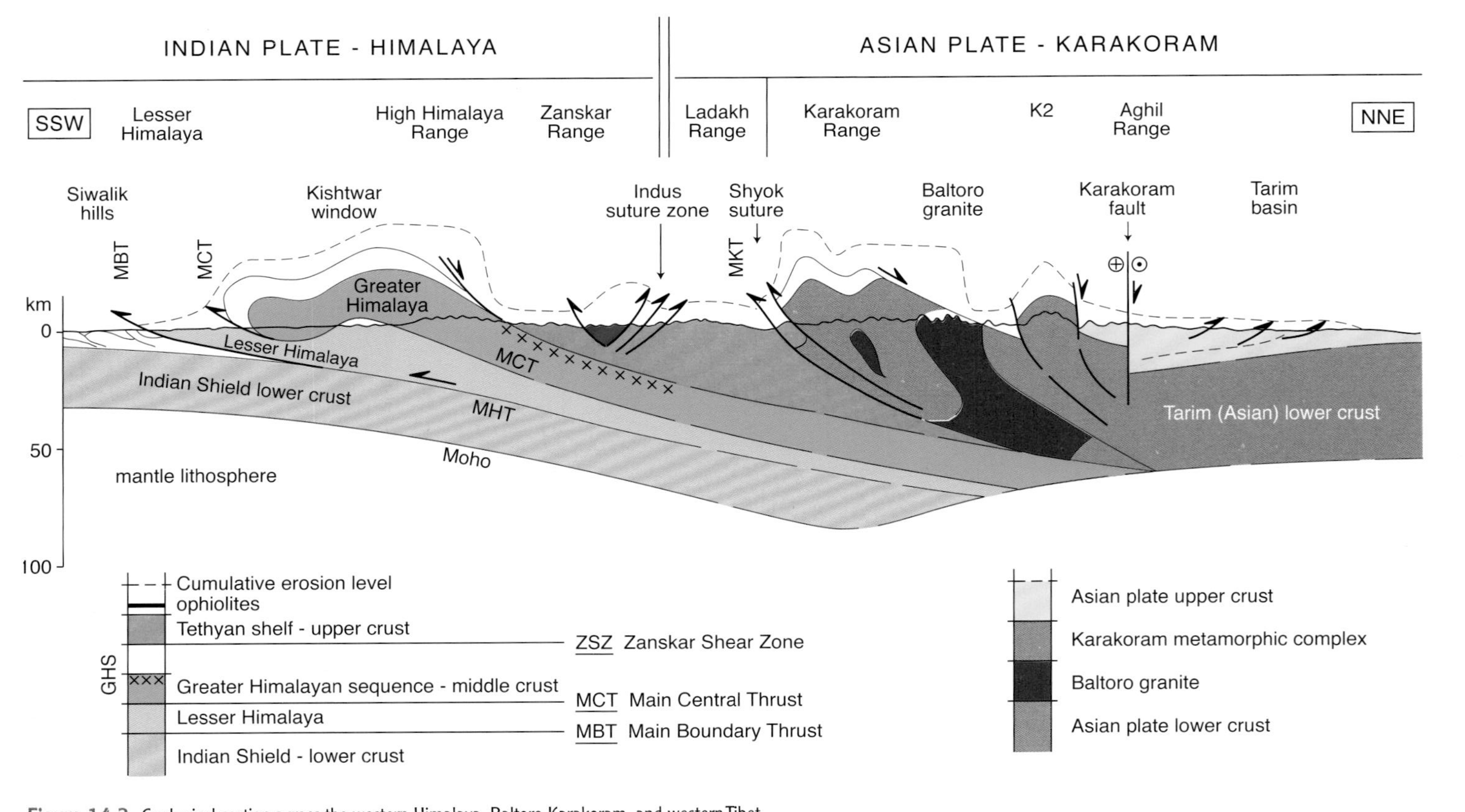

Figure 14.3 Geological section across the western Himalaya, Baltoro Karakoram, and western Tibet.

burial of up to 20 or 30 kilometres in the crust, between 1.7 and 1 million years ago, and are now at elevations of 6 to 7 kilometres on Nanga Parbat. The youngest U-Pb ages ever measured are monazite grains from a garnet, biotite, sillimanite melt vein that records ages as young as 790,000 years, and four xenotime crystals that have ages between 720,000 and 690,000 years old. These rocks record the highest exhumation and erosion rates known anywhere on Earth—more than 11–13 mm per year.

Crustal Melting and Channel Flow

The Channel Flow model was developed for the Himalaya using accurate geological constraints taken from mapping structures, thermobarometric profiles across the Himalaya, and the age dating of granites and metamorphic rocks.[1] The model also relied on linking the geology of the Himalaya with the deep crustal seismic profiles taken from seismic reflection and refraction.[2] Combining structural sections constructed from surface geology with these seismic sections in Tibet showed that the two major detachments mapped along the Himalaya—the Main Central Thrust and the South Tibetan Detachment—were continuous beneath southern Tibet and that these detachments sandwiched a mid-crustal zone that was partially molten with pockets of fluids (granites) at depths as shallow as 17–20 kilometres beneath south Tibet. Mapping isograds along Himalayan profiles has shown that inverted metamorphism—that is, decreasing metamorphic grade with depth—occurred along the Main Central Thrust zone, and right-way-up metamorphism along the South Tibetan Detachment. Structural studies showed that the mid-crust layer was moving south relative to the underthrusting Indian lower crust below and the passively riding, already shortened, Himalayan upper crust above. In one area in the far western part of the Himalaya in western Zanskar and eastern Kashmir, the metamorphic isograds above the Main Central Thrust could actually be mapped to join up with the isograds beneath the low-angle normal fault, the South Tibetan Detachment, proving that southward expulsion of the partially molten mid-crust layer was real. Dating the granite melts has shown that Channel Flow was operating during the Miocene between about 24 and 15 million years ago. The link between the Miocene channel frozen in along the crest of the High Himalayan peaks and the present-day partially molten middle layer beneath south Tibet makes for an exciting model. Maybe the pockets of melts imaged today beneath southern Tibet are Himalayan leucogranites in the making, formed at similar temperatures, pressures, and depths in the crust. The southward flow of these rocks might be transporting these young crustal rocks to the Himalaya of the future, in about 10 or 20 million years' time (Figures 14.4 and 14.5).

Figure 14.4 Kangchenjunga (8,586 m), the world's third highest mountain on the border of Nepal and Sikkim. The entire mountain is made of a series of young granite sills, the thickest granite sheets along the Himalaya. Granite melts like these lubricated the southward flow of the Greater Himalaya during the Miocene between about 21 and 13 million years ago. Photo courtesy of Steve Razzetti.

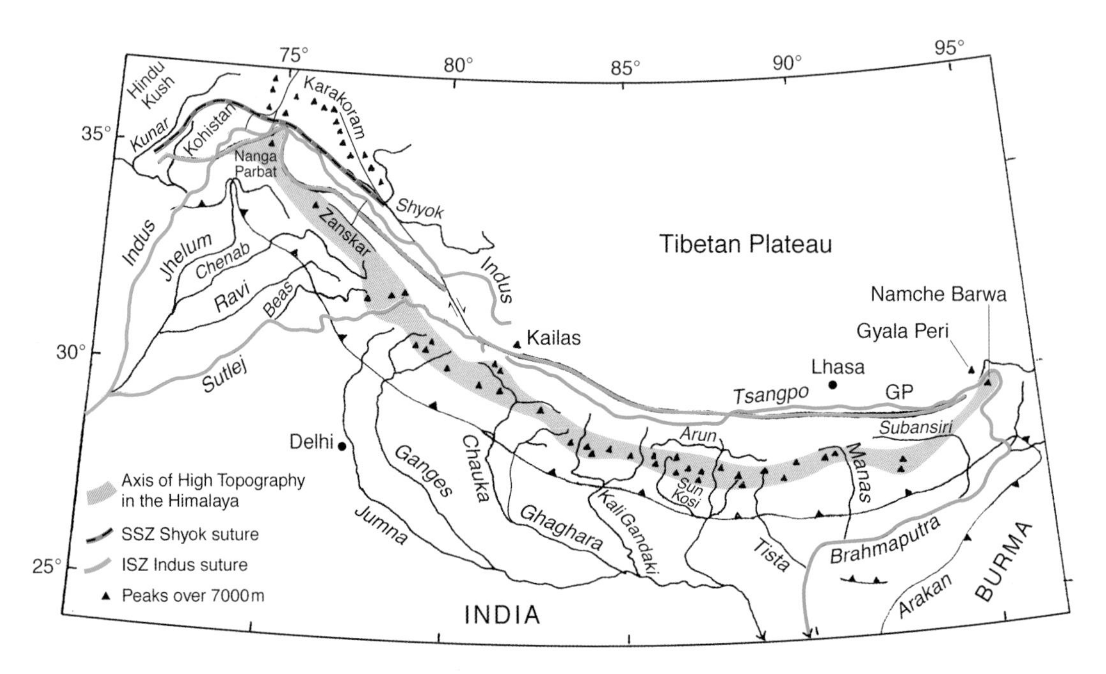

Figure 14.5 Map of the Himalaya showing the axis of high topography, prominent mountains, and courses of the three major antecedent rivers, the Indus, Sutlej, and Tsangpo Rivers.

Colliding Continents: the Tibetan View

Every model that has been proposed for the Himalaya, Karakoram, and Tibet is almost certainly wrong; some may be slightly useful, many are wildly inaccurate. Most geologists working in Tibet have found it difficult to understand how a feature like Tibet, with its double thickness crust and inherently hot, weak, mechanical strength could be maintained for a long period of time. Hence most workers thought that Tibetan uplift was a relatively young phenomenon, starting possibly 7 or as much as 15 million years ago.[3] The model involved progressive homogeneous shortening and thickening of the entire lithosphere, until the mantle became thermally unstable and dropped off to be replaced by hotter asthenosphere from below, presumably from around the margins of Tibet. This model was largely imagination and had no geological data to support it. As more and more ages of volcanic rocks across Tibet were published it was clear that hot mantle underlay the plateau from the time of collision 50 million years ago until about 11 million years ago, and after that only in the north along the Kun Lun. The underthusting of Indian cold lithosphere progressively shunted the hotter mantle northwards until only in the far north, the Kun Lun was underlain by hotter mantle. The volcanoes in the Kun Lun are the youngest in the plateau, and there is also one igneous intrusion of interesting crustal melt granite at the isolated large mountain of Ulugh Muztagh.

The two extreme models invented to describe the structure of Tibet are both incorrect, or at least are far too generalized. The homogeneous crustal thickening model[4] to explain the double thickness of crust should involve large amounts (about 1000 kilometres) of post-collision (younger than 50 million years) crustal shortening by the folding and thrusting of Tibetan crust. In fact almost all deformation and metamorphism in Tibet is old, predating the collision, mainly Triassic in age, and evidence of Tertiary folding and thickening is almost completely lacking. The high plateau can be compensated by a thick crustal root (Airy isostasy) or by a hot upper mantle (Pratt isostasy). We know from seismic evidence that Tibet has a thick crustal root (between 70 and 90 kilometres' thickness), and we now know that the upper mantle under most of the plateau is cold, probably as a result of underthrusting of cold Indian plate lithosphere beneath.[5]

The continental extrusion model for Tibet[6] is also far too extreme. We know from the distribution of earthquakes that Tibet does not behave as a rigid block, or even as a number of smaller blocks. We know that the bounding strike-slip faults of Tibet do not accommodate thousands of kilometres of relative motion, but more like 100–150 kilometres. It is extremely unlikely that these faults extend down through the entire crust into the mantle, and it is certainly not possible for the faults to produce large granite melts or metamorphism by frictional heating. Probably the most contentious sorts of models are not models at all, but simulations dressed up as models.[7] If all the bounding constraints from geological mapping, structure, P–T conditions, and ages are input into the model, then the model simply proves what the author tells it to. There is no point in simulating in a numerically simplistic way what we already know from geology and field constraints. Some of these numerical models might be mathematically clever, but they are mostly useless for interpreting the real world. As Ernest Rutherford put it, 'If your experiment needs statistics, you ought to have done a better experiment', or a quotation attributed to Fred Vine, 'I never touch statistics, I just deal with the facts.'

Contrary to the model of sudden uplift of the Tibetan Plateau 7 million years ago, our studies in the Karakoram and western Tibet suggest that the southern margin of Asia had thick crust, resulting in high-grade metamorphism, and was topographically high as far back as 65 million years ago, before the collision.[8] Initially the thickening was a result of Andean-type crustal thickening processes, and after about 50 million years ago, the result of Indian lower crust underthrusting the plateau. In eastern and northern Tibet there was good evidence of even older, Triassic–Jurassic crustal thickening during the Indosinian mountain-building event, with widespread intense folding and thickening of the Triassic sedimentary rocks. Andean-type processes along southern Tibet involved magmatic thickening of the crust by the intrusion of subduction-related granite batholiths and eruption of andesitic volcanic rocks. Contrary to the continental extrusion model, our field-based studies suggested that the large strike-slip faults of Tibet cut across all earlier formed granites and metamorphic rocks, had only limited geological offsets, and could not extend all the way down to the upper mantle.[9]

Jelly Sandwiches, Crème Brûlées, and Custard Pies

Having now criticized most models presented to explain the Himalaya and Tibet, it must be admitted that some models are useful, even as pictures to imagine whilst carrying out all-important mapping. Geophysicists have a sweet tooth, and the two competing models to explain the rheology and strength of the lithosphere use Tibet as a prime example. In one camp are those who think that the strength of the lithosphere resides in the upper crust 'floating' on a ductile, deforming, weak lower crust, with a strong, rigid, upper mantle driving the plate motion, the so-called 'jelly sandwich model.'[10] In the other camp are the people who believe the strength of the lithosphere resides completely in the upper crust, which rides on a weak lower crust and weak mantle, the so-called 'crème brûlée model.'[11] Under particular contention is the location of several deep earthquakes along the Himalayan region where the Indian plate buckles down beneath south Tibet. How temperatures deep under parts of the Himalaya can be cold enough to allow earthquakes to occur remains a big mystery. Earthquakes are generally restricted to the brittle deforming upper crust, where temperatures are below about 450°C and where rocks deform by folding and faulting, or brittle fracture. When temperatures are higher than 500–550°C, feldspars and quartz deform by plastic or fluid-like flow, and rocks will deform by ductile deformation. Any geologist who has mapped metamorphic belts across the world knows that rocks at high temperature and pressure deform in a ductile fashion, but treating the entire Tibetan lithosphere like this is a wild over-simplification and completely misleading.

Whereas the rheological arguments between those who prefer the jelly sandwich and those who prefer crème brûlée are destined to go on for a long time, the Himalaya Channel Flow model is seeded in mapping along the Himalaya combined with pressure–temperature and time calibrations. Here the exhumed Miocene mid-crustal partially molten channel can be seen and mapped and sampled along numerous Himalayan valleys. The Channel Flow 'custard pie model' is rooted in reality so is less contentious.[1]

Capturing Rivers

Some of the more intriguing models proposed seek to explain the geomorphology and river systems of the Himalaya, Tibet, and South East Asia. Rivers can be diverted along active faults and can capture drainage by headwall incision, cutting progressively back into the highlands until they meet another river system. As the Tibetan Plateau rose, rivers were formed flowing off the high plateau in all directions. As we saw in Chapter 11, the three great rivers of south Asia—the Indus, Sutlej, and Yarlung Tsangpo—all of which rise from south-west Tibet near Mount Kailas, flow across the plateau for some distance before cutting right through the high peaks of the Himalaya. These are antecedent rivers, ones whose origins pre-date the rise of the Himalaya. The Indus 'molasse' includes conglomerates deposited by rivers and silt deposits in lakes along a basin, the Indus basin that follows the suture zone in Ladakh and across south Tibet. These rocks are the earliest deposits formed by great river systems, the ancestral drainages of the Indus and Yarlung Tsangpo River systems. In Ladakh, the dating of the youngest granites beneath the Indus molasse basin shows that the basin initiated about 46 million years ago.[12] We have already seen in Chapter 5 how the Karakoram fault has captured and diverted the antecedent Indus River for over 120 kilometres (Figure 14.6).

Most other Himalayan rivers were initiated later, during the initial rise of the Himalaya, probably at least 35–45 million years ago. These rivers, including the

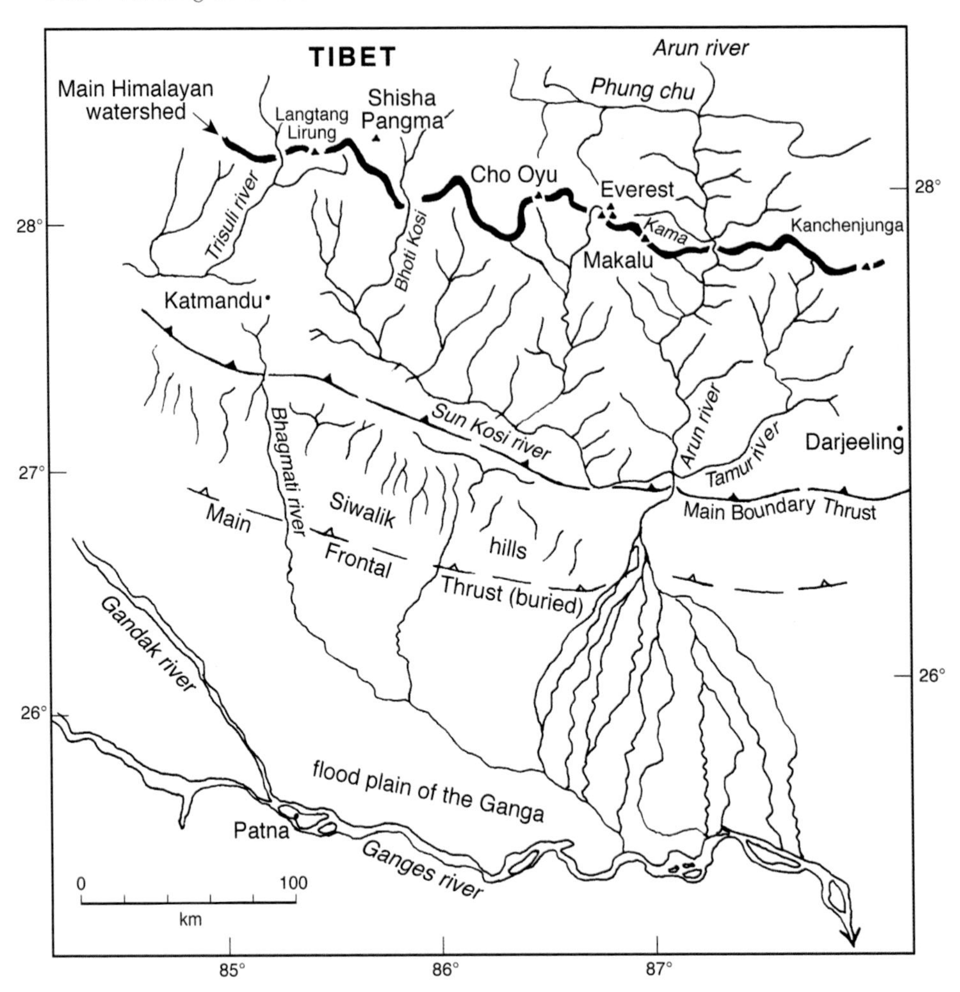

Figure 14.6 Map of the river drainage of eastern Nepal. The Arun River is an antecedent river that cuts through the main Himalayan divide (thick line) between Makalu and Kangchenjunga. The Arun was initiated prior to the uplift of the main Himalaya and has incised progressively northwards into Tibet. The Trisuli and Bjote Kosi Rivers also cut through the main axis of uplift. Younger river systems such as the Tamur and Sun Kosi were formed after the initial rise of the Himalaya.

Beas and Ganges in India, the Karnali, Kali Gandaki, Burhi Gandaki, Trisuli, and Sun Kosi Rivers in Nepal, and the Manas River in Bhutan, were all initiated during the rise of the Himalaya. As the mountains rose higher, the rivers gradually cut back into their headwalls, incising northwards with time. Many of these rivers actually cut across the axis of the highest mountains today, suggesting that they were all present at least for the past 20–30 million years. Many of these Himalayan rivers are very close, in geological time, to cutting all the way north to the Yarlung Tsangpo River which flows west to east across the southern Tibet trans-Himalaya. Lawrence Wager first noted that some large Himalayan rivers such as the Arun may have been antecedent, formed prior to the uplift of the Himalaya.[13] The Arun River rises in south Tibet and cuts right through the axis of the highest peaks to the east of Makalu. The headwaters of the Arun are only 10 kilometres or so from the Yarlung Tsangpo today. When the Arun River incision does break through to the Yarlung Tsangpo there will be a sudden and drastic reorganization as the entire western Tibet drainage will pour south through Nepal, diverting well over half the drainage of the Yarlung Tsangpo away from the deep gorge of the Namche Barwa Big Bend, the east Himalayan syntaxis region discussed in Chapter 10.

The three great rivers of East Asia—the Yangtze, Mekong, and Salween—all rise on the Tibetan Plateau and run close together, only tens of kilometres apart for over 300 kilometres in deeply incised gorges, before diverging and flowing into different seas. It is possible that the courses of these rivers actually represent strain markers, squeezed close together during the northward indentation of India into Asia.[14] We have seen how many Himalayan rivers are incising rapidly northwards, and in the near future will most likely cut north all the way to capture the east-flowing Yarlung Tsangpo drainage. It has been proposed that the Yarlung Tsangpo River originally flowed east into the Yangtse, before it was captured first by the Red River then by the Mekong and possibly also the Salween River, before finally being captured by the Brahmaputra River (Figure 14.7). This intriguing new story was first proposed by Marin Clark and her colleagues from MIT in the US, from a regional compilation of drainage histories.[15]

The Red River has a relatively short course today, rising from the peaks around Erhai Lake in Yunnan and flowing south-east across North Vietnam to the Gulf

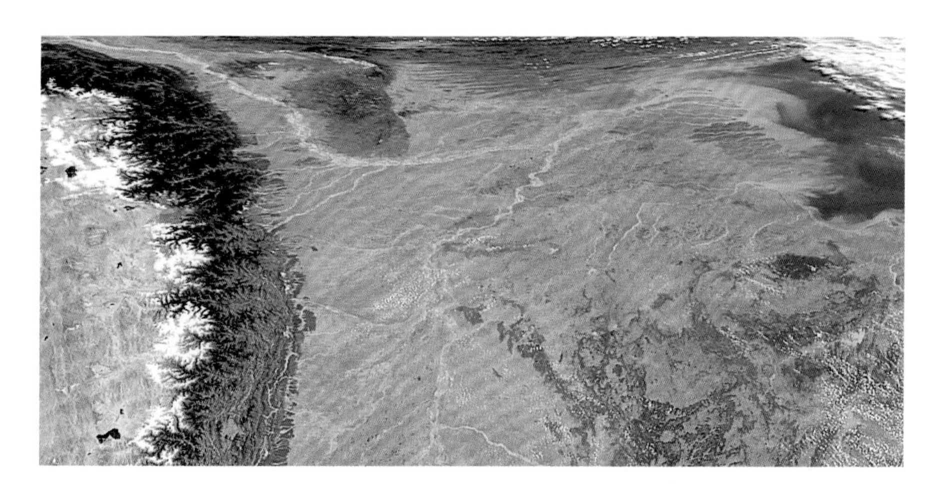

Figure 14.7 Oblique view of the eastern Himalaya and Indo-Burma ranges showing the Ganges River (below) and the Bhramaputra River (top left) merging into the delta region of Bangladesh and flowing south into the Bay of Bengal. Photo courtesy of NASA.

Figure 14.8 View westwards across the mountains of Burma and the eastern Himalaya. The Sagaing fault runs from the east Himalayan syntaxis region south to the Andaman Sea. The south-flowing Irrawaddy River has captured the Chindwin River as the fault was dragging it southwards. Photo courtesy of NASA.

of Tonkin and the South China Sea. The delta region near Hanoi is more than 50 kilometres wide, and the amount of sediment deposited in the delta is far more than one would expect from the drainage basin today. The sediment was mostly derived from eastern Tibet and was transported along the palaeo-Yangtze and the palaeo-Mekong Rivers, which both originally drained into the Red River. At some point the upper part of the Yangtze River was captured away from the Red River, shutting off the sediment source from Tibet. Similar river capture events probably occurred along the Mekong, Salween, and the Irrawaddy Rivers, but the most spectacular capture was the Brahmaputra which incised north through the Namche Barwa massif and captured flow from the Yarlung Tsangpo (Figure 14.8).

Indian Ocean Monsoon

With the rapid rise of climate change in public perception, it became fashionable to link mountain belts that showed rapid uplift, erosion, or exhumation rates to periods of inferred climate change in the geological record. Accurate climate records in Asia go back only a few decades. Further back in time we are reliant on sedimentary and isotope data collected from cored sediments in lakes or ocean basins. Relating geological events such as the India–Asia collision or the uplift of the Tibetan Plateau to periods of past climate change involves numerous assumptions, and thus is highly controversial. The greatest climatic changes in Asia arose from the onset of the Indian Ocean summer monsoon, the annual winds that blow northwards across India during the summer months, bringing torrential rains and making the sub-continent habitable. Equatorial winds flow from east to west across the Indian Ocean and are then diverted northwards along the south coast of Arabia and across India. The air is laden with moisture that precipitates as rainfall across India. The ground-hugging winds rise along the southern slopes of the Himalaya where most of the rain is dumped. North of the Himalayan watershed, southern Tibet lies in the rain-shadow of the monsoon winds. Whereas the monsoon used to be very predictable, beginning in south India around mid-June, reaching the eastern Himalaya by early July, and the western Himalaya by late July, recent years have shown increasingly unpredictable patterns.

In normal Hadley Cell circulation, equatorial heating causes warm air to rise, and then flow north towards the pole at altitude, where it becomes deflected to the east. This is known as the Coriolis Effect and is caused by the spin of the Earth. Dry air subsides and is returned southwards towards the equator. When the Tibetan Plateau was thrust up it created a strong summer heat source causing air to rise over the plateau and reverse the Hadley Cell. Rising air over Tibet sucks warm, moist air in from the Indian Ocean resulting in the summer monsoon. The onset of the Indian Ocean summer monsoon winds is imprecisely known, but the monsoon is known to have strengthened considerably approximately 7 million years ago.[3] Maureen Raymo from MIT and Bill Ruddiman from the University of Virginia suggested that global cooling in the Cenozoic which led to the growth of large continental ice sheets may have been caused by the uplift of the Tibetan Plateau.[16] The theory proposes that a tectonically driven increase in chemical weathering, which involves carbon dioxide from the atmosphere combining with water to form carbonic acid, caused a drawdown of atmospheric CO_2, leading to global cooling. Tectonic uplift of Tibet leads to a positive feedback effect: the higher the plateau rises, the greater the monsoon rains, the more chemical weathering, and the greater the global cooling. The reduction in atmospheric CO_2 led directly to the Pleistocene Ice Ages.

Evidence from oxygen, strontium, and osmium isotopes appears to corroborate this model. Oxygen isotope ratios ($^{18}O/^{16}O$) are directly related to global cooling. As global cooling occurs, oxygen isotope ratios in marine limestones rise. The oxygen isotope record shows a general increase over the last 55 million years, since the India–Asia collision, with the most rapid changes occurring around 36 and 15 million years ago, corresponding very closely to the timing of Himalayan metamorphism and increased erosion along the Himalaya. Silicate weathering results in high strontium ratios ($^{87}Sr/^{86}Sr$), which are a good proxy for erosion rates. Himalayan granites have some of the highest $^{87}Sr/^{86}Sr$ ratios of any rock in

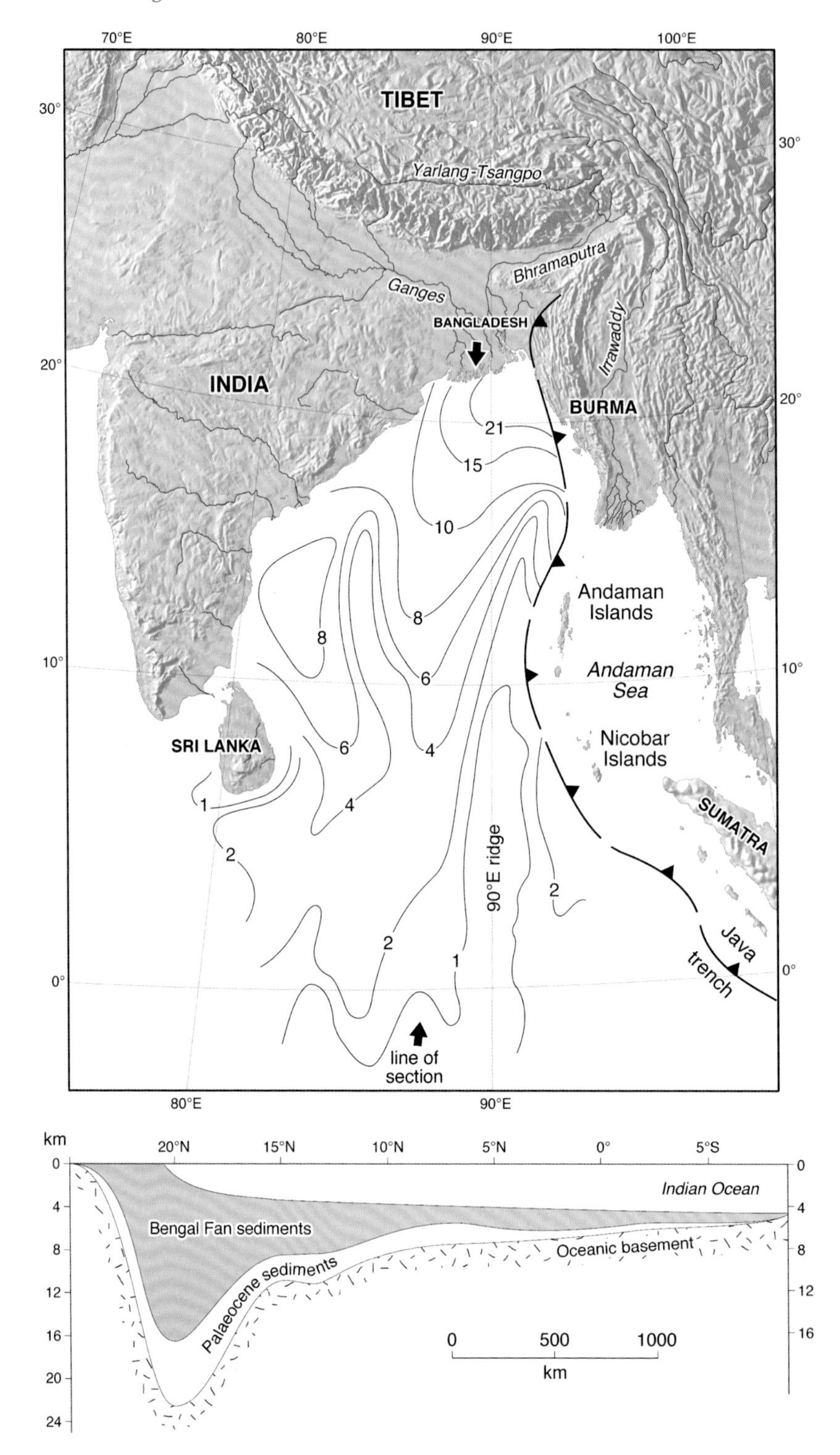

70°E
80°E
90°E
100°E
30°
20°
10°
0°
TIBET
Yarlang-Tsangpo
Bhramaputra
Ganges
BANGLADESH
Irrawaddy
INDIA
BURMA
21
15
10
8
6
4
2
1
Andaman Islands
Andaman Sea
Nicobar Islands
SRI LANKA
SUMATRA
90°E ridge
Java trench
line of section
km
20°N
15°N
10°N
5°N
0°
5°S
Indian Ocean
Bengal Fan sediments
Oceanic basement
Palaeocene sediments
0
4
8
12
16
20
24
0 500 1000
km

the world. Strontium ratios in the oceans began to increase rapidly about 40 million years ago, corresponding closely to the timing of Himalayan metamorphism and melting, the creation of thick crust and hence higher erosion.[17] Rapid erosion is coupled with high rainfall along the Himalaya. The highest rainfall recorded is 4,000 mm per 160 inches per year in the Assam–Meghalaya region of north-east India. An incredible 18 metres of rain has been recorded in one year in Meghalaya, almost all of it falling in the three months of the summer monsoon (June–July–August).

Accompanying all this monsoonal rain is high erosion. Material eroded from the high peaks along the Himalaya is transported via rivers to the Ganges delta and then dumped into the Bay of Bengal. The Bengal Fan extends from the coast of Bengal at least 2,500 kilometres south into the Indian Ocean. Offshore Calcutta the sediments derived from the Himalaya reach over 18 km thickness, and must be undergoing some form of burial metamorphism today. The cycle of sedimentation, tectonics, metamorphism, uplift, erosion, deposition, and burial continues in a continuous geological rhythm (Figure 14.9).

The Cosmic Mandala

'In a hundred ages of the gods I could not tell thee of the glories of the Himalaya', say the Puranas, the 'ancient stories', myths, and legends of India, transmitted orally and first recorded in texts dating back before the Mahabharata. It is not surprising that all the Asian religions worship and revere the mountains. The Himalaya, the 'abode of snow', was where the unknown forces lived that controlled everything on Earth. The old myths and legends are brought back to life in the masked demon dances and festivals of the Buddhist Himalaya. The ancient Bon religion still lives on in remote desert valleys of western Tibet. Every summer Hindu and Buddhist pilgrims walk the sacred '*kora*', the six-day circumvention of the holy Mount Kailas in Tibet. Hot springs throughout the Himalaya and Tibet are held in reverence and worshipped. *Saddhus* still meditate cross-legged on icy boulders at Gaumukh, the snout of the Gangotri Glacier, and the source of the sacred Ganges.

The Cosmic mandala, that spiralling *thanka* hanging in *dzongs* and *gompas* across the Himalaya, records 'the becoming, being and vanishing of all worlds and the impermanence of all existing things' (Figure 14.10). How appropriate for the Himalaya, a mountain range born out of the Tethys Ocean, subducted, collided,

Figure 14.9 (*opposite*) Map of the Bay of Bengal showing contours (in kilometres) on the depth of the Bengal fan sediments. The erosional debris from the Himalaya extends south as far as the Equator and reaches a depth of over 18 km, south of the Ganges delta.

Figure 14.10 *Thanka* or painting of the Cosmic mandala from Paro *dzong* in Bhutan.

buried, metamorphosed, melted, uplifted, eroded, and transported back to the ocean. The endless cycle of life, death, and rebirth is evident from the geology. Rocks from the summit of Everest still preserve tiny crinoid ossicles, the fossilized stems of sea-lilies that lived in a shallow, warm, tropical sea 400 million years ago. Fossil remains of tiny primitive horses, hippos, and palm trees are preserved in the 5-kilometre-high barren desert plateau of Tibet, showing that major climate and altitude changes have occurred. Ancient human remains have been found in the Siwalik deposits along the southernmost ranges of the Himalaya. There is even a suggestion that the migration of Tibetan people south into the Sherpa lands of Nepal, Sikkim, and Bhutan may have preceded the final stage of the rise of the Himalaya during the Holocene.

Certainly the climate is changing drastically today. Himalayan glaciers are almost all in retreat and have been since the past century at least. The monsoon is becoming increasingly erratic. Violent floods occur, particularly in the Indus and Ganges basins, with increasing regularity. Low-lying delta regions such as in Bangladesh are doomed to continuous flooding. Earthquakes have always occurred along the active margin of the Himalaya, but the huge increase in population of most Asian countries makes tragedies like the Kashmir earthquake or Sumatra–Andaman earthquake more likely in the future. Without the monsoon, food production in the Indian subcontinent would be impossible. If the monsoon is directly linked to the rise of the Himalaya and Tibet then surely it is important to study the relationship in far more detail.

It has been a great privilege to work and travel along the length and breadth of the Himalaya, Karakoram, and Tibet for the last thirty years. I have seen astonishing geological wonders of every description: towering snowy mountains, canyons 6 or 7 kilometres deep, active faults pulling rivers apart, volcanoes erupting, earthquakes pushing land up by metres instantly, new islands suddenly appearing out of the sea. I have climbed on the rock spires of the Baltoro, sampled rocks from high on K2 and Masherbrum, stood on unclimbed summits in the Karakoram and Pamirs, across Kulu, Ladakh, Zanskar, and Nepal, crossed the frozen icecap of Snow Lake, and almost reached the summit of fabulous mountains like Shivling. I have trekked along the frozen waters of the Zanskar River, and crossed the Himalaya about six times on foot, and trekked the length and breadth of the Hunza, Baltoro, and Hushe Karakoram. In the Karakoram and Hindu Kush I have found exquisite crystals of tourmaline, aquamarine, spinel, rubies, andalusite, kyanite, and garnet. I have seen donkey loads of beautiful lapis lazuli and turquoise hauled across the Hindu Kush from the lapis mines of Badakshan. In Indochina I have panned for gold in the Shan States of Burma, hunted for rubies and spinel from the marble tracts of Mogok and Vietnam, found sapphires in young basalts in remote corners of Thailand and Laos, and seen the fabulous imperial jade from the jade mines of Burma. It has been a wonderful journey, and I have shared it with many remarkable people—geologists, climbers, trekkers, local folk, wandering *saddhus,* and the occasional crazed militants. Along the way I hope I have been able to add a little more to our understanding of how the Earth works, and how mountains are made, and how the rocks have been coaxed into giving up their secrets. *Tashe delek*!

APPENDIX A

BASIC CONCEPTS AND DEFINITIONS OF ROCK DEFORMATION, METAMORPHISM, AND MAGMATISM

Deformation of rocks

Newton's Third Law of Motion states: 'For a body at rest or in uniform motion, to every action there is an equal and opposite reaction.'

Stress is a pair of equal and opposite forces acting on a unit area of rock.

Strain is the change in size and shape of a body resulting from the action of an applied stress field.

In the brittle upper crust, rocks deform by folding and faulting. ***Earthquakes*** occur when rocks fail by fracturing or faulting. Earthquakes are usually confined to the brittle upper crust, the ***seismogenic thickness*** (*Ts*). In old, cold, stable shield regions (e.g. peninsula India) earthquakes can occur throughout the crust and temperatures at the ***Mohorovicic discontinuity*** or ***Moho*** for short (base of crust) can be as low as 400–500°C. In active mountain belts (e.g. Tibetan plateau) the brittle-ductile transition can be at high levels (10–15 km depth) in the crust due to high heat flow and metamorphism.

Brittle deformation occurs at lower temperatures (<300–400°C) in the crust when elastic deformation leads to failure, faulting, and jointing. A ***fault*** is a planar fracture plane in the rock, across which there has been some displacement.

Ductile deformation occurs at higher temperatures deeper in the crust, where rocks undergo permanent deformation without fracturing. In the ductile lower crust, below the brittle-ductile transition, rocks deform by viscous, plastic flow. ***Mylonites*** are finely laminated rocks formed under ductile conditions by recrystallization and flow. They occur in zones of high strain along large faults. Strain may be concentrated along ***ductile shear zones.***

Folds are formed when an original planar feature becomes bent or curved as a result of deformation. ***Anticlines*** are bowed up and ***synclines*** are bowed down. ***Recumbent folds*** are where one limb is completely overturned and sometimes referred to as ***nappes. Foliations*** are a set of new planar surfaces produced as a result of deformation. These may include ***cleavage***, such as formed in slates, and ***schistosity*** as produced in metamorphic rocks at increased pressure and temperature. ***Lineations*** are linear structures produced as a result of deformation (e.g. intersections of bedding and cleavage, slickensides, minerals elongated in the direction of stretching, etc.) ***Fabric*** or ***texture*** is a general term used to define planar and linear features in a rock. These can be used for determining ***sense of shear*** in a rock. ***Vergence*** direction is the direction of movement as recorded by folds. ***Facing*** direction is the way-up along a fold axial plane. 'Way-up' can be determined from stratigraphy (younger above older) and gravitational structures in rocks (e.g. graded bedding, cross-bedding in sediments, shape of pillow lavas, etc.)

Thrust faults or reverse faults result from compression. Thrusting effectively thickens the crust in convergent mountain belts by repeated thrusting of one slice of rock on top of another. Folds and faults in the brittle upper crust can be restored to determine the amounts of crustal shortening or extension across mountain belts. ***Balanced cross-sections*** are sections where the deformed and restored sections show the same cross-sectional area, that is, no rock material has been lost from the section. Balanced cross-sections may not always be correct (because some material may have been lost through lateral motion, subduction etc.), but unbalanced cross-sections are always incorrect.

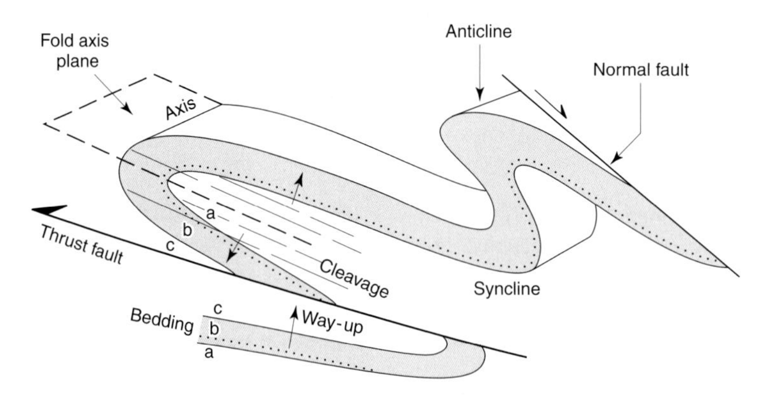

Figure A1 Sketch of a fold in sedimentary rocks (showing right 'way-up' with fold axis and axial plane cleavage, and a thrust fault.

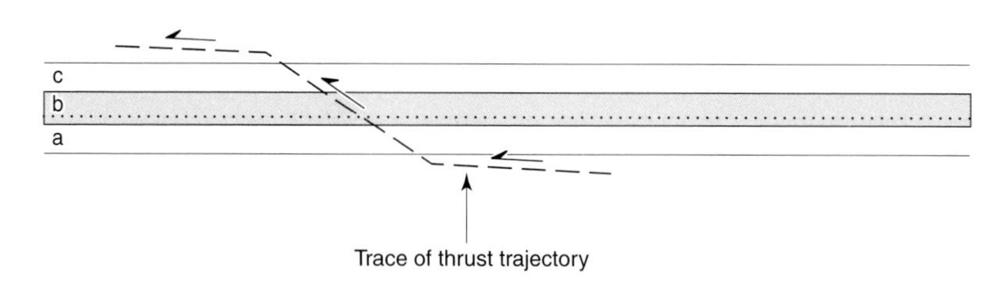

Figure A2 A restored version of this fold structure with the trajectory of the future thrust fault cutting up-section towards the transport direction.

Normal faults result from extension. ***Grabens*** are down-faulting blocks forming rift valleys; ***horsts*** are up-faulted blocks. Normal faults can be at high-angles in regions of crustal extension, or low-angled when they are sometimes called ***detachments***.

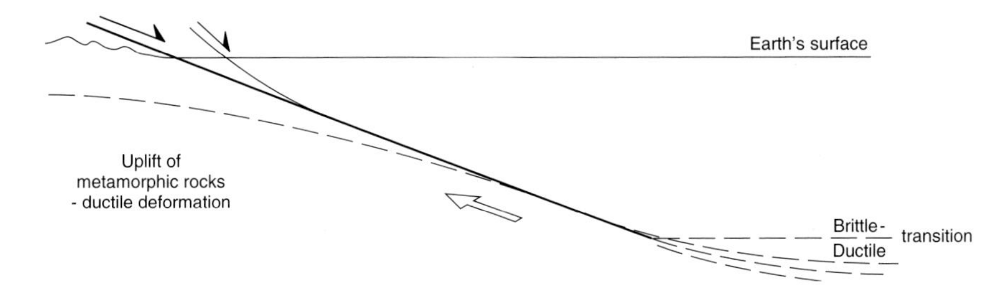

Figure A3 Sketch section across a low-angle normal fault, the transition between upper level brittle fault with deeper level ductile shear zone.

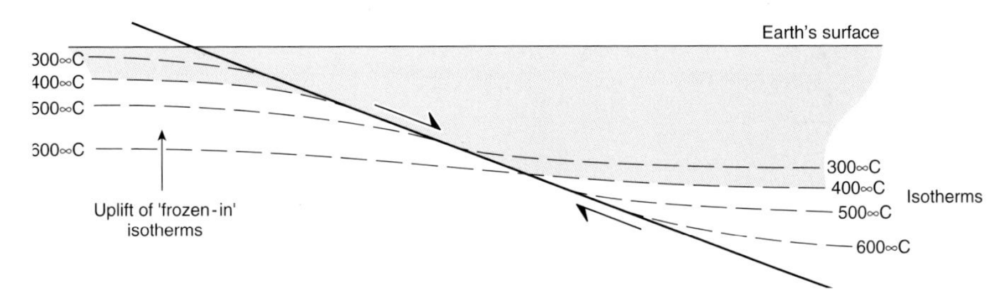

Figure A4 Uplifted isotherms during development of a metamorphic core complex.

Strike-slip faults or transcurrent faults show lateral motion; ***dextral*** motion associated with clockwise movement, ***sinistral*** motion associated with anti-clockwise movement in map view. A combination of strike-slip faulting and compression may result in ***transpression*** and uplift (e.g. K2—Gasherbrum Range, Karakoram). A combination of strike-slip faulting and extension may result in ***transtension*** and subsidence (e.g. Dead Sea, Turfan depression).

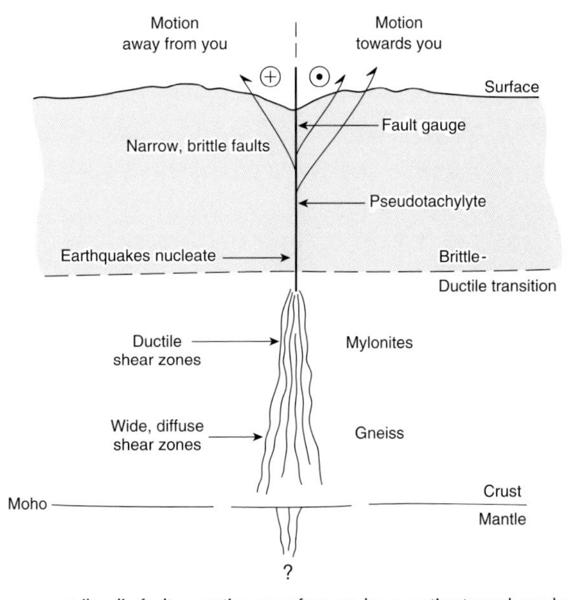

Figure A5 Profile across a strike-slip fault. + motion away from reader; • motion towards reader.

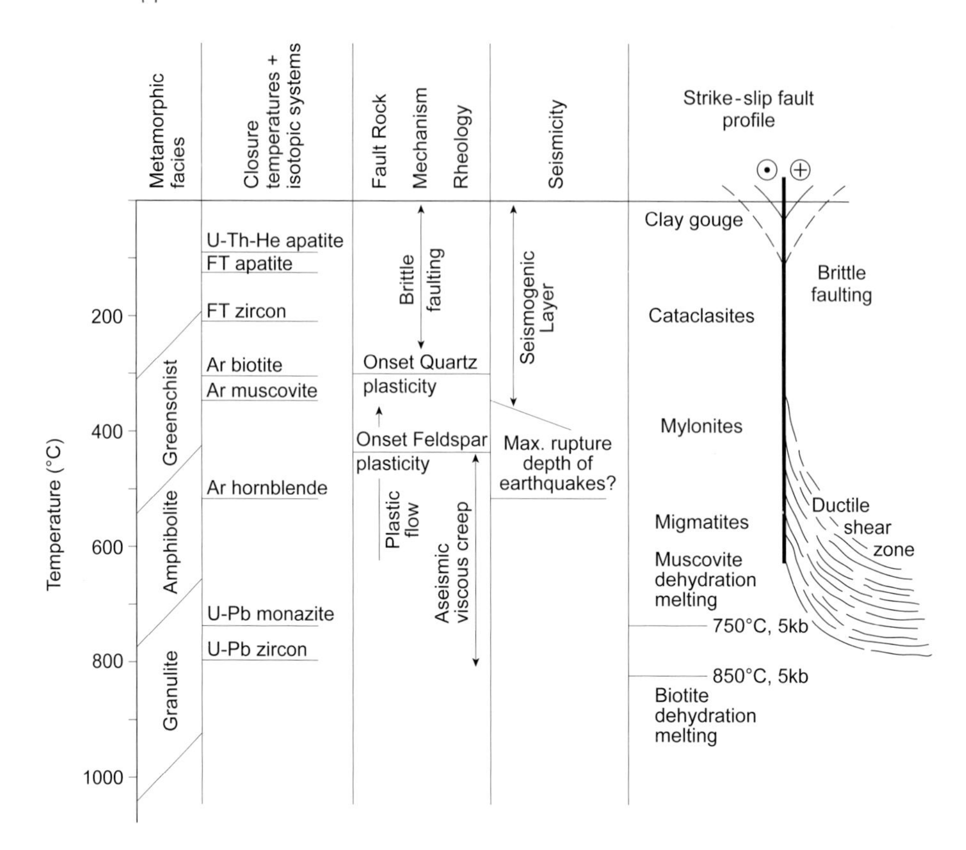

Figure A6 Temperature profile across a strike-slip fault showing fault rocks and the temperature and depth recorded by dating of minerals in various isotopic systems; FT—Fission Track; Ar-Ar—Argon 39/40; U-Pb—Uranium-Lead.

Metamorphism of rocks: temperature and pressure

Metamorphic rocks are crystalline rocks formed from their original sedimentary or igneous origins by chemical and mineralogical changes caused by the increase in temperature and pressure.

Temperature (T) is measured in degrees Celsius (°C) or the absolute Kelvin scale (K).

$$TK = T3°C + 273.$$

Geothermal gradient is the rate at which temperature increases with depth, and is usually within the range 15–30 °C/km. A ***geotherm*** is the curve on a graph of temperature versus depth.

Pressure (P) is the force per unit area to which a rock is subjected, and is measured in bar or kilobar (kbar). 1 bar* = 0.987 atmospheres = 14.5 pounds per square inch.

*1 bar = I megapascal (Mpa) or 0.1 gigapascal (Gpa).

Lithostatic pressure is the pressure exerted on a rock from the weight of the overlying column of rock; the lithostatic pressure gradient is approximately 3.5—3.7 km/kbar or 0.285 kbar/km for continental crustal rocks, and 3.9 km/kbar for denser oceanic rocks.

Metamorphic facies:

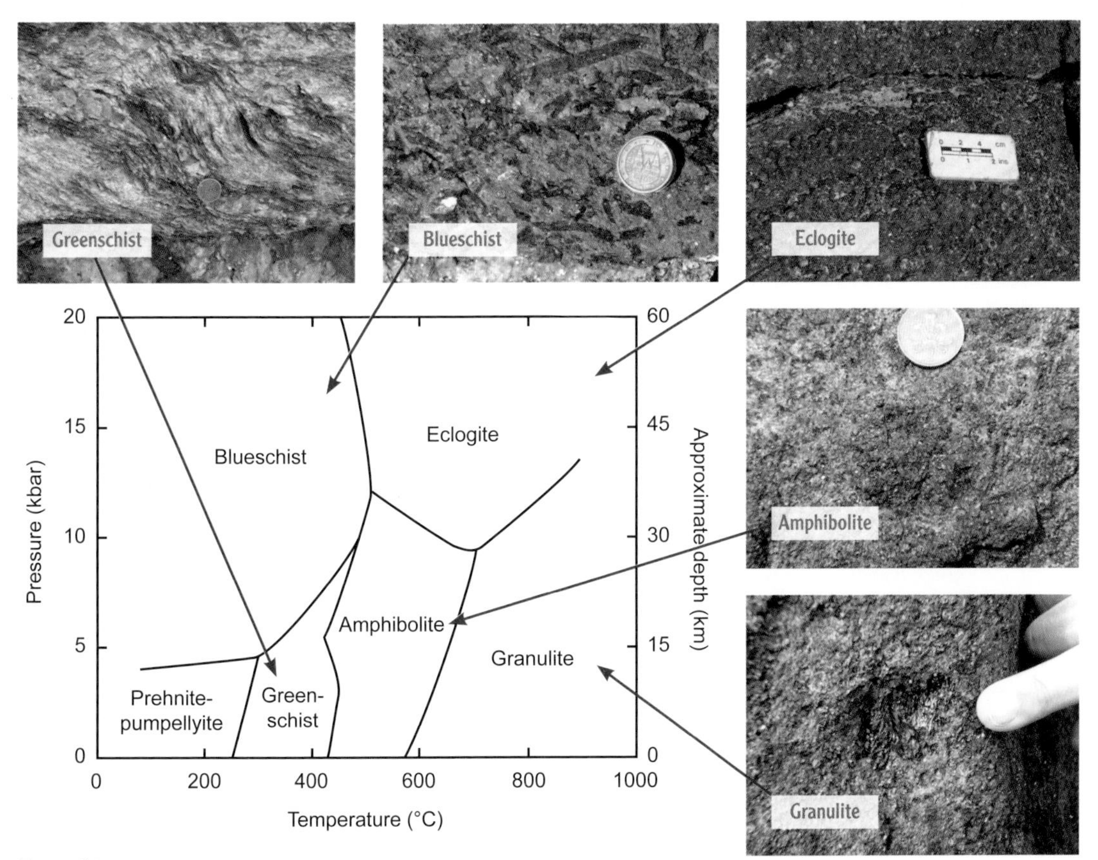

Figure A7 Metamorphic facies diagram.

Regional metamorphism is caused by increased P-T as a result of crustal thickening in mountain ranges. In shaley rocks, the normal succession of ***index minerals***, from low to high grade conditions is:

chlorite → biotite → garnet → staurolite → kyanite → sillimanite

During metamorphism, shales and mudstones convert to pelites or schists, limestones convert to marbles, and basalts turn to metabasites or eclogites. ***Gneisses*** are high-grade metamorphic rocks. At the highest temperatures, rocks will begin to melt producing a ***migmatite*** (part granite melt, part gneissic restite).

Contact metamorphism is caused by the localized thermal effects around an igneous intrusion, characterized by High-Temperature Low-Pressure minerals. An example is a metamorphic aureole surrounding a granite intrusion. Temperatures decrease away from the granite heat source.

Isograds are lines of constant metamorphic grade, and can be used for mapping purposes. The garnet-in isograd for example represents the first appearance of garnet in rocks.

Isotherms are lines of equal temperature.

Isobars are lines of equal pressure.

Thermobarometry is the determination of P-T conditions of the rock, providing that the minerals in the rock formed at ***equilibrium***, that is, at the same time.

CHARACTERISTIC MINERALS FOUND IN DIFFERING METAMORPHIC FACIES

Original rock type

Facies	Shale	Basalt
Granulite [high T med P]	K-feldspar, biotite, andalusite	pyroxene, garnet, plagioclase
Amphibolite [high P-T]	garnet, biotite, muscovite, sillimanite, kyanite	amphibole, plagioclase, garnet
Greenschist [med P-T]	chlorite, muscovite, plagioclase	chlorite, amphibole, plagioclase, epidote
Blueschist [high P low T]	glaucophane (blue amphibole), chloritoid, garnet, chlorite	glaucophane, lawsonite, garnet, epidote
Eclogite [high P med T]	pyroxene (jadeite), garnet, kyanite, phengite, glaucophane, chloritoid	jadeite, garnet, glaucophane
Hornfels [high T low P]	andalusite, biotite, K-feldspar	pyroxene, plagioclase
Zeolite [low P-T]	zeolite, pyrophyllite, micas	calcite, chlorite, zeolite

Characteristics of granites

	Andean-type e.g. Transhimalayan granites	**Himalayan-type** e.g. Everest granites
Minerals	quartz, K-feldspar, plagioclase, hornblende, biotite	quartz, K-feldspar, plagioclase muscovite, biotite, garnet, tourmaline
Accessory minerals	zircon, monazite, sphene	sillimanite, cordierite, uranium-bearing minerals (eg: zircon)
Mineralization	copper, gold, silver	tin, tungsten, lithium REE gems: aquamarine, topaz, tourmaline
Source	melting above subducted oceanic slab; hot, dry tonalitic magma	*in situ* melting from sedimentary source; associated with regional metamorphic rocks, migmatites
Mantle component	yes	none, purely crustal origin
Heat source	mantle-derived	regional metamorphic heat from crustal thickening
Radioactivity	small crustal source	high uranium contents from sedimentary source rocks
Melting temperature	ca 800°C	650–750°C
Volcanism	andesite volcanoes above	no volcanic rocks associated
Emplacement	vertical intrusions, large-scale plutons forming linear batholiths	segregation from migmatites, lateral intrusion along sills
Metamorphism	high-temperature thermal aureoles in surrounding rocks	intrusive into regional high-grade metamorphic rocks
Isotopes	low $^{87}Sr/^{86}Sr$ ratios	high $^{87}Sr/^{86}Sr$ ratios
Tectonic setting	continental margins adjacent to subducting oceanic plate	continent–continent collision belts, regional crustal thickening
Asian examples	Kohistan–Ladakh–Gangdese granites	High-Himalayan leucogranites (eg: Shivling, Manaslu, Shisha Pangma, Makalu, Jannu, etc.)

GEOCHRONOLOGY

Isotopes and decay systems

An isotope is a chemical element in which the atoms have a common number of protons and electrons (i.e. their atomic number is the same), but have a variable number of neutrons in their nucleus. Isotopes may be produced by various nuclear reactions and the products are usually radioactive. Radioactive decay is the process when an unstable nucleus ejects or captures particles, transforming the radioactive nuclide into an isotope of another element.

Isotopic decay systems:

U-Pb = uranium – lead: dating system based on the radioactive decay of ^{238}U to ^{206}Pb (half life 4510 million years) and decay of ^{235}U to ^{207}Pb (half life 713 million years). The amount of radiogenic lead must be distinguished from naturally occurring lead, calculated by using the ratio with the stable isotope ^{204}Pb. If the mineral has remained as a closed system, after correcting for original lead, the ratios of $^{235}U : {}^{207}Pb$ and $^{238}U : {}^{206}Pb$ should agree. If so they are 'concordant' and can be plotted on a Concordia curve, with the age determined being the true age. If the two ages do not agree, they are 'discordant' and some lead loss must have occurred, possibly during a subsequent heating event.

Th-Pb = thorium – lead: radioactive decay of ^{232}Th decays to $^{208}Pb + 6He^4$ with a half-life of 13 900 million years.

Rb-Sr = rubidium – strontium: dating system based on the radioactive decay of ^{87}Rb to ^{87}Sr. ^{87}Rb decays in a single step to ^{87}Sr by the emission of a low-energy beta-particle. $^{87}Sr/^{86}Sr$ is plotted against $^{87}Rb/^{86}Sr$ to produce a straight line 'isochron' from which the age can be determined.

K-Ar = potassium – argon: dating system based on the radioactive decay of ^{40}K to ^{40}Ar. ^{40}K decays by two modes, by electron capture to ^{40}Ar and by beta-emission to ^{40}Ca. ^{40}Ca is the most abundant isotope of Ca and almost ubiquitous in rocks and minerals so it is difficult to determine the amount of ^{40}Ca present initially, and therefore not used for dating. ^{40}Ar is an inert gas, which does not combine chemically with any other element and so escapes from rocks when they are heated. When the rock cools, the radiogenic ^{40}Ar is trapped within the crystal structure and accumulates with time. If the rock is re-heated later then some of the ^{40}Ar may escape and the clock will be partially or completely reset.

Ar-Ar = argon 40 – argon 39: a more modern variant of K-Ar dating, where a sample is bombarded by neutrons in a nuclear reactor resulting in some ^{40}K being converted to ^{39}Ar. After irradiation the samples are step-heated to progressively higher temperatures, which incrementally releases Ar gas. An age is typically calculated from a number of consistent $^{40}Ar/^{39}Ar$ measurements obtained from several individual heating steps.

Common mineral phases used for geochronology

Mineral	Formula	Closure temperature (°C)	Decay system
Zircon	$ZrSiO_4$	>800°C	U-Pb
Monazite	$CePO_4$	750-720°C	U-Pb, Th-Pb
Xenotime	YPO_4	~750°C	U-Pb
Allanite	$CaFeAl_2Si_3O_{12}$	800-750°C	U-Pb, Th-Pb
Sphene	$CaTiO_3$	670-500°C	U-Pb
Hornblende	$Ca_2(Mg,Fe)_4Al_2Si_7O_{22}(OH)_2$	550-450°C	K-Ar, Ar-Ar
Muscovite	$KAl_3Si_3O_{10}(OH)_2$	550-500°C 400-300°C	Rb-Sr K-Ar, Ar-Ar
Biotite	$K(Mg,Fe)_3Si_3AlO_{10}(OH)_2$	350-250°C 350-250°C	Rb-Sr K-Ar, Ar-Ar

Closure temperature: Isotopic closure occurs at a time when a mineral will no longer exchange isotopes with its surroundings. When the rock cools, the closure temperature is the temperature at which the mineral becomes stable. The isotopic closure temperature of minerals can be combined with pressure–temperature (P-T) conditions of formation to determine 'peak' metamorphic timing or timing of granite crystallisation.

Cooling histories: If a single rock contained several of the minerals listed above which were dated, it would be possible to plot the ages on a temperature–time diagram. Minerals with successively lower closure temperatures should plot as successively younger ages, as the rock cools during its exhumation from deep levels in the crust towards the Earth's surface. The line joining all the ages from this single rock would plot on a cooling path.

Dating instruments and methods

ICP-MS – Inductively Coupled Plasma Mass Spectrometer
TIMS – Thermal Ionization Mass Spectrometer
SIMS – Secondary Ionization Mass Spectrometer
SHRIMP – Sensitive High-Resolution Ion Microprobe
LA – ICPMS – Laser-Ablation Inductively Coupled Plasma Mass Spectrometer
LA-PIMMS – Laser-Ablation Plasma Ionization Multicollector Mass Spectrometer
SEM – Scanning Electron Microscope

Fission-track dating

Fission tracks, or radiation damage tracks, form in uranium-bearing minerals from the spontaneous fission of atoms of ^{238}U. Numbers of fission tracks increase with time, allowing the calculation of fission-track ages. Heating anneals the tracks causing reductions in track length and in the apparent fission-track age by amounts controlled mainly by maximum temperature experienced. Uranium concentration can be measured by counting tracks produced by fission of ^{235}U as a result of the irradiation of the sample with thermal neutrons in a nuclear reactor.

MINERALS USED

The main minerals used in Fission-track dating are apatite ($Ca_5(PO_4)_3(F,Cl,OH)$) and zircon ($ZrSiO_4$).

CLOSURE TEMPERATURES

Closure temperatures for fission tracks in zircon is 200 ± 50°C and for apatite is around 120°C. During the motion of the rock towards the Earth's surface (the exhumation path), the sample will pass through the partial annealing zone, where fission tracks are formed. As erosion removes material from above, the rock cools and retains its 'frozen-in' fission tracks. The elevation at the surface prior to uplift of the rock sample can be geologically constrained, so the change in mean surface elevation can be calculated.

Surface uplift = rock uplift minus denudation.

FISSION-TRACK AGE VERSUS TOPOGRAPHIC HEIGHT

In high mountains, the fission-track ages of apatite and zircon will normally increase with increased elevation. The youngest ages will be at the base of the mountain where the sample has only recently been exhumed, whilst the mountain summit should have the oldest age. Where samples over a wide altitude range have similar ages, it must mean that uplift of the rocks and erosion has been very rapid.

INTERPRETATION OF DATA

Fission-track dating can be used to interpret low-temperature cooling of rocks as they rise towards the Earth's surface during exhumation. These data are most useful for constraining timing and amounts of erosion or denudation over the upper ~5–10 km of crust, and particularly for dating landscape evolution, glacial unroofing and periods of rapid mountain uplift.

Uranium–Thorium–Helium dating

Uranium–thorium–helium (^{238}U-^{234}U-^{230}Th) dating is an even lower temperature thermo-chronometer, which can be used in conjunction with fission-track dating. The accumulation of radiogenic He from the decay of U and Th in apatite crystals has a closure temperature of only 75 ± 5°C. In a normal geothermal gradient the zone of partial He retention occurs in a zone ~2 km thick centred at a depth of only 2 km beneath the surface. It is analogous to the fission-track partial annealing zone, but lies at ~2 km shallower depth.

Obtaining erosion rates from rocks

Long-term cooling rates can be derived from dating several minerals with different closure temperatures from the same rock (e.g. U-Pb dating of monazites, zircons; K-Ar dating of micas; fission-track dating of apatite and zircon). This gives a number of points on a temperature versus time diagram which can be used for constructing a 'cooling curve'.

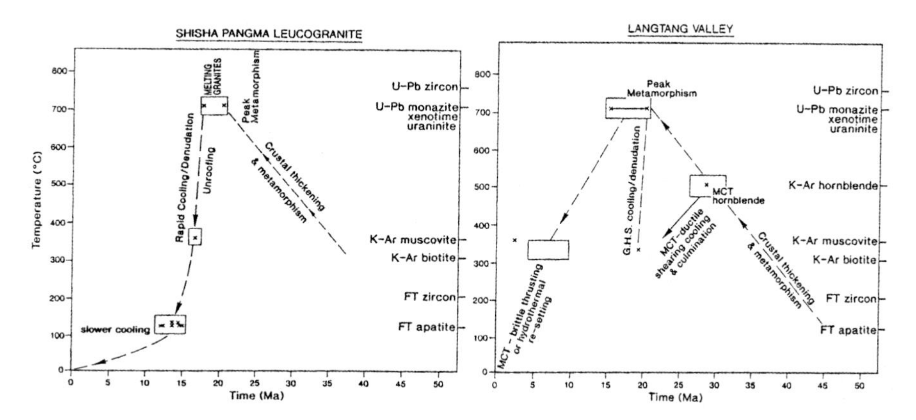

Figure A8 Cooling curves for the Shisha Pangma leucogranite, South Tibet (left) and the Langtang Valley, Nepal (right), showing ages of minerals with different closure temperatures. Periods of time when the cooling curve is steep (e.g. 17–15 million years ago for Shisha Pangma) means that rapid rock uplift and rapid erosion has removed a lot of material in a short time, resulting in rapid cooling of the rocks. This can be interpreted as the timing of increased mountain uplift and higher topography along the Himalaya.

Short-term cooling rates can be derived from fission-track dating of apatite and zircon, U-He dating of apatite. These low-temperature geochronometers can be used for dating landscape evolution, assuming a geothermal gradient.

Caveats:

- Low-temperature isotherms (lines of equal temperature) may be warped up beneath very high mountain massifs, like Everest, and down beneath deeply incised rivers, like the Arun River in Nepal which cuts 4 kilometres deep into the Makalu Himalaya.
- Geothermal gradients (temperature versus depth) may not be uniform under large mountain massifs, or where hot fluids are circulating in the crust.
- Climatic controls have a positive feedback mechanism. Increased precipitation results in increased weathering, which results in increased uplift at the climatic front (e.g. High Himalaya).

Uplift, erosion, and exhumation

DEFINITIONS

Uplift is displacement in the direction opposite to the gravity vector.

Surface uplift is the displacement of the Earth's surface with respect to mean sea level.

Uplift of rocks means that the rocks are moving up in relation to sea level, or the geoid, but not necessarily with respect to the Earth's surface.

Exhumation is the unroofing of rocks, relative to the Earth's surface. The rate at which a rock rises towards the Earth's surface (at whatever elevation that might be) is referred to as the 'exhumation rate'.

Erosion encompasses all the processes of denudation, including chemical weathering and mechanical weathering, such as glacial, wind, and water erosion that has the net effect of lowering surface topography.

METHODS OF INFERRING RATES OF EROSION

Long-term erosion rates can be inferred by:

- Radiometric ages used cooling curves. Periods of rapid erosion can be inferred when the cooling curves are steep, for example in the 20–17 m.y. time interval for the Greater Himalaya (see Everest Temp–Time fig.). The inference is that periods of rapid exhumation (uplift of rocks) correlate with rapid erosion and greater relief.
- Sedimentary fluxes in rivers and foreland basins. Periods of rapid erosion will result in higher fluxes in the rivers and thicker sedimentary deposits in the basins.

Short-term erosion rates can be inferred from:

- Fission-track (FT) dating of rocks; vertical sampling up high mountains. Fission tracks will fade in zircon crystals at temperatures of ~200°C, and are completely annealed in apatite crystals at temperatures above ~100 ± 10°C. Normally FT ages increase with increasing altitude.
- Bedrock incision rates. River incision determines hillslope gradients. Abandoned straths, or terraces, are left when the river cuts deeper into the valley. Rates of river incision can be calculated using age of the terrace and the height above the river.
- Cosmogenic radionuclide (e.g. ^{10}Be) dating of terraces for river incision rates.

Methods of inferring rates of uplift and palaeo-altitude

- Palaeobotany. The shape and physiognomy of leaves is dependent on climate and altitude. Small, simple leaves occur in dry, cold conditions above 3,000 m; moderate size leaves with toothed margins occur in cool, wet climates at 2–3,000 m; large leaves with smooth margins and pointed tips are characteristic of hot, wet conditions below 1,000 m.
- Oxygen isotopes. The isotopic composition of meteoric water decreases with increasing elevation; ratios of $^{18}O/^{16}O$ decrease with altitude.

- Cosmogenic radionuclide dating. Production rates vary strongly with altitude (3 times higher for every 1,700 m increase in altitude)
- River profiles can be used to infer bedrock uplift through time. Knick points in rivers might indicate location of active faults and uplift boundaries.

Glacial cycles, ice ages, and climate change

The Earth's orbit through time is calculated by:

a) the precession of equinoxes (when the Earth is nearest the Sun) from January to July;
b) the tilt of the Earth's rotation axis (which varies from 22° to 24.5° over 41,000-year timescale); and
c) changes in the eccentricity of the Earth's orbit around the Sun (over a period of 100,000 years). As the Earth moves around the Sun, it is also spinning obliquely about its own axis, like the motion of a spinning top.

The orbital variations are known as ***Milankovitch cycles***, after the Yugoslav astronomer who first discovered them, and are responsible for the periodic glaciations that have occurred on roughly 100,000 year cycles during the Quaternary period. The amount and intensity of solar radiation that falls on the Earth varies with the changing position of the planet in the cycle, and affects the changes in climate. Milankovitch cycles correspond very closely to climatic oscillations, including the size of ice sheets, sea level and mean surface temperature. It is a worrying fact that very small changes in the elliptical orbit of the Earth can result in the Earth being forced into, or out of, a glacial period.

GREENHOUSE EFFECT

Greenhouse gasses, such as carbon dioxide, methane, ozone, nitrous oxides, and chlorofluorocarbons, form a protective blanket around the Earth's atmosphere that keeps damaging solar radiation away from the Earth's surface. CO_2 in the atmosphere can be increased by sudden emissions from volcanic eruptions, or by large-scale metamorphism of carbonate rocks that turn limestone into marble and would release CO_2 into the atmosphere. There has been a progressive increase in CO_2 since the Industrial Revolution and the large-scale burning of fossil fuels, amounting to about 25% increase since 1850. The rapid increase of CO_2 is the major cause of global warming today. The results of global warming include large-scale environmental problems such as the break-up of the Antarctic ice shelves, the desertification of large parts of Africa and the Middle East, and inundation of low-lying coastal areas, where a large percentage of the world's population live, by rising sea level.

ICEHOUSE EFFECT

Decreased atmospheric CO_2 levels lead to global cooling as polar ice sheets spread, and biological productivity in the oceans may become reduced. Continental margins are major repositories of organic carbon in modern oceans, so periods of '***snowball earth***' usually follow periods of continental break-up in the rock record. Major snowball earth episodes of global glaciation occurred during the Neoproterozoic, approximately 600 million years ago, and during the Permo-Carboniferous, approximately 300–280 million years ago.

Methods used for dating Quaternary features

1. ***Optically stimulated luminescence dating.*** This method can be used to determine the time elapsed since a sample (quartz silt) was exposed to daylight. It relies on the interaction of ionizing radiation with electrons within semiconducting crystals, resulting in the accumulation of charge within the grains. This charge can be determined by heating (thermoluminescence) or illuminating (optically stimulated luminescence) the sample.
2. ***Cosmogenic radionuclide dating.*** This method measures the isotopes of beryllium (^{10}Be), aluminium (^{26}Al), and chlorine (^{36}Cl) in quartz boulders. When the boulder is exhumed from depth, the production rate of cosmogenic isotopes increases exponentially from cosmic rays coming in from the Earth's atmosphere. Corrections have to be made for altitude and geomagnetic latitude.
3. ***Carbon-14 dating.*** This method uses the radioactive decay of ^{14}C to its daughter ^{14}N through emission of an electron: a beta particle. Any plant or animal organic matter, buried for example in a moraine, will begin to decay upon death, changing the ratio of carbon isotopes in the cells. Widely used for archaeological dating, ^{14}C dating is good for ages younger than ~35,000 years.
4. ***Uranium-Thorium-Helium dating.*** This precise method of dating relies on the radioactive decay of parent isotopes ^{238}U, ^{235}U, and ^{232}Th to daughter isotopes ^{206}Pb, ^{207}Pb, and ^{208}Pb respectively. The U-Th series can be used for calibrating the ^{14}C clock where both methods can be used to date a single sample (e.g. corals).
5. ***Dendrochronology.*** Certain trees grow new wood seasonally in rings, which can be counted to reveal the age. Trees grow rapidly in the wet season in the summer and slowly during the winter cold season. The age of petrified fossil trees, preserved in sediments or ash layers, can be derived from the number of rings, combined with a knowledge of the climate.

The Indian summer monsoon

FACTORS INFLUENCING THE MONSOON

Four major boundary conditions affect the development of the Indian monsoon:

1. Orbital forcing parameters such as solar radiation, which directly affects monsoon precipitation.
2. Height of the Tibetan plateau, which affects atmospheric pressure conditions.
3. Glacial conditions such as extent of surface ice, albedo (reflectance level) of the land surface, which will affect the seasonal timing and strength of the monsoon.
4. Atmospheric CO_2 levels, which are lower in interglacial periods.

The uplift of the vast plateau of Tibet results in the development of a low-atmospheric pressure system over northern India during the northern hemisphere summer, which effectively draws in the warm, moist air from the Indian Ocean. Warm air rises above the plateau, atmospheric pressure falls, and air is sucked into the void. When the moisture-laden clouds from the tropical Indian Ocean reach the land, the clouds spontaneously release their moisture as rainfall over India and the Himalaya.

The Indian monsoon lasts usually from late June to September and brings torrential rains to the Indian sub-continent. The monsoon starts in the south and spreads north towards the Himalaya. The Himalaya acts as a physical barrier to the northward-flowing monsoon winds. The southern slopes of the Himalaya are lush and green whereas the northern slopes are barren and dry. The Khumbu Valley draining south from Everest passes through coniferous and rhododendron forests to the jungles of the Terai lowlands. The Tibetan plateau is in the rain-shadow of the Indian monsoon. The Rongbuk Glacier draining north of Mount Everest flows through barren desert with practically no vegetation at all.

EFFECTS OF THE MONSOON

Increased rainfall in the Himalaya results in increased erosion levels. Increased erosion results in increased rates of both mechanical and chemical weathering. ***Mechanical weathering*** is the mechanism whereby water, wind, or ice physically break up the rock, and the debris is transported down rivers to the lowlands. ***Chemical weathering*** produces dissolved carbonate minerals that are transported down rivers to the oceans. Chemical weathering removes CO_2 from the atmosphere leading to cooling (the '***icehouse effect***'). Cooler climate leads to growth of glaciers and a general increase in storminess of the weather, which can exacerbate the cooling trend and lead to a 'runaway icehouse effect'.

Periods of major climate change in Asia

Periods of major climate change that can be ascertained from the geological record are:

- Enhanced aridity occurred across most of the Asian interior, 9–8 million years ago.
- Strengthening of the Indian monsoon about 7–8 million years ago. Climate change can be determined from marine cores in oceanic sediments. The relative abundance of the planktonic foraminifera *Globigerina bulloides* is an indicator of coastal upwelling during the monsoon.
- Changes in vegetation from C3 (forests) to C4 (grasses) in northern Pakistan and India from 8 million years ago.
- Change from mixed broad-leaf forests to grasslands in north-east Tibet from 8.5 million years ago.
- Intensification of the monsoon occurred about 3 million years ago.
- Increased variability of the monsoon in the last 2 million years.

GLOBAL CENOZOIC CHANGES

Warmest period in the Cenozoic	~55 million years ago.
Periods of global cooling	50–40 m.yr and 35–25 m.yr
Warm period	20–15 m.yr.
Ocean deep-water cooling (4–5°C)	15–14 m.yr
First ice sheets appear in Antarctica	ca. 33 m.yr.
Large Antarctic ice sheet appears	15 m.yr.
Sea-level falling (growth of ice sheets)	11–10 m.yr.
Intensification of the Indian monsoon	8–7 m.yr.
Development of the Greenland ice sheet	4–3 m.yr.
Initiation of northern hemisphere glacial oscillations	~4.5 m.yr.
Major northern hemisphere glaciation starts	2 m.yr.

NB.

- The Earth's climate was relatively warm when the Himalaya and Tibetan plateau were at or near their maximum elevation.
- Over the past 13 million years, the Earth has followed a cooling trend, as the Himalaya and Tibet have remained at roughly the same high elevation.

A brief history of climate change and Himalayan uplift

- 40 million years ago: Sudden rise in $^{87}Sr/^{86}Sr$ ratios in seawater indicating increase in weathering and erosion interpreted as the beginning of topographic elevation in the Himalaya.
- 35–30 million years ago: Peak metamorphism along the Himalaya, indicating crustal thickening, high topography and high erosion–exhumation rates.
- 21–16 million years ago: High Himalayan granites formed; possible climax of Greater Himalaya crustal thickening and topographic elevation. Timing of Himalayan Channel Flow.
- 21–18 million years ago: Extremely rapid increase in $^{87}Sr/^{86}Sr$ ratios in seawater, correlating with the timing of maximum exhumation, cooling, and probably high topography along the Himalaya.
- 20–16 million years ago: Major faunal changes in northern India and Pakistan.
- 17–15 million years ago: Rapid cooling rates, rapid erosion and exhumation; high topography along the Himalaya.
- 8 million years ago: Increased aridity and major change in vegetation from C3 (forests) to C4 (grasses) in northern Pakistan and India.
- 8–7 million years ago: Strengthening of the Indian summer monsoon.
- 7.4 million years ago: Increased chemical weathering rates from Sr isotope records and reduced sediment loads, despite an intensifying monsoon.
- 3.6–2.6 million years ago: Intensification of the summer monsoon.
- 2.5 million years ago: Start of Quaternary glaciations in the Everest area, increased dust in the atmosphere leading to the beginning of Central China loess deposition, global cooling, increased CO_2 in the atmosphere, increased variability of the monsoon.
- 2 million years ago: Start of a second rapid phase of exhumation of Himalayan granites on Mount Everest due to increased erosion associated with glaciation.
- 2–0.7 million years ago: High-temperature metamorphism and anatexis in Nanga Parbat syntaxis. Rapid exhumation, high erosion rates and extreme surface uplift of syntaxis region.
- 20,000 years ago: Maximum advance of the glaciers south of Mount Everest.
- 18,000 years ago–today: Global warming, retreat of Himalayan glaciers. Increased monsoonal precipitation along Himalaya, increased aridity in Tibet. Rapid uplift of Karakoram.

APPENDIX B

GLOSSARY

Ablation valley—valley formed along the sides of a glacier by melting of the ice.

Acid rock—igneous rock composed dominantly of quartz and feldspars.

Adakite—intrusive dykes or volcanic rocks requiring a garnet-bearing eclogite or amphibolite deep crustal source.

Apatite—$Ca_5(PO_4)_3(OH,F,Cl)$; common accessory mineral used in fission-track dating.

Alpine Himalayan mountain belt—a chain of mountains stretching across Europe and Asia formed as a result of the closing of the Tethys Ocean and collision of two continental land masses.

Amphibolite—a high-grade metamorphic rock characterized by hornblende amphibole and plagioclase feldspar. Amphibolite facies indicates a range of Pressure–Temperature conditions between greenschist facies and granulite facies.

Andalusite—Al_2SiO_5; aluminium silicate mineral associated with high-temperature low-pressure contact metamorphism.

Andesite—volcanic rock composed of feldspar, hornblende, pyroxene, quartz, named after the volcanoes of the Andes.

Asthenosphere—deeper region of the mantle directly beneath the lithosphere.

Aquamarine (beryl)—$Be_3Al_2Si_6O_{18}$; pale blue gem mineral frequently found in pegmatites.

Balanced cross-section—a geological section across a mountain range, balanced for known stratigraphic thicknesses and therefore potentially restorable to its pre-deformed state.

Barrovian metamorphism—regional metamorphism as recorded by the progressive change of minerals in meta-pelite index minerals: chlorite-biotite-garnet-staurolite-kyanite-sillimanite; named after George Barrow (1853–1932).

Basalt—fine-grained basic volcanic rock composed of pyroxene, olivine, and feldspar, the common rock of the oceanic crust.

Batholith—large igneous, granitic intrusion, sometimes composed of several smaller plutons.

Benioff zone—subduction zone where one tectonic plate descends beneath another, marked by a narrow, inclined zone of deep earthquakes.

Benthic—deep ocean environment.

Biotite—dark-coloured mica, a sheet silicate mineral common in metamorphic rocks.

Blueschist—an intermediate high-pressure metamorphic rock characterized by the presence of glaucophane, a blue amphibole.

Brittle deformation—fracturing and folding characteristic of the upper levels of the crust. Brittle fractures result in faulting and earthquakes.

Cataclasites—fault rocks (e.g. breccias, fault gouge).

Cenozoic era—the last sixty-five million years of Earth history.

Channel Flow model—model developed to explain metamorphism of the Greater Himalaya involving the southward ductile extrusion of a partially molten layer of middle crust. Distinct from lower crustal flow.

Coesite—a high-pressure polymorph of quartz (SiO_2) indicative of ultrahigh-pressure metamorphism.

Conglomerate—coarse-grained sedimentary rock composed of rounded boulders formed in a high-energy fluvial environment.

Constructive plate margins—mid-ocean ridges along which new oceanic crust is formed.

Contact metamorphism—passive high-temperature metamorphism formed around the margins of granitic igneous intrusions.

Continental crust—the outer layer of the Earth dominated by quartz and feldspar on average about 35 km thick, and up to 70–80 km beneath high plateaux areas, e.g. Tibet.

Core—central part of the Earth composed mainly of iron and nickel where the Earth's magnetic field is generated; innermost core is solid, whereas the outer core is liquid.

Coriolis force—forces acting on moving objects as a result of the Earth's rotation; causes oceanic and atmospheric currents to be deflected to the east in the northern hemisphere and to the west in the southern hemisphere. Force varies from zero at the equator to a maximum at the poles.

Corundum—mineral Al_2O_3 including the gemstones sapphire, ruby, and emerald; hardness 9.

Craton—old, stable part of the continental crust.

Crevasse—deep icy fissure in the surface of a glacier.

Crust—the upper level of the lithosphere comprising about 3–7 km thickness of basalts and gabbros in the oceans, and ca. 35 km thickness of granites, metamorphic and sedimentary rocks in the continents. The continental crust reaches a maximum thickness of 75–90 km in parts of Tibet.

Denudation—erosional removal of rock.

Destructive plate margins—subduction zone margins where lithospheric plates are returned into the mantle.

Diamond—a high-pressure mineral polymorph of carbon, a characteristic constituent of kimberlites or ultrahigh-pressure metamorphic rocks; hardness 10.

Ductile deformation—plastic flow folding and shearing occurring below the brittle upper crust at high temperatures deep in the crust usually concomitant with metamorphism.

Dyke—a vertical igneous intrusion usually cross-cutting surrounding country rocks.

Earthquake—shaking of the Earth caused by sudden release of stress during movement on a fault; magnitude measured on the Richter scale.

Eclogite—a high-pressure metamorphic rock composed of garnet and pyroxene indicative of deep burial in a subduction zone type setting.

Erosion—removal of rock by chemical or mechanical weathering.

Exhumation—the movement of a rock towards the Earth's surface (at whatever height that may be) by a combination of tectonic uplift of rock and removal of overburden.

Fault—fracture line in the Earth's crust; sudden movement along a fault line triggers earthquakes.

Feldspar—most common mineral constituent of igneous rocks, includes alkali feldspar (K,Na) $AlSi_3O_8$ and plagioclase (ranging from sodic albite to calcic anorthite).

Felsic—a term applied to light-coloured igneous minerals, such as quartz and feldspar.

Fission-track dating—dating method using radiation damage tracks in uranium-bearing minerals from the spontaneous fission of atoms of ^{238}U. Numbers of fission tracks increase with time, allowing the calculation of fission-track ages.

Foliation—a penetrative planar fabric (e.g. cleavage, schistosity) in a rock.

Foraminifera—amoeba-like protozoan (single-celled) organism formed of calcium carbonate, widely used for stratigraphic dating of sedimentary rocks.

Garnet—$[Mg, Fe,Mn, Ca]_3Al_2Si_3O_{12}$; metamorphic mineral associated with upper greenschist, amphibolite, eclogite or granulite facies rocks; also can be an igneous mineral component of a crustal melt granite.

Geochronology—the study of age dating rocks, commonly using radioactive decay (e.g. U-Th-Pb dating).

Geomorphology—the study of surface processes and landscapes; physical geography.

Geothermal gradient—the rate at which temperature increases with depth. The geotherm is usually within the range 15–30°C per kilometre depth.

Granite—coarse-grained acidic intrusive igneous rock formed by melting of the continental crust, composed mainly of quartz and feldspars with lesser amounts of micas, hornblende, etc.

Granulite—a high-grade metamorphic rock composed mainly of pyroxene, amphibole, garnet, and feldspar indicative of high temperature and intermediate-high pressure, typically forming the lower continental crust.

Greenhouse effect—warming of the Earth's atmosphere due to the build-up of greenhouse gases such as carbon dioxide and methane.

Greenschist—a low-grade metamorphic rock composed of a variety of micaceous mainly meta-sedimentary rocks.

Gneiss—a high-grade metamorphic rock showing strong foliation and compositional layering.

Hornblende—$NaCa(Mg,Fe,Al)_5(Al,Si)_8O_{22}(OH)_2$; green or black amphibole group mineral.

Hornfels—a rock derived during contact metamorphism in the aureole around a hot igneous (granitic) intrusion, generally high-temperature and low-pressure.

Hotspot—localized zone of asthenospheric upwelling, typically with an alkali volcanic edifice (e.g. Hawaii, Réunion, Comores islands).

Igneous rock—a crystalline rock, such as granite or gabbro that has cooled from a molten state.

Isobar—line of equal pressure.

Isograd—line of constant metamorphic grade used for mapping purposes (e.g. garnet-in isograd is first appearance of garnet in rocks).

Isotherm—line of equal temperature.

Isotope—one of two or more forms of an element differing in atomic weight.

Isostasy—the principle of a buoyant crust floating on a fluid-like substrate (mantle) according to Archimedes principle.

Jet stream—concentrated high-level high-speed air flow usually from west to east in the northern hemisphere, as a result of the Coriolis effect.

Kyanite—Al_2SiO_5; blue aluminium silicate mineral associated with high-pressure regional metamorphism.

Leucogranite—an igneous rock composed dominantly of quartz and feldspars with characteristic minerals such as biotite, muscovite, garnet, tourmaline, etc.

Lineation—a linear rock fabric on a foliation plane usually indicating a direction of motion during deformation.

Lithosphere—strong outer part of the Earth forming 'plates' composed of the Earth's crust and upper mantle, usually about 100 km thick.

Lithostatic pressure—pressure exerted on a rock from the weight of the overlying column of rock (about 3.5 kilometres per kilobar).

Mafic rock—a dark-coloured rock composed dominantly of minerals, such as olivine, pyroxene, micas, etc.

Main Boundary Thrust—the large-scale active thrust fault along the southern boundary of the Himalaya.

Main Central Thrust—a large-scale south-vergent thrust fault along the base of the Greater Himalaya.

Mantle—part of the Earth beneath the crust and above the core, composed almost exclusively of ultramafic rocks made up of olivine and pyroxenes.

Magma—molten form of an igneous rock e.g. granite.

Mélange—an association of different rock types showing a mixture of lithologies usually either in a sedimentary matrix or a serpentinite matrix.

Metamorphism—process whereby an original sedimentary or igneous rock has been changed to a crystalline rock by the increase of pressure (P) and temperature (T) during burial and heating.

Metasomatism—process of alteration of a metamorphic rock by hot fluids.

Mid-Ocean Ridge Basalt (MORB)—tholeiitic lava erupted from ocean-spreading centres.

Migmatite—a high-grade partially molten metamorphic rock composed of a melt phase (leucosome) and a restite phase (melanosome).

Mesozoic era—period of Earth history between 250 million years and 65 million years ago.

Mid-ocean ridge—linear zone of submarine mountains along which new basaltic oceanic crust is formed by volcanic eruptions.

Milankovich cycle—cyclic variations in climate as a result of the Earth's rotation and orbit, generally on 20,000, 40,000, and 100,000 year periodicity cycles.

Moho—short for Mohorovicic, disconformity a geophysical line separating the crust above from the mantle below.

Mohs' scale of mineral hardness—from 1 (talc) through gypsum (2), calcite (3), fluorite (4), apatite (5), orthoclase (6), quartz (7), topaz (8), corundum (9) to diamond (10).

Molasse—continental sediments deposited by rivers (conglomerates) or lakes (silts) following continental collision.

Monazite—$(Ce,La,Y,Th)PO_4$; very small uranium-bearing mineral, commonly used for uranium-lead geochronology and dating of metamorphic and igneous rocks.

Monsoon—regional wind system associated with heavy precipitation associated with tropical wind systems, from the Arabic '*mausim*'.
Moraine—linear mounds of loose rock debris formed along sides (lateral moraines) or termination (terminal moraines) of glaciers.
Muscovite—white mica, a sheet silicate mineral common in the Himalaya.
Mylonite—finely laminated rock formed under ductile conditions by recrystallization and flow, associated with large faults.
Nappe—large-scale recumbent fold sometimes associated with thrust fault along the base, formed as a result of horizontal compression in Alpine-type mountain belts.
Normal fault—extensional fault placing younger (or shallower) rocks onto older (or deeper) rocks.
Obduction—the process of emplacement of ophiolite complexes onto continental margins.
Oceanic crust—upper 5–7 km of the oceans composed of deep igneous rocks (gabbros), vertical sheeted dykes and upper pillow basalts with deep-sea sedimentary rocks, e.g. cherts, umbers, etc.
Olivine—silicate mineral rich in Fe and Mg, the common component of the mantle.
Ophiolites—a sequence of oceanic crustal rocks (pillow lavas, sheeted dykes, gabbros) and upper mantle rocks (peridotites) formed in the oceans but emplaced onto continental margins by a process called obduction (as opposed to subduction).
Orogeny—a period of mountain building (e.g. Himalayan orogeny).
Pahoehoe—ropy lava.
Palaeozoic—period of Earth history between 550 and 250 million years ago.
Passive continental margin—aseismic margins typically dominated by carbonate sedimentation.
Pelagic—open ocean environment, far away from a continental margin.
Pegmatite—coarse-graining granitic intrusive dyke rock.
Peridotite—an ultrabasic rock composed almost entirely of olivine and pyroxene with no quartz or feldspar; main rock type of the Earth's mantle.
Plagioclase feldspar—important rock-forming silicate mineral showing solid solution between the sodic end-member albite ($NaAlSi_3O_8$), through oligoclase, andesine, labradorite, bytownite to the calcic end-member anorthite ($CaAlSi_3O_5$).
Plate tectonics—the theory that Earth's relatively rigid lithospheric plates move independently driven by mantle convection.
Pluton—a large body of igneous rock (e.g. granite) intruded upwards in the crust.
Potassium-Argon (K-Ar)—dating system based on the radioactive decay of ^{40}K to ^{40}Ar.
Precambrian—period of Earth history older than 550 million years going back to the formation of the Earth ~4.5 billion years ago (includes Archaean and Proterozoic periods).
Pressure-Temperature-Time (PTt) path—the evolution of a metamorphic rock during its heating and burial (prograde path) and its subsequent exhumation (retrograde path).
Pseudotachylyte—glassy rock formed by frictional heating along a fault during an earthquake.

Pyroxene—silicate mineral commonly found in basic igneous rocks and a common component of the mantle. Monoclinic clinopyroxenes include augite, diopside, jadeite; orthorhombic orthopyroxenes include enstatite, hypersthene.
Quartz—SiO_2 silicate mineral composed entirely of silicon and oxygen.
Radioactive decay—the spontaneous decay of an atom to an atom of a different element by emission of one or more particles or photons from its nucleus (alpha, beta, gamma decay).
Rare earth element (REE)—Elements with atomic number between 57 and 71 occurring in tiny amounts in minerals.
Rayleigh number—fluid parameter that determines whether convection and flow will occur.
Red bed—an iron-stained sedimentary rock sequence formed in a continental setting.
Rheology—the study of flow and ductile deformation of rocks, including viscosity, plasticity and elasticity.
Ruby—Al_2SiO_3 corundum, a bright red precious gemstone found in marbles.
Schist—metamorphic rock showing a strong foliation (schistosity).
Sea-floor spreading—the process by which oceanic crust is formed along mid-ocean ridges and tectonic plates spread apart.
Sedimentary rock—bedded rock formed by surface processes of deposition in oceans (e.g. shallow marine limestones, deep marine turbidites, etc., or on continents by fresh-water deposition (e.g. lacustrine, fluvial) or wind (aeolian) processes.
Seismic velocity—the speed with which an elastic wave propagates through a medium.
Serac—a tower of ice formed as a glacier moves over a slope.
Serpentinite—rock composed of hydrated olivine or serpentine ($Mg_3Si_2O_5(OH)_4$).
Shale—a fine-grained, fissile sedimentary rock formed from compaction of mudstones.
Shear zone—a narrow zone of intense deformation, usually of ductile strain.
Shoshonite—potassium- or sodium-rich volcanic rock, requiring a deep hot mantle source.
Sill—a horizontal or layer-parallel igneous intrusion.
Sillimanite—Al_2SiO_5; aluminium silicate mineral typically associated with high-temperature regional metamorphism.
Solar radiation—electromagnetic radiated energy from the Sun.
South Tibetan Detachment—a large-scale north-dipping low-angle normal fault bounding the top of the Greater Himalayan metamorphic rocks.
Stratigraphy—the branch of geology that relates age ranges of sedimentary rocks to fossils and geochronology.
Strike-slip fault—a vertical fault line along which the blocks move in opposite directions on the horizontal plane.
Subduction zone—arcuate line along which one lithospheric plate slides beneath another associated with zone of deep earthquakes, high-pressure metamorphic rocks and deep trench; analogous with Benioff zone.
Surface uplift—uplift of the topographic surface relative to mean sea level (geoid).
Suture zone—a zone of collision between two continental plates usually containing rocks that were formed in that ocean (e.g. remnant ophiolites, deep-sea sedimentary rocks, mélanges, etc.).
Syntaxis—a 90° bend in an orogenic belt (e.g. West Himalayan Nanga Parbat syntaxis; East Himalayan Namche Barwa syntaxis).

Tethys Ocean—a large ocean spanning Late Palaeozoic and Mesozoic times that separated the Gondwana continents to the south (Africa, Arabia, India) from the European and Asian continental masses to the north. A northern Palaeo-Tethys ocean and a southern Neo-Tethys ocean were separated by a series of Cimmerian continental fragments.

Thermobarometry—determination of Pressure (P) and Temperature (T) conditions of a metamorphic rock using equilibrium thermodynamics.

Thrust fault—a shallow-dipping contractional fault where older (or deeper) rocks are emplaced above younger (or shallower) rocks.

Tourmaline—igneous or metamorphic boron silicate mineral typical component of Himalayan leucogranites, usually black in colour but can be pink, green, or red.

Trench—a long, narrow, and deep bathymetric depression forming above a subduction zone, characteristic of oceanic plate boundary (e.g. Mariana trench).

Tsunami—a large ocean surface wave formed as a result of uplifted ocean floor due to an earthquake.

Turbidite—sedimentary deposit typically showing graded bedding deposited as result of sediment transport down a continental margin slope.

Ultrabasic rock—rock composed almost entirely of mafic minerals (e.g. olivine, pyroxene) commonly called peridotite.

Unconformity—the irregular contact between two rock formations usually indicative of a period of erosion between the two.

Uplift—the upward motion of a rock relative to the Earth's surface or geoid (exhumation) or the surface uplift of the ground.

Uranium-Lead dating—dating system based on the radioactive decay of ^{238}U to ^{206}Pb (half life 4510 million years) and decay of ^{235}U to ^{207}Pb (half life 713 million years).

Viscosity—a measure of the strength of a fluid measured in poises, that determines whether it is capable of flow.

Volcanic arc—a chain of volcanic islands formed above a subduction zone characterized by basalts and andesite lavas.

Volcanic rock—a rock erupted from a volcano such as a basalt (basic), andesite, dacite, or rhyolite (acid).

Wadi—dry river valley (Arabic).

Xenolith—fragment of country rock preserved in volcanic feeder dykes or granites.

Zircon—$ZrSiO_4$; very small uranium-bearing mineral widely used in U-Pb geochronology for dating timing of crystallization of granite

APPENDIX C

LOCAL GLOSSARY

Allah-u-Akbar— 'God is Great' (Arabic)
Bhanjyang—pass (Nepali)
Bhotia—Tibetan (Nepali)
Bhote—from Tibet
Brahmin—upper caste (Hindu)
Caka—salt lake (Tibetan)
Caste system—social hierarchy class system (Hindu)
Chaddur—frozen Zanskar River in winter (Ladakh)
Chahara—waterfall (Nepali)
Chai—tea (Hindi)
Chang—local brew beer made from barley
Chomolungma—Mount Everest (Sherpa)
Chu—water, river (Tibetan)
Chutsen—hot spring (Tibetan)
Co or Tso—lake (Tibetan)
Cwm—top end of a valley (Welsh)
Dacoit—highway robber (India, Nepal)
Dalai Lama—spiritual leader of Tibet, Tenzin Gyatso
Dara—river (Tadjik, Khirghiz)
Deorali—ridge top (Nepali)
Dhal bhat—lentils and rice, staple food of Asia
Doonga wallah—geologist, rock wallah (Nepalese)
Dun—inter-montane valley in Siwalik hills (Hindi)
Drokpa—nomads of the high plateau (Tibet)
Druk Yul—Land of the Thunder Dragon (Bhutan)
Dun—valley in Siwalik hills (Hindi)
Dzo—cross between a yak and a cow
Dzong—fort, hilltop monastic town (Bhutanese, Tibetan)
Feng—peak (Chinese)
Gandaki—river (Nepali)
Gangri—snow peak (Tibetan)
Giapo—King (Ladakh, Zanskar)
Gol—stream (Khowar)
Gompa—monastery (Ladakhi, Tibetan, Nepali)
Haj—pilgrimage to Mecca (Arabic)
Himal—mountain range (Nepali)
Jebel—mountain (Arabic)
Kham—province in eastern Tibet
Khampa—people from Kham (east Tibet); bandits
Khola—stream, river (Nepali)
Khrebet—mountains (Russian)
Khlong—canal, stream (Thai)
Kosi—river (Nepali)
Kuh—mountains (Dari, Persian)
Kul—lake (Khirghiz)
Kund—holy lake (Nepali)
Kyang—wild ass (Tibetan)
La—pass (Tibetan)
Lama—Tibetan Buddhist monk
Lambadar—headman (Wakkhi)
Lekh—mountain range (Nepali)
Mandala—Tibetan wheel of life
Mani stones—Buddhist carved stones, sometimes in long walls
Muang—town (Burmese)
Mujahadeen—'holy warrior', freedom fighter (Arabic)
Mustagh—ice mountain (Turkic, Wakhi)
Nala—stream, valley (Hindi)

Namaste—hello (Nepali, Hindi)
Nirvana—Buddhist state of mind free from birth, suffering, and death
Nur—lake (Mongol)
Om mani padme hum—sacred Buddhist chant, 'hail to the jewel in the lotus'
Ozero—lake (Russian)
Pamir—high-altitude grassy pasture (Wakhi)
Phedi—foot of a hill (Nepali)
Phu—upper part of a valley (Tibetan)
Pokhari—tank, lake (Nepali)
Pundits—local Indian surveyors
Rai—Nepali people who live in lowlands, south of Everest
Rakshi—local brew spirit made from rice
Ri—peak (Nepali, Tibetan)
Sadhu—Hindu holy man (Hindi)
Sagamatha—Mount Everest (Nepali)
Salaam m'aleikum—'peace be with you' greeting (Arabic)
Saligram—black mud balls enclosing ammonite fossil (Nepali)
Serac—tower of ice on a glacier
Shar—east (Tibetan)
Shan—mountain (Chinese)
Sherpa—Buddhist hill people living around Everest; 'people from the east'
Shikara—small boat (Kashmiri)
Sirdar—porter leader (Nepali, Hindi, Urdu)
Sola Khumbu—district of Nepal around Mount Everest
Stupa—Buddhist circular shrine
Tal—lake (Indian, Nepalese)
Tamang—Tibeto-Burman people who live in foothill regions, Nepal
Tang—plain (Tibetan)
Tau—mountain (Kazakh)
Terai—flat plains south of the Himalayan front (Nepal)
Thanka—Buddhist religious painting, scroll
Tsampa—roasted barley flour, staple food of Tibet
Tulku—reincarnate lama (Tibet)
Tsang—province of west-central Tibet, region around Shigatse
Tso—or 'Co'; lake (Tibetan, Ladakhi)
U—province of central Tibet, around Lhasa
Wadi—dry river valley (Arabic)
Wallah—person; e.g. *doud* wallah—milk-man; *chai* wallah—tea person
Xia—gorge (Chinese)
Xian (Shian)—county (Chinese)
Yatra—pilgrimage (Hindi)
Yeti—mythical abominable snowman
Zangpo (Tsangpo)—river (Tibetan)
Zhunglam—'middle way' trekking route (Zanskari)
Zindabad—'long live' (Arabic, Urdu)
Zuk—coracle made from yak skins (Balti)

APPENDIX D

CHRONOLOGY OF EXPLORATION

1976 Expedition to Patagonian Andes, Chile. First ascents on Cerro Castillo (2,675 m), Cerro Puntudo, attempt on Cerro Hyades. Ascents of several volcanoes in Chile, including Volcan Osorno, Volcan Villarica. Trek across Cordillera Vilcabamba and Cordillera Blanca, Peru. Ascent of Volcan Cotopaxi (6,000 m) in Ecuador. Visit Galapagos Islands.

1978 Mountaineering expedition to East Kulu Himalaya, Tos Glacier region, India. Ascent of White Sail (6,730 m) and several other peaks on the Kulu–Lahoul border.

1980 Mountaineering expedition to south face of Langtang Lirung (7,234 m), central Nepal Himalaya. Trek from Trisuli bazaar to Langtang, up Langtang Glacier across Helambu-Gosainkund back to Kathmandu.
Move to Memorial University of Newfoundland, St.Johns, Canada.

1981 Geological exploration of Ladakh and Zanskar. Trek Kargil–Padam–Lamayuru. Climb peaks around the Kashmir Valley.

1982 Traverse of the western Himalaya. Leh–Marka–*'zhung lam'*–Padam–Manali–Kulu. Return trek to Kashmir from Manali to Kishtwar along the Chenab Valley.

1984 First trip to Tibet. Field-trip around Lhasa, Shigatse, Yamdock Tso, NyenchenTanggla Yarlung Tsangpo Valley. *Move to Leicester University, UK.*
First trip to the Karakoram, Pakistan. Trek from Skardu to Askole and along the Biafo Glacier to the Latok–Ogre Range, Snow Lake region.

1985 Expedition to the Baltoro Glacier region, Pakistan Karakoram. Climb part of the Abruzzi Ridge K2 (8,611 m), Gasherbrum Range, attempt to climb Masherbrum (7,821 m), and exploration of the Trango, Muztagh, Abruzzi, Vigne, Baltoro Kangri Glacier systems.

1986 Mountaineering expedition to Shivling (6,543 m), Garhwal Himalaya, India, and mapping along the Gangotri Glacier region.

1987 Mountaineering expedition to the Hushe Valley, Gondogoro-la, Layla Peak, Masherbrum south, North Pakistan.

1988 Mountaineering expedition to Biale (6,730 m), Lobsang Spires and Trango Towers, Baltoro Glacier, Pakistan. Crossing of the Karakoram from the Hunza Valley, via Shimshal, Lukpe-la, Snow Lake, Biafo Glacier, to Askole and Skardu.

1989 Mountaineering expedition to the Hispar Glacier, Snow Lake, and ascent of Mikeron Chhish (6,000 m). Exploration of the Hunza Valley, Rakhiot, and Diamir faces of Nanga Parbat. *Move to Oxford University, UK.*

1990 Mountaineering expedition to Shivling (6,543 m) Garhwal Himalaya, India, and mapping along the Gangotri Glacier.

1991	Ladakh, Zanskar, Kashmir. Geological mapping along the Indus suture zone and Zanskar Himalaya. Geological mapping along Langtang Himalaya, Nepal.
1992	Field excursion to Pamirs, drive from Pakistan, Hunza Xalley, Tashkurgan; Karakoram fault in Xinjiang; Gez defile (Kongur, Muztagh Ata peaks) Kashgar, southern Tien Shan, Urumchi.
1993	Geological explorations of Ladakh, Kishtwar, Chamba, Kashmir.
1994	Kathmandu HKT. Annapurna Base Camp trek, Nepal. Mountaineering expedition to Shisha Pangma (8,013 m); overland from Kathmandu to south Tibet. Geological mapping along the Hunza Valley and around Nanga Parbat, Pakistan.
1995	Ladakh Winter expedition along the *Chaddur*, the frozen Zanskar River and ski traverse Zanskar; Suru, Indus, Nubra valleys. Summer trek to the Spontang ophiolite region, Ladakh and Zanskar. First trip to Karakoram fault, Tangste Valley, and Pangong Lake. Tibet: drive across the plateau from Lhasa to the Karakoram, return to Kathmandu.
1996	Geological exploration of Kohistan, Pakistan, Hunza, Hindu Kush, Chitral, Kalash valleys Kafiristan. Geological mapping in Ladakh, Zanskar, Pangong Lake, Kashmir.
1997	Geological exploration of Sola Khumbu, Nepal–Everest Base Camp and cross Cho-la to Cho Oyu. Return to film Everest, Ama Dablam (*Earth Story*).
1998	Peshawar HKT, Pakistan. Geological exploration Kohistan, Hunza Karakoram, Skardu. Ladakh: Karakoram fault in Nubra Valley, Pangong Lake. Nepal: trek to Everest Base Camp.
1999	Thailand, Similan islands diving expedition.
2000	Thailand, Burma, Mergui islands, diving; North Thailand. Chengdu HKT, Yunnan, field trip to eastern Tibetan plateau, Gongga Shan. Tibet: Everest north side, Rongbuk Glacier, Cho Oyu Base Camp. Karakoram, Kohistan. Ladakh: mapping Karakoram fault in Nubra, Pangong Lake. Burma: Mogok belt, Mandalay, Myektina, Bagan, Rangoon.
2001	Karakoram; Return to the Baltoro Glacier, K2–Gondogoro-la–Hushe. Bhutan, Chomolhari–Lingshi trek.
2002	Sikkim, Gangtok HKT. Trek to Kangchenjunga, geology around Darjeeling. Nepal: Annapurna circuit, Kali Gandaki. Indonesia; Bali, Flores, Komodo Islands diving trip.
2003	Geological mapping around Everest, Nepal. Thailand–Laos, Mekong River exploration. Geological mapping around Everest, from Tibet, Rongbuk Glacier.
2004	Tibet: Kailas–Shiquanhe–central Tibet, Lhasa. Tibet: Everest, trek into Kangshung Valley.
2005	Geological trip to the Andaman Islands following 26 Dec. 2004 earthquake. North Vietnam; mapping the Red River fault, climbed Mt. FanSiPan, Sapa. Phuket and south Thailand, Malaysia, Taman Negara, Tioman Island. Nepal: Arun Valley–Makalu trek.
2006	China–field excursion North and NE Tibet–Golmud–Dunhuang–Xining. India–Ladakh, HKT (Tso Morari, Tangtse). Nepal; round Manaslu trek.

2007	Hong Kong HKT, Red River fault, Ailao Shan, and Diancang Shan, Yunnan.
	Thailand, Lansang, Wang Chao fault Chiang Mai.
2008	Andaman islands.
	Nepal; Geological mapping along Langtang Valley.
	North Vietnam; Red River fault, Ha Long Bay.
	Ladakh, Leh HKT: Tso Morari, Tangtse, Pangong, Zanskar, Kargil to Padam.
2009	Nepal, Pokhara Kali Gandaki filming (*How the Earth was Made*).
	Beijing HKT; Tibet: Linzhi, Basum Tso to Lhasa.
2010	Nepal—Everest Sola Khumbu.
	Tibet: Nyenchen Tangla Range, Lhasa to Basum Tso, Bayi, Linzhi.
2011	Malaysia granite trip. Kuala Lumpur, Cameron Highlands, Penang, Langkawi, Tuba Island, Kuala Terrenganu, Perhentian Islands, Kuantan.
	Nepal: Kali Gandaki, Dhaulagiri, and Pokhara Valley region.
	Sichuan: Chengdu to Baoxing, Barmie, Danba, Kangding, Moxi, Gongga Shan, east Tibet.
2012	Nepal: Modi khola, Annapurna Base Camp, Seti khola and filming with BBC (*Continents series*).
	Phuket Island granites; Malay granite fieldwork, Johore, Pahang, Singapore, Tioman Island, Kuala Lumpur.

HKT—Himalayan–Karakoram–Tibet Workshop meetings

NOTES

Chapter 1. Oceans and Continents

1. The most exhaustive account of the Survey of India is contained within the five volumes of R. H. Phillimore, *Historical Records of the Survey of India* (1950–1968, Dehra Dun). Excellent accounts of the Great Trigonometrical Survey are found in John Keay's books, *The Honourable Company: A History of the English East India Company* (1991, London, HarperCollins), *The Great Arc* (2000, London, HarperCollins), and *When Men and Mountains Meet* (1977, London, John Murray).
2. A. B. Watts (2001) *Isostasy and Flexure of the Lithosphere*, Cambridge; New York; Melbourne: Cambridge University Press.
3. S. G. Burrard and H. H. Hayden (1907) *A Sketch of the Geography and Geology of the Himalaya Mountains and Tibet*, Calcutta: Government of India (price two rupees!)
4. Arnold Heim and Augusto Gansser (1939) *Central Himalaya: Geological Observations of the Swiss Expedition 1936*, Zurich.
5. Augusto Gansser (1964) *Geology of the Himalayas*, Chichester: John Wiley & Sons.
6. One of the best summaries of the internal structure of the Earth, sea-floor spreading, continental drift, and all aspects of tectonics can be found in P. Kearey and F. J. Vine's *Global Tectonics* (1990), Blackwell Science.
7. Arthur Holmes (1944) *Principles of Physical Geology*, Edinburgh: Thomas Nelson & Sons.
8. E. Argand (1924) *La Tectonique de l'Asie*. 13th International Geological Congress, Brussels 1924, 171–372.
9. J. F. Dewey and J. M. Bird (1970) 'Mountain Belts and the New Global Tectonics', *Journal of Geophysical Research*, 75, 2625–47.
10. F. J. Vine and D. H. Matthews (1963) 'Magnetic Anomalies over Oceanic Ridges', *Nature*, 199, 947–9.
11. A. G. Smith, A. H. Hurley, and J. C. Briden (1981) *Phanerozoic Palaeocontinental World Maps*, Cambridge: Cambridge University Press. The most detailed maps of the break-up of Gondwana can be found in M. J. DeWit, M. Jeffery, H. Bergh, and L. Nicolaysen (1988) *Geological Map of Sectors of Gondwana Reconstructed to their Disposition~150 Ma*. American Association of Petroleum Geologists and University of Witwatersrand.
12. Dorrik Stow (2010) *Vanishing Ocean: How Tethys Reshaped the World*, Oxford; New York: Oxford University Press.
13. O. R. Green, M. P. Searle, R. I. Corfield, and R. M. Corfield (2008) 'Cretaceous-Tertiary Carbonate Platform Evolution and the Age of the India-Asia Collision along the Ladakh Himalaya (Northwest India), *Journal of Geology*, 116, 331–53.

14. Two of the great pioneers of ophiolite research were Ian Gass and Bob Coleman, who both organized the first detailed research projects in Oman following their early pioneering work in Cyprus and North America respectively. An excellent summary of the ophiolite concept and the evolution of thought has been published in Y. Dilek and S. Newcomb (2003) *Ophiolite Concept and the Evolution of Geological Thought*, Geological Society of America, Special Paper 373.
15. The pioneering and most significant early research on the Oman Mountains was the superb work published by Ken Glennie and his team of Shell geologists who spent five years mapping the whole Oman mountain region from 1969 to 1974. K. Glennie, M. G. A. Boeuf, M. W. Hughes-Clarke, M. Moody-Stuart, W. F. H. Pilaar, and B. M. Reinhardt (1974) *Geology of the Oman Mountains*, 2 volumes and geological map, Verhandelingen van het Koninklijk Nederlands geologisch mijnbouwkundig Genootschap.
16. M. P. Searle and J. Malpas (1980) 'Structure and Metamorphism of Rocks beneath the Semail Ophiolite of Oman and their Significance in Ophiolite Obduction', *Philosophical Transactions of the Royal Society of Edinburgh*, 71, 213–28. A more recent summary of the tectonic setting of the Oman Ophiolite can be found in M. P. Searle and J. S. Cox (1999) 'Tectonic Setting, Origin and Obduction of the Oman Ophiolite', *Geological Society of America Bulletin*, 111, 104–22.
17. J. A. Pearce, T. Alabaster, A. W. Shelton, and M. P. Searle (1981) 'The Oman Ophiolite as a Cretaceous Arc-Basin Complex: Evidence and Implications', *Philosophical Transactions of the Royal Society of London* Series A, 300, 299–317.

Chapter 2. Continents in Collision: Kashmir, Ladakh, Zanskar

1. There are many excellent books on the history and exploration of Western Tibet, Ladakh, Baltistan, and Kashmir, such as John Keay's two books *When Men and Mountains Meet* (1977) and *The Gilgit Game* (1979), both published by John Murray.
2. A selection of the best photo books on the scenery and culture of Ladakh are the following: Heinrich Harrer (1980) *Ladakh*, Penguin-Verlag, Insbruck, Austria; Rajesh Bedi and Ramesh Bedi (1986) *Ladakh*, Delhi: Brijbasi Press; Siddiq Wahid and Kenneth Storm Jr. (1989) *Ladakh Between Earth and Sky*, Bombay: B.I. Publications; Kenneth Storm Jr., Joanna van Gruisen, and Nina Rao (1989) *Ladakh: The Secret Land beyond the Himalaya*, New Delhi: Lustre Press. For a thorough, detailed description of the culture of Ladakh, see D. L. Snellgrove and T. Skorupsky, (1980) *Cultural Heritage of Ladakh*, Warminster: Aris & Philips.
3. For a good description of the Kashmir conflict, see Victoria Schofield (2000) *Kashmir in Conflict*, London; New York: I.B. Tauris. Also see Eric S. Margolis (2000) *War at the Top of the World*, New York: Routledge.
4. For descriptions of the folds and thrusts of the Zanskar Range and the southward extrusion of the middle crust, see M. P. Searle (1986) 'Structural Evolution and Sequence of Thrusting in the High Himalayan, Tibetan Tethys and Indus Suture Zones of Zanskar and Ladakh, Western Himalaya', *Journal of Structural Geology*, 8, 923–36; M. P. Searle, D. J. W. Cooper, and A. J. Rex (1988) 'Collision Tectonics of the Ladakh-Zanskar Himalaya', Philosophical Transactions of the Royal Society, London, A326, 117–150, and M. P. Searle and A. J. Rex (1989) 'Thermal Model for the Zanskar Himalaya', *Journal of Metamorphic Geology*, 7, 127–34.

5. See B. J. Stephenson, D. J. Waters, and M. P. Searle (2000) 'Inverted Metamorphism and the Main Central Thrust: Field Relations and Thermobarometric Constraints from the Kishtwar Window, NW Indian Himalaya', *Journal of Metamorphic Geology*, 18, 571–90, and B. J. Stephenson, M. P. Searle, D. J. Waters, and D. C. Rex (2001) 'Structure of the Main Central Thrust Zone and Extrusion of the High Himalayan Deep Crustal Wedge, Kishtwar-Zanskar Himalaya', *Journal of the Geological Society of London*, 158: 637–52.
6. A. Heim and A. Gansser (1939) and A. Gansser (1964) *Geology of the Himalaya*, Chichester: John Wiley & Sons. See also the nice photo book by Blanche Olschak, Augusto Gansser, and Emil Bührer (1987) *Himalaya*, Switzerland: Motovun.
7. J. F. Dewey, R. M. Shackleton, Chengfa Chang, and Yiyin Sun (editors) 1990 *The Geological Evolution of Tibet*, Royal Society of London Memoir.
8. P. LeFort (1981) 'Manaslu Leucogranite: A Collision Signature of the Himalaya, A Model for its Genesis and Emplacement', *Journal of Geophysical Research*, 86: 10545–68.

Chapter 3. The Dreaming Spires of the Karakoram

1. The best books on the history of exploration and mountaineering on K2 are Jim Curran (1995) *K2: The Story of the Savage Mountain*, London: Hodder and Stoughton; Galen Rowell's (1977) *In the Throne Room of the Mountain Gods*, London: George Allen & Unwin; and Roberto Mantovani and Kurt Diemberger (1997) *K2: Challenging the Sky*, Seattle. MA: Mountaimeers Books.
2. Eric Shipton was an icon of true exploration and early mountain climbing. His book on the extensive exploration around K2 and the Karakoram is told in Eric Shipton (1938) *Blank on the Map*, London: Hodder and Stoughton; Eric Shipton (1969) *That Untravelled World*, London: Hodder and Stoughton.
3. Charles Houston and Bob Bates wrote two mountaineering classics (1939) *Five Miles High*, New York: Dodd, Mead, & Co., and (1954) *K2: The Savage Mountain*, New York: McGraw-Hill.
4. A. Desio and B. Zanettin (1974) *Geology of the Baltoro Basin: Italian Expeditions to the Karakoram and Hindu Kush*, Leiden: Brill. The story of the first ascent of K2 is told in Lino Lacedelli and Giovanni Caenacchi (2006) *K2: The Price of Conquest*, Ross-on-Wye: Carreg Ltd.
5. Two lavishly illustrated books on mountaineering exploits are Doug Scott's *Himalayan Climber: A Life Time's Quest to the World's Greater Ranges*, London: Bâton Wickes, and Chris Bonnington's (1989) *Mountaineer: Thirty Years of Climbing on the World's Great Peaks*, London: Diadem Books.
6. Our first papers on the Baltoro and Hushe Karakoram work were published in:
   A. J. Rex, M. P. Searle, R. Tirrul, M. B. Crawford, D. J. Prior, D. C. Rex, and A. Barnicoat (1988) 'The Geochemical and Tectonic Evolution of the Central Karakoram, N. Pakistan', *Philosophical Transactions of the Royal Society*, London, A326, 229–55.
   M. P. Searle, A. J. Rex, R. Tirrul, D. C. Rex, A. Barnicoat, and B. F. Windley (1989) 'Metamorphic, Magmatic and Tectonic Evolution of the Central Karakoram in the Biafo-Baltoro-Hushe Regions of N. Pakistan', *Geological Society America Special Paper* 232: 47–73.
   M. P. Searle, R. R. Parrish, R. Tirrul, D. C. Rex (1990) 'Age of Crystallization and Cooling of the K2 Gneiss in the Baltoro Karakoram', *Journal Geological Society of London*, 147: 603–6.

M. P. Searle (1991) *Geology and Tectonics of the Karakoram Mountains*, Chichester: John Wiley & Sons. This book includes a geological map of the Central Karakoram Mountains at scale 1:250,000.

7. Cameron Wake and Mike Searle (1993) 'Rapid Advance of the Pumarikish Glacier, Hispar Glacier Basin, Karakoram Himalaya', *Journal of Glaciology*, 39(131): 204–206.

Chapter 4. Pressure, Temperature, Time, and Space

1. One of the best accounts of the evolution of geological ideas is given in Bill Bryson's superb book (2003) *A Short History of Nearly Everything*, London: Doubleday Transworld Publishers.
2. A thorough account of the Everest expedition including Wyn Harris and Wager's attempt on the summit is given in Hugh Ruttledge (1934) *Everest 1933*, London: Hodder & Stouton.
3. An entertaining account of William Smith, the man who made the first geological map of the British Isles in 1815, and the birth of geology as a science is given in Simon Winchester (2001) *The Map that Changed the World*, London: Penguin, Viking.
4. An excellent account of the Highland Controversy in Scotland and the history of geological exploration and ideas is given in Malcolm Rider (2005) *Hutton's Arse: 3 Billion Years of Extraordinary Geology in Scotland's Northern Highlands*, Sutherland: Rider-French.
5. B. N. Peach, J. Horne, W. Gunn, C. T. Clough, L. W. Hinxman, and J. J. H. Teall (1907) *The Geological Structure of the North-West Highlands of Scotland*, Memoir of the Geological Survey of Great Britain, Glasgow: HMSO.
6. A. Heim and A. Gansser (1939) *Himalaya: Geological Results of the Swiss Expedition, 1933*; and A. Gansser (1964) *Geology of the Himalaya*, Chichester: John Wiley & Sons.
7. Bruce Yardley (1989) *Introduction to Metamorphic Petrology*, Harlow: Longman Earth Science Series.
8. A recent discussion of the Himalayan inverted metamorphism along the Main Central Thrust can be found in M. P. Searle, R. D. Law, L. Godin, K. P. Larson, M. J. Streule, J. M. Cottle, and M. J. Jessup (2008) 'Defining the Himalayan Main Central Thrust in Nepal', *Journal of the Geological Society of London*, 164: 523–34.
9. B. J. Stephenson, D. J. Waters, and M. P. Searle (2000) 'Inverted Metamorphism and the Main Central Thrust: Field Relations and Thermobarometric Constraints from the Kishtwar Window, NW Indian Himalaya', *Journal of Metamorphic Geology*, 18: 571–90.
   J. D. Walker, M. W. Martin, S. Bowring, M. P. Searle, D. J., Waters, and K. V. Hodges, (1999) 'Metamorphism, Melting and Extension: Age Constraints from the High Himalayan Slab of SE Zanskar and NW Lahoul, Indian Himalaya', *Journal of Geology*, 107: 473–95.
   C. B. Walker, M. P. Searle, and D. J. Waters (2001) 'An Integrated Tectono-Thermal Model for the Evolution of the High Himalayas in Western Zanskar, with Constraints from Thermobarometry Metamorphic Modelling', *Tectonics*, 20: 810–33.10.
10. M. J. Jessup, M. P. Searle, J. M. Cottle, R. D. Law, D. L. Newell, R. J. Tracy, and D. J. Waters (2008) 'P-T-t-D Paths of Everest Series Schist, Nepal', *Journal of Metamorphic Petrology*, doi:10.1111/j.1525-1314.2008.00784.x.
11. M. J. Streule, M. P. Searle, D. J. Waters, and M. Horstwood (2010) 'Metamorphism, Melting and Channel Flow in the Greater Himalayan Sequence and Makalu Leucogranite: Constraints from Thermobarometry, Metamorphic Modeling and U-Pb Geochronology, *Tectonics*, 29, TC5011 doi:10.1029/2009TC002533.

12. G. Brent Dalrymple (1991) *The Age of the Earth*, Stanford: Stanford University Press.
13. An excellent account of the science of dating rocks in particular following the career of Arthur Holmes is given in Cherry Lewis (2000) *The Dating Game*, Cambridge: Cambridge University Press.
14. Randall Parrish and Stephen Noble (2003) 'Zircon U-Th-Pb Geochronology by Isotope Dilution – Thermal Ionization Mass Spectrometry (ID-TIMS)', in J. Hanchar and P. Hoskin (eds) *Zircon: Reviews of Mineralogy and Geochemistry*, Mineralogical Society of America, 183–213.
15. G. A. Wagner and P. Van den Haute (1992) *Fission-Track Dating*, Dordrecht: Kluwer Academic Publishers.
16. T. Dunai (2010). *Cosmogenic Nuclides: Principles, Concepts and Applications in the Earth Surface Science*, Cambridge: Cambridge University Press.
17. Douglas Burbank and Robert Anderson (2005) *Tectonic Geomorphology*, Chichester: Wiley-Blackwell.
18. M. P. Searle, S. R. Noble, A. J. Hurford, and D. C. Rex (1999) 'Age of Crustal Melting, Emplacement and Exhumation History of the Shivling Leucogranite', *Geological Magazine*, 136: 513–25.
19. M. P. Searle, P. R. Parrish, K. V. Hodges, A. Hurford, M. W. Ayres, and M. J. Whitehouse (1997) 'Shisha Pangma Leucogranite, South Tibet: Field Relations, Geochemistry, Age, Origin and Emplacement', *Journal of Geology*, 105: 295–317.
20. S. R. Noble and M. P. Searle (1995) 'Age of Crustal Melting and Leucogranite Formation from U-Pb Zircon and Monazite Dating in the Western Himalaya, Zanskar, India', *Geology*, 23: 1135–8.

Chapter 5. Frozen Rivers and Fault Lines

1. Mike Searle, Richard Corfield, Ben Stephenson, and Joe McCarron (1997) 'Structure of the North Indian Continental Margin in the Ladakh-Zanskar Himalaya: Implications for the Timing of Obduction of the Spontang Ophiolite, India-Asia Collision and Deformation Events in the Himalaya', *Geological Magazine*, 134: 297–316.
2. Zanskar Ski School in Padam: <http://zanskarskischool.org/>.
3. P. Molnar and P. Tapponnier (1975) 'Cenozoic Tectonics of Asia: Effects of a Continental Collision', *Science*, 189: 419–26.
4. M. P. Searle (1996) 'Geological Evidence Against Large-Scale Pre-Holocene Offsets along the Karakoram Fault: Implications for the Limited Extrusion of the Tibetan Plateau', *Tectonics*, 15: 171–86.
5. R. J. Phillips, R. R. Parrish, and M. P. Searle (2004) 'Age Constraints on Ductile Deformation and Long-Term Slip Rates along the Karakoram Fault', *Earth and Planetary Science Letters*, 226: 305–19.
   R. J. Phillips and M. P. Searle (2007) 'Macrostructural and Microstructural Architecture of the Karkoram Fault: Relationship between Magmatism and Strike-Slip Faulting', *Tectonics*, 26 TC3017, doi:10.029/2006TC001946.
6. M. P. Searle (1991) *Geology and Tectonics of the Karakoram Mountains*, with geological map of the central Karakoram 1:250,000 scale, Chichester: John Wiley & Sons.

7. M. P. Searle and R. J. Phillips (2007) 'Relationships between Right-Lateral Shear along the Karakoram Fault Metamorphism, Magmatism Exhumation and Uplift: Evidence from the K2-Gasherbrum-Pangong Ranges, North Pakistan and Ladakh', *Journal of Geological Society of London*, 164: 439–50.
8. R. Laccasin, F. Valli et al. (2004) 'Large-Scale Geometry, Offset and Kinematic Evolution of the Karakoram Fault, Tibet', *Earth and Planetary Science Letters*, 219: 255–69.
9. M. J. Streule, R. J. Phillips, M. P. Searle, D. J. Waters, and M. Horstwood, (2009) 'Evolution and Chronology of the Pangong Metamorphic Complex adjacent to the Karakoram Fault, Ladakh: Constraints from Thermobarometry, Metamorphic Modelling and U-Pb Geochronology', *Journal of the Geological Society of London*, 166: 919–32.
10. The arguments for both models are discussed fully in the review paper: M. P. Searle, J. R. Elliott, R. J. Phillips, and Sun-Lin Chung (2011) 'Crustal-Lithospheric Structure and Continental Extrusion of Tibet', *Journal of the Geological Society of London*, 168: 633–72, doi: 10.1144/0016-76492010-139.
11. M. P. Searle, R. R. Parrish, A. V. Thow, S. R. Noble, R .J. Phillips, and D. J. Waters (2010) 'Anatomy, Age and Evolution of a Collisional Mountain Belt: The Baltoro Granite Batholith and Karakoram Metamorphic Complex, Pakistani Karakoram', *Journal of the Geological Society of London*, 167: 183–202, doi: 10.1144/0016-76492009-043.

Chapter 6. North-West Frontier: Kohistan, Hindu Kush, Pamirs

1. The most readable account of the rise of the Taliban and conflicts in Afghanistan and Pakistan is Ahmed Rashid's (2001) *Taliban, The Story of the Afghan Warlords*, London: Pan Books.
2. M. P. Searle (1991) *Geology and Tectonics of the Karakoram Mountains*, with geological map of the Central Karakoram Range, scale 1:250,000, Chichester: John Wiley & Sons.
3. P. R. Hildebrand, S. R. Noble, M. P. Searle, and R. R. Parrish (2001) 'Old Origin for an Active Mountain Range: Geology and Geochronology of the Eastern Hindu Kush, Pakistan', *Geological Society of America Bulletin*, 113, 625–39. See also the geological map of the Hindu Kush in P. R. Hildebrand, M. P. Searle, Zafarali Khan Shakirullah, and H. J. van Heijst (2000) 'Geological Evolution of the Hindu Kush, NW Frontier Pakistan: Active Margin to Continent–continent Collision Zone', in M. Asif Khan, P. J. Treloar, M. P. Searle, and M. Qasim Jan, (eds) 'Tectonics of the Nanga Parbat Syntaxis and the Western Himalayas', *Geological Society of London Special Publication*, 170: 277–93.
4. P. R. Hildebrand, S. R. Noble, M. P. Searle, R. R. Parrish, and Z. K. Shakirullah (1998) 'Tectonic Significance of 24 Ma Crustal Melting in the Eastern Hindu Kush, Pakistan', *Geology*, 26: 871–4.
5. M. G. Pettersen (2010) 'A Review of the Geology and Tectonics of the Kohistan Island Arc', *Geological Society of London Special Publication* 338: 287–327.
6. R. G. Coleman and X. Wang (eds) (1995) *Ultrahigh Pressure Metamorphism*, New York: Cambridge University Press.
7. The graphite–diamond transition has a positive slope in P-T space so at lower temperatures of about 500°C it occurs at 30 kilobars but at 900°C it occurs at 40 kilobars, equivalent to about 150 kilometres depth.
8. P. O'Brien, N. Zotov, R. Law, M. A. Khan, and M. Q. Jan (2001) 'Coesite in Himalayan Eclogite and Implications for Models of India-Asia Collision', *Geology*, 29: 435–8.

9. R. R. Parrish, S. J. Gough, M. P. Searle, and D. J. Waters (2006) 'Plate Velocity Exhumation of Ultrahigh-Pressure Eclogites in the Pakistan Himalaya', *Geology*, 34: 989–92.
10. G. Pegler and S. Das (1998) 'An Enhanced Image of the Pamir–Hindu Kush Seismic Zone from Relocated Earthquake Hypocenters', *Geophysical Journal International*, 134: 573–95.
11. M. P. Searle, B. R. Hacker, and R. Bilham (2001) 'The Hindu Kush Seismic Zone as a Paradigm or the Creation of Ultrahigh-Pressure Diamond- and Coesite-Bearing Continental Rocks', *Journal of Geology*, 109: 143–53.

Chapter 7. Faces of Everest

1. Toni Hagen (1960) *Nepal*, Bern: Kümmerly and Frey.
2. Tony Astill (2005) *Mount Everest—The Reconnaisance 1935: The Forgotten Adventure*, privately published.
   Michael Ward (2003) *Everest: A Thousand Years of Exploration*, Glasgow: The Ernest Press.
3. Stephen Venables (2000) *Everest: Alone at the Summit*, Bath: Odyssey Books.
4. Jon Krakauer (1998) *Into Thin Air: A Personal Account of the Mt. Everest Disaster*, New York: Anchor Books Doubleday.
5. Tim McCartney-Snape (1991) *Everest: From Sea to Summit*, Australian Geographic Society.
6. Anatoli Boukreev and G. Weston DeWald (1999) *The Climb: Tragic Ambitions on Everest*, New York: St. Martin's Griffin.
7. David Breashers (1999) *An Enduring Passion for Everest and Unforgiving Places*, New York: Simon & Schuster.
8. Reinhold Messner (1998) *The Crystal Horizon: Everest—The First Solo Ascent*, and (1999) *Everest: Expedition to the Ultimate*, both published by Mountaineers Books, Seattle, WA, US.
9. Roberto Mantovani (1997). *Everest: The History of the Himalayan Giant*, New Delhi: Om Book Service.
10. John Hunt (1953) *The Ascent of Everest*, London: Hodder and Stoughton; Edmund Hillary (1955) *High Adventure*, London: Hodder and Stoughton, reprinted by Oxford University Press; Jan Morris, (2000) *Coronation Everest*, London: Faber & Faber.
11. Jamling Tenzin Norgay (2002) *Touching My Father's Soul: In the Footsteps of Tenzin Norgay*, London: Wisdom Books.

Chapter 8. Mapping the Geology of Everest and Makalu

1. P. F. M. Fellowes (1934) *First Over Everest: The Houston Mount Everest Expedition 1933*, New York: Robert M. McBride.
2. A. M. Heron (1922) 'Geological Results of the Mount Everest Reconnaissance Expedition', *Records of the Geological Survey of India*, LIV: 215–34; N. E. Odell (1925) 'Observations on the Rocks and Glaciers of Mount Everest', *Geographical Journal*, LXVI: 289–315.
3. L. R. Wager (1934) 'A Review of Geology and Some New Observations', in: H. Ruttledge, *Everest 1933*, 312–36, Hodder and Stoughton; L. R. Wager (1939) 'The Lachi Series of North Sikkim and the Age of the Rocks Forming Mount Everest', *Records of the Geological Survey of India*, LXXIV: 171–88; L. R. Wager (1965) 'Injected Granite Sheets of the Rongbuk Valley and the North Face of Mount Everest', D. N. Wadia Commemorative volume, Mining and Metallurgical Institute of India.

4. National Geographic Map of Mount Everest, Boston Museum of Science.
5. Leo Dickinson, (1993) *Ballooning over Everest*, Jonathan Cape Ltd.
6. M. P. Searle (2003) Geological Map of the Mount Everest Region Nepal and South Tibet, scale 1:50,000. Oxford University. Second edition published 2007.
7. M. P. Searle (1999) 'Extensional and Compressional Faults in the Everest–Lhotse Massif, Khumbu Himalaya, Nepal', *Journal of the Geological Society of London*, 156: 227–40; M. P. Searle (1999) 'Emplacement of Himalayan Leucogranites by Magma Injecion Along Giant Sill Complexes: Examples from the Cho Oyu, Gyachung Kang and Everest Leucogranites (Nepal Himalaya), *Journal of Asian Earth Sciences*, 17: 773–83.
8. Simon Lamb and David Sington (2003) *Earth Story*, London: BBC Books.
9. J. M. Cottle, M. J. Jessup, D. L. Newell, M. P. Searle, R. D. Law, and M. S. A. Horstwood (2007) 'Structural Insights into the Early Stages of Exhumation along an Orogen-Scale Detachment: The South Tibetan Detachment System, Dzakaa Chu Section, Eastern Himalaya', *Journal of Structural Geology*, 29: 1781–97, doi:10.1016/j.jsg.2007.08.007.
10. Augusto Gansser (1964) *Geology of the Himalaya*, Chichester: John Wiley & Sons.
11. John McPhee (1982) *Basin and Range*, New York: Farrar, Straus and Giroux, 240.
12. P. M. Myrow, N. C. Hughes, M. P. Searle, C. M. Fanning, S-C. Peng, and S. K. Parcha (2008) 'Stratigraphic Correlation of Cambrian-Ordovician Deposits along the Himalaya: Implications for the Age and Nature of Rocks in the Mount Everest Region', *Bulletin of the Geological Society of America*, 120: 323–32.
13. R. D. Law, M. P. Searle, and R. L. Simpson (2004) 'Strain, Deformation Temperatures and Vorticity of Flow at the Top of the Greater Himalayan Slab, Everest Massif, Tibet', *Journal of Geological Society of London*, 161: 305–20.
    M. Jessup, R. D. Law, M. P. Searle, and M. S. Hubbard (2006) 'Structural Evolution and Vorticity of Flow during Extrusion and Exhumation of the Greater Himalayan Slab, Mount Everest Massif, Tibet/Nepal: Implications for Orogen-Scale Flow Partitioning', in: R. D. Law, M. P. Searle, and L. Godin (eds) 'Channel Flow, Ductile Extrusion and Exhumation in Continental Collision Zone', *Geological Society of London Special Publication*, 268: 379–413.
14. *How The Earth Was Made*, History Channel TV documentary.
15. B. C. Burchfiel, C. Zhiliang, K. V. Hodges, L. Yuping, L. Royden, D. Changrong, and X. Jiene, (1992) 'The South Tibetan Detachment System, Himalayan Orogen: Extension Contemporaneous with and Parallel to Shortening in a Collisional Mountain Belt', *Geological Society of America Special Paper*, 269.
16. M. P. Searle, R. L. Simpson, R. D. Law, R. R. Parrish, and D. J. Waters (2003) 'The Structural Geometry, Metamorphic and Magmatic Evolution of the Everest Massif, High Himalaya of Nepal–South Tibet', *Journal of Geological Society, London*, 160: 345–66.
    M. P. Searle, R. D. Law, M. Jessup, and R. L. Simpson (2006) 'Crustal Structure and Evolution of the Greater Himalaya in Nepal–South Tibet: Implications for Channel Flow and Ductile Extrusion of the Middle Crust', in R. D. Law, M. P. Searle, and L. Godin (eds) 'Channel Flow, Ductile Extrusion and Exhumation in Continental Collision Zones', *Geological Society of London Special Publication*, 268: 355–78.
17. J. M. Cottle, M. P. Searle, M. S. A. Horstwood, and D. J. Waters (2009) 'Timing of Mid-Crustal Metamorphism, Melting and Deformation in the Mount Everest Region of South Tibet Revealed by U-Th-Pb Geochronology', *Journal of Geology*, 117: 643–64, doi: 10.1086/60599.

18. M. J. Streule, M. P. Searle, D. J. Waters, and M. Horstwood (2010) 'Metamorphism, Melting and Channel Flow in the Greater Himlayan Sequence and Makalu Leucogranite: Constraints from Thermobarometry, Metamorphic Modeling and U-Pb Geochronology', Tectonics, 29, TC5011, doi:10.1029/2009TC002533.
19. M. P. Searle and A. J. Rex (1989) 'Thermal Model for the Zanskar Himalaya', *Journal of Metamorphic Geology*, 7: 127–34.
20. C. Beaumont, R. A. Jamieson, M. H. Nguyen, and B. Lee (2001) 'Himalayan Tectonics Explained by Extrusion of a Low-Viscosity Crustal Channel Coupled to Focused Surface Denudation. *Nature*, 414: 738–42.
21. K. D. Nelson, W. Zhao, and project INDEPTH members (1996) 'Partially Molten Middle Crust beneath Southern Tibet: Synthesis of Project INDEPTH Results', *Science*, 274: 1684–96.

Chapter 9. Mountains and Maoists: Annapurna, Manaslu

1. M. Colchen, A. Pêcher, and P. LeFort (1986) *Annapurna—Manaslu—Ganesh Himalaya*, Paris: CNRS. Map dated 1980.
2. The first geologists to propose a low-angle normal fault along the northern Himalaya were R. Caby et al. (1984) working in Annapurna region and J-P. Burg working in southern Tibet.
3. K. V. Hodges, R. R. Parrish, and M. P. Searle (1996) 'Tectonic Evolution of the Central Annapurna Range, Nepalese Himalayas', *Tectonics*, 15(6): 1264–91.
4. M. P. Searle and L. Godin (2003) 'The South Tibetan Detachment and the Manaslu Leucogranite: A Structural Reinterpretation and Restoration of the Annapurna-Manaslu Himalaya, Nepal', *Journal of Geology*, 111: 505–23.
5. Prakash A. Raj (2004) *Maoists in the Land of Buddha*. Nirala, Delhi and Kathmandu.
6. Jonathan Gregson (2002) *Massacre at the Palace—The Doomed Royal Dynasty of Nepal*, New York: Miramax.
7. M. P. Searle (2010) 'Low-Angle Normal Faults in the Compressional Himalayan Orogen: Evidence from the Annapurna–Dhaulagiri Himalaya, Nepal', *Geosphere*, 6(4): 296–315.

Chapter 10. Around the Bend: Nanga Parbat, Namche Barwa

1. Heinrich Harrer (1953) *Seven Years in Tibet*, London: Rupert Hart-Davis.
2. Herman Buhl (1956) *Nanga Parbat Pilgrimage*, London: Hodder and Stoughton.
3. Reinhold Messner (1980). *Solo, Nanga Parbat*, New York: Oxford University Press.
4. D. N. Wadia (1931) 'The Syntaxis of the Northwest Himalaya: Its Rocks, Tectonics and Orogeny, *Records of the Geological Survey of India*, 65: 189–220.
5. Peter Misch (1949) 'Metasomatic Granitization of Batholithic Dimensions', *American Journal of Science*, 247: 209–45.
6. Bowen's reaction series was proposed in 1928 to explain how minerals respond to changing equilibrium conditions during cooling of magma. The series demonstrates the evolving minerals in a discontinuous series from the first to crystallize: olivine—pyroxene—amphibole—biotite—muscovite—quartz, and the progression from calcium-rich to sodium- to potassium-rich feldspar in the continuous series.

7. An entertaining account of the granite controversy is given in Davis A. Young (2003) *Mind over Magma: The Story of Igneous Petrology*, Princeton, NJ: Princeton University Press.
8. M. P. Searle, J. M. Cottle, M. J. Streule, and D. J. Waters (2010) 'Crustal Melt Granites and Migmatites along the Himalaya: Melt Source, Segregation, Transport and Granite Emplacement Mechanisms', *Transactions of the Royal Society*, Edinburgh, 100: 219–33.
9. P. K. Zeitler et al. (2001) 'Crustal Reworking at Nanga Parbat, Pakistan: Metamorphic Consequences of Thermal-Mechanical Coupling Facilitated by Erosion', *Tectonics*, 20: 712–28.
10. J. L. Crowley, D. J. Waters, M. P. Searle, and S. A. Bowring (2009) 'Pleistocene Melting and Rapid Exhumation of the Nanga Parbat Massif, Pakistan: Age and P-T Conditions of Accessory Mineral Growth in Migmatite and Leucogranite', *Earth and Planetary Science Letters*, 288: 408–20.
11. F. Drew (1875) *Jummo and Kashmir*, London: Edward Stanford.
12. R. W. H. Butler and D. J. Prior (1988) 'Tectonic Controls on the Uplift of the Nanga Parbat Massif, Pakistan Himalaya', *Nature*, 333: 247–50.
13. Frank Kingdon-Ward (1926) *The Riddle of the Tsangpo Gorges*, London: Edward Arnold & Co.
14. A. L. Booth, C. P. Chamberlain, W. S. F. Kidd, and P. K. Zeitler (2009) 'Constraints on the Metamorphic Evolution of the Eastern Himalayan Syntaxis from Geochronologic and Petrologic Studies of Namche Barwa', *Geological Society of America Bulletin*, 121: 385–407.
15. P. K. Zeitler et al. (2001) 'Erosion, Himalayan Geodynamics and the Geomorphology of Metamorphism', *GSA Today*, 11: 4–9.
16. F. M. Bailey (1945) *China, Tibet, Assam*, London: Cape; F. M. Bailey (1957) *No Passport to Tibet*, London: Rupert Hart-Davis,.
17. Frank Kingdon-Ward wrote many books about his plant-hunting expeditions amongst which are: (1913) *The Land of the Blue Poppy*, Cambridge: Cambridge University Press; (1934) *A Plant-hunter in Tibet*, London: Jonathan Cape; (1941) *Assam Adventure*, London: Jonathan Cape; (1949) *Burma's Icy Mountain*, London: Jonathan Cape.
18. K. Cox, K. Storm, and L. Baker, I. (eds) (1999) *Frank Kingdon Ward's Riddle of the Tsangpo Gorges: Retracing the Epic Journey of 1924–25 in Southeast Tibet*, Woodbridge, Suffolk: Antique Collectors Club.
19. I. Baker (2004) *The Heart of the World: A Journey to the Last Secret Place*, New York: Souvenir Press.

Chapter 11. Roof of the World: Tibet, Pamirs

1. Charles Allen (1982) *A Mountain in Tibet: The Search for Mount Kailas and the Sources of the Great Rivers of India*, London: Deutsch.
   B. Olschak, A. Gansser, and E. M. Bührer (1987) *The Himalayas*, New York and Oxford: Facts on File.
2. Frank Kingdon-Ward (1913) *Land of the Blue Poppy*.
   Frank Kingdon-Ward (1926) *Riddle of the Tsangpo Gorges*.
   Ian Baker (2006) *Heart of the Himalaya, a Journey to Tibet's Lost Paradise*. (Penguin)
3. Probably the best books on the subject are John Keay's (1999) *The Great Game*, London: Jonathan Cape, and four outstanding and very readable accounts of early travellers in Tibet and Central Asia by Peter Hopkirk: (1984) *Foreign Devils on the Silk Road: The Search for the Lost Cities and Treasures of Chinese Central Asia*, London: John Murray; (1995) *Setting the East Ablaze:*

Lenin's Dream of an Empire in Asia, London: Kodansha International; (1982) *Trespassers on the Roof of the World: The Secret Exploration of Tibet*, London: John Murray; (2006) *The Great Game*, London: John Murray.

4. Derek Waller (1988) *The Pundits: British Exploration of Tibet and Central Asia*, Lexington: University Press of Kentucky.
5. F. M. Bailey (1945) *China, Tibet, Assam*, London: Cape.
   F. M. Bailey (1957) *No Passport to Tibet*, London: Rupert Hart-Davies.
6. Sven Hedin (1898) *Through Asia*, London: Methuen & Co, 2 volumes; Sven Hedin (1910) *Trans-Himalaya Discoveries and Adventures in Tibet*, 4 volumes. MacMillan, London.
7. Francis Younghusband (1896) *The Heart of a Continent*, London: John Murray, reprinted 1984 Oxford University Press.
8. A wonderful biography of Francis Younghusband is by Patrick French (1994) *Younghusband: The Last Great Imperial Adventurer*, London: HarperCollins.
9. C. J. Allégre and 34 others (1984) 'Structure and Evolution of the Himalaya—Tibet Orogenic Belt, *Nature*, 307: 17–22.
10. J-P. Burg (1984) Geological map of South Tibet, CNRS, Paris.
11. Heinrich Harrer (1953) *Seven Years in Tibet*.
12. Chang Chengfa, R. M. Shackleton, and J. F. Dewey (1988) *Geological Evolution of Tibet*, Report on the 1985 Royal Society—Academia Sinica Geotraverse of the Qinghai–Xizang Plateau, The Royal Society, London, with map.
13. W. Gan, P. Zhang et al. (2007) 'Present-Day Crustal Motion within the Tibetan Plateau Inferred from GPS measurements', *Journal of Geophysical Research*, 112, B0846.
14. P. Molnar and P. Tapponnier (1975) 'Cenozoic Tectonics of Asia: Effects of a Continental Collision', *Science*, 189: 419–426; P. Molnar and P. Tapponnier (1978) Active Tectonics of Tibet, *Journal of Geophysical Research*, 83: 5361–75.
15. L. Royden, C. Burchfield, R. King, E. Wang, Z. Chen, F. She, and Y. Liu (1997) 'Surface Deformations and Lower Crustal Flow in Eastern Tibet, *Science*, 276: 788–90.
16. L. Hoke, S. Lamb, D. Hilton, and R. Pored (2000) 'Southern Limit of Mantle-Derived Geothermal Helium Emissions in Tibet: Implications for Lithospheric Structure', *Earth and Planetary Science Letters*, 180: 297–308.
17. S-L. Chung et al. (2005) 'Tibetan Tectonic Evolution Inferred from Spatial and Temporal Variations in Post-Collisional Magmatism', *Earth Science Reviews*, 68: 173–96.
18. G. H. Chan, D. J. Waters, M. P. Searle, J. Aitchison, M. Horstwood, Q. Crowley, C-H. Lo, and J. Chan (2009) 'Probing the Basement of Southern Tibet: Evidence from Crustal Xenoliths Entrained in a Miocene Ultrapotassic Dyke', *Journal of Geological Society of London*, 166: 45–52.
19. K. D. Nelson, W. Zhao, and project INDEPTH members (1996) 'Partially Molten Middle Crust Beneath Southern Tibet: Synthesis of Project INDEPTH results', *Science*, 274: 1684–96.
20. Several papers linking the INDEPTH seismic lines with Himalayan geology are found in M. Hauck, K. D. Nelson, L. D. Brown, W. Zhao, and A. R. Ross (1998) 'Crustal Structure of the Himalayan Orogen at 90° East Longitude from Project INDEPTH Deep Reflection Profiles', *Tectonics*, 17: 481–500.

M. P. Searle and A. G. Szulc (2005) 'Channel Low and Ductile Extrusion of the High Himalayan Slab—the Kangchenjunga–Darjeeling Profile, Sikkim Himalaya', *Journal of Asian Earth Sciences*, 25: 173–85.

M. P. Searle, R, D. Law, and M. J. Jessup (2006) 'Crustal Structure, Restoration and Evolution of the Greater Himalaya in Nepal—South Tibet: Implications for Channel Flow and Ductile Extrusion of the Middle Crust', *Geological Society* of *London*, Special Publication 268: 355–78.

21. E. Argand (1924) La tectonique de l'Asie. International Geological Congress, report session 13: 170–372.
22. J. F. Dewey and K. C. A. Burke (1973) 'Tibetan, Variscan and Precambrian Basement Reactivation: Products of a Continental Collision', *Journal of Geology*, 81: 683–92.
23. R. Armijo, P. Tapponnier, and T. Han (1989) 'Late Cenozoic Right-Lateral Strike-Slip Faulting in Southern Tibet', *Journal of Geophysical Research*, 94: 2787–838; G. Peltzer and P. Tapponnier (1988) 'Formation and Evolution of Strike-Slip Faults, Rifts and Basins during the India–Asia Collision: An Experimental Approach', *Journal of Geophysical Research*, 93: 15085–117
24. P. England and G. Houseman (1988) 'The Mechanics of the Tibetan Plateau', *Philosophical Transactions of the Royal Society*, London, 326: 301–19.
25. P. Molnar, P. England, and J. Martinod (1993) 'Mantle Dynamics, Uplift of the Tibetan Plateau and the Indian Monsoon', *Reviews of Geophysics*, 31: 357–96.
26. K. Preistley and D. McKenzie (2006) 'The Thermal Structure of the Lithosphere from Shear Wave Velocities', *Earth and Planetary Science Letters*, 244: 285–301.
27. P. Kapp, P. G. DeCelles, G. Gehrels, M. Heizler, and L. Ding (2005) 'Geological Records of the Lhasa-Qiantang and Indo-Asian Collisions in the Nima Area Central Tibet', *Geological Society of America Bulletin*, 119: 917–32.
28. R, A. Spicer et al. (2003) 'Constant Elevation of South Tibet over the Past 15 Million Years', *Nature*, 421: 622–4.
29. M. P. Searle, R. Parrish, A. V. Thow, S. R. Noble, R. J. Phillips, and D. J. Waters (2010) 'Anatomy, Age and Evolution of a Collisional Mountain Belt: The Baltoro Granite Batholith and the Karakoram Metamorphic Complex, Pakistani Karakoram', *Journal of the Geological Society of London*, 167: 183–202.
30. P. Kapp, P. DeCelles, G. Gehrels, M. Heizler, and L. Ding (2007) 'Geological Records of the Lhasa-Qiangtang and Indo-Asian Collisions in the Nima Area, Central Tibet', *Geological Society of America Bulletin*, 119: 917–32.
31. See numerous articles and photographs by Tom Nakamura in the Japanese Alpine Journals.
32. Chris Bonnington and Charles Clarke (1999) *Tibet's Secret Mountain: Sepu Kangri*, London: Trafalgar Square Publishing.
33. M. R. St-Onge, M. P. Searle, and N. Wodicka (2006) 'Trans-Hudson Orogen of North America and Himalaya-Karakoram-Tibetan Orogen of Asia: Structural and Thermal Characteristics of the Lower and Upper Plates', *Tectonics*, 25, doi:10.1029/2005TC001907.2006.
34. R. J. Phillips, R. R. Parrish, and M. P. Searle (2004) 'Age Constraints on Ductile Deformation and Long-Term Slip Rates along the Karakoram Fault Zone, Ladakh', *Earth and Planetary Science Letters*, 226: 305–19.

35. P. H. Leloup, R. Lacassin, et al. (1995) The Ailao Shan–Red River Shear Zone (Yunnan, China): Tertiary Transform Boundary of Indochina. *Tectonophysics*, 251: 3–84.

Chapter 12. Extruding Indochina: Burma, Thailand, Vietnam, Yunnan

1. J. R. Curray (2005) 'Tectonics and History of the Andaman Sea Region', *Journal of Asian Earth Sciences*, 25: 1–42.
2. L.A.N. Iyer (1953) 'The Geology and Gemstones of the Mogok Stone Tract, Burma', *Memoir of the Geological Survey of India*, 82: 1–100.
   R. E. Kane and R. C. Kammerling (1992) 'Status of Ruby and Sapphire Mining in the Mogok Stone Tract', *Gems and Gemology*, 28: 152–74.
   Tony Waltham (1999) 'The ruby mines of Mogok. *Geology Today*, 143–9.
3. M. P. Searle, S. R. Noble, J. M. Cottle, D. J. Waters, A. H. G. Mitchell, T. Hlaing, and M. S. A. Horstwood (2007) 'Tectonic Evolution of the Mogok Metamorphic Belt, Burma (Myanmar) Constrained by U-Th-Pb Dating of Metamorphic and Magmatic Rocks, *Tectonics*, 26, TC3014, doi: 10.1029/2006TC002083.
4. A. L. Stork, N. D. Selby, R. Heyburn, and M. P. Searle (2008) 'Accurate Relative Earthquake Hypocenters Reveal Structure of the Burma Seismic Zone', *Bulletin of the Seismological Society of America*, 98: 2815–927.
5. M. P. Searle and C. K. Morley (2010) 'Tectonics and Thermal Evolution of Thailand in the Regional Context of Southeast Asia', *Geological Society Memoir, Geology of Thailand*, Ch. 20: 540–71.
6. P. H. Leloup, T. M. Harrison, F. J. Ryerson, Li Qi Chen Wenji, P. Tapponnier, and R. Lacassin (1993) 'Structural, Petrological and Thermal Evolution of a Tertiary Ductile Strike-Slip Shear Zone, Diancang Shan, Yunnan', *Journal of Geophysical Research*, 98: 6715–43.
   P. H. Leloup, R. Lacassin, P. Tapponnier, U. Schärer, Z. Dalai, L. Xiaohan, Z. Liangshang, J. Shaocheng, and P. T. Trinh (1995) 'The Ailao Shan–Red River Shear Zone (Yunnan, China), Tertiary Transform Boundary of Indochina', *Tectonophysics*, 251: 3–84.
   P. H. Leloup, N. Arnaud, R. Lacassin, J. R. Kienast, T. M. Harrison, T. P. Trong, A. Replumaz, and P. Tapponnier, (2001) 'New Constraints on the Structure, Thermochronology and Timing of the Ailao Shan-Red River Shear Zone, SE Asia', *Journal of Geophysical Research*, 106: 6657–71.
7. M. P. Searle (2006) 'Role of the Red River Fault, Yunnan and Vietnam, in the Continental Extrusion of Southeast Asia', *Journal of the Geological Society of London*, 163: 1–12.
8. S-L. Chung, M. Chu, Y. Zhang, Y. Xie, C. Lo, T. Lee, C. Lan, X. Li, Q. Zhang, and Y. Wang (2005) 'Tibetan Tectonic Evolution Inferred from Spatial and Temporal Variations in Post-Collisional Magmatism, *Earth Science Reviews*, 68: 173–96.
9. M. P. Searle, M-W. Yeh, T-H. Lin, and S-L. Chung, (2010). Structural constraints on the timing of left-lateral shear along the Red River shear zone in the Ailao Shan and Diancang Shan Ranges, Yunnan, SW China. *Geosphere*, 6, no. 4, 1-23, doi: 10.1130/GES00580.1.
10. L. H. Royden, B. C. Burchfiel, R. W. King, E. Wang, Z. Chen, F. Shen, and L. Lupine (1998) 'Surface Deformation and Lower Crustal Flow in Eastern Tibet', *Science*, 276: 788–90.
    M. K. Clark and L. H. Royden (2000) 'Topographic Ooze: Building the Eastern Margin of Tibet by Lower Crustal Flow, *Geology*, 28: 703–6.

11. E. Kirby, (2000) 'Neotectonics of the Min Shan, China: Implications for Mechanisms Driving Quaternary Deformation Along the Eastern Margin of the Tibetan Plateau', *Geological Society of America Bulletin*, 112: 375–93.
    E. Kirby et al. (2002) 'Late Cenozoic Evolution of the Eastern Margin of the Tibetan Plateau: Inferences from ^{40}Ar/^{39}Ar and (U-Th)/He Thermochronology', *Tectonics*, 21, doi: 10.1029/2000TC001246.
12. B. C. Burchfiel, L. H. Royden, R. D. van der Hilst, B. H. Hager, Z. Chen, R. W. King, C. Li, J. Lu, H. Yao, and E. Kirby (2008) 'A Geological and Geophysical Context for the Wenchuan Earthquake of 12 May 2008, Sichuan, People's Republic of China', *GSA Today*: 18: 4–11.
13. C. J. L. Wilson, M.J. Harrowfield, and A.J. Reid (2006) 'Brittle Modification of Triassic Architecture in Eastern Tibet: Implications for the Construction of the Cenozoic Plateau', *Journal of Asian Earth Sciences*, 27, 341–57.
14. F. Roger, S. Calasou, J. Lancelot, J. Malavielle, M. Mattauer, X. Zhiqin, H. Ziwen, and H. Liwei (1995) 'Miocene Emplacement and Deformation of the Konga Shan Granite (Xianshui He Fault Zone, West Sichuan, China): Geodynamic Implications', *Earth and Planetary Science Letters*, 130: 201–216.
15. W. Gan, et al. (2007) 'Present-Day Crustal Motion within the Tibetan Plateau Inferred from GPS Measurements', *Journal of Geophysical Research*, 112, B08416.
16. R. Lacassin, A. Replumaz, A. and P. H. Leloup (1998) 'Hairpin River Loops and Slip Sense Inversion on Southeast Asian Strike-Slip Faults', *Geology*, 26: 703–6.

Chapter 13. The Day the Earth Shook: the Sumatra–Andaman Earthquake, December 2004

1. C. J. Ammon et al. (2005) 'Rupture Process of the 2004 Sumatra–Andaman Earthquake', *Science*, 308: 1133–9.
   R. Bilham (2005) 'A Flying Start then a Slow Slip', *Science*, 308: 1126–7.
   R. Bilham, E. R. Engdahl, N. Feldl, S. P. Satyabala (2005) 'Partial and Complete Rupture of the Indo-Andaman Plate Boundary 1847–2004, *Seismol. Res. Lett.* 76, 299–311 (2005).
   S. Stein and E. A. Okal (2005) 'Speed and Size of the Sumatra Earthquake', *Nature*, 434: 581–2.
2. NASA news January 2005.
3. K. Monecke et al. (2008) 'A 1,000 Year Sediment Record of Tsunami Recurrence in Northern Sumatra, *Nature*, 455: 1232–4.
4. J. R. Curray (2005) 'Tectonics and History of the Andaman Sea Region', *Journal of Asian Earth Sciences*, 25: 187–232.
5. T. N. Pandit (1990) *The Sentinelese*, Kolkata: Seagull Books.
   Adam Goodheart (2000) *The Lost Island of the Savages*, The American Scholar, 69, 13–44.
6. M. P. Searle (2006) 'Co-seismic Uplift of Coral Reefs along the Western Andaman Islands during the December 26th 2004 Earthquake', *Coral Reefs*, 25(2), doi: 10.1007/s00338-005-0051-z.
7. NASA Jet Propulsion Laboratory. Website <http://www.jpl.nasa.gov/>.
8. R. W. Briggs, K. Sieh, et al. (2006) 'Deformation and Slip along the Sunda Megathrust in the Great 2005 Nias-Simeulue Earthquake', *Science*, 311: 1897–901.

9. M. Tobita, H. Suito, T. Imakire, M. Kato, S. Fujiwara, and M. Murakami (2006) 'Outline of Vertical Displacement of the 2004 and 2005 Sumatra Earthquakes Revealed by Satellite Radar Imagery', *Earth Planets Space*, 58: e1–14.
10. Simon Winchester (2003) *Krakatao: The Day the World Exploded: August 27, 1883*, Ontario: Harper.
11. C. A. Chesner, W. I. Rose, A. Deino, R. Drake, and J. A. Westgate (1991) 'Eruptive History of Earth's largest Quaternary Caldera (Toba, Indonesia) Clarified', *Geology*, 19: 200–3.
W. I. Rose and C. A. Chesner (1987) 'Dispersal of Ash in the great Toba Eruption, 75 ka', *Geology*, 15: 913–7.
12. M. R. Rampino and S. Self (1992) 'Volcanic Winter and Accelerated Glaciation Following the Toba Super-Eruption', *Nature*, 359: 50–2.
M. R. Rampino, and S. Self (1993) 'Climate-Volcanism Feedback and the Toba Eruption ~74,000 Years Ago', *Quaternary Research*, 40: 269–80.
13. R. B. Stothers (1984) 'The Great Tambora Eruption in 1815 and its Aftermath', *Science*, 224: 1191–8.
14. Clive Oppenheimer (2011) *Eruptions that Shook the World*, Cambridge, UK; New York: Cambridge University Press.
15. T. Pal (2010) 'The 2005–2006 Eruption of the Barren Volcano, Andaman Sea: Evolution of Basaltic Magmatism in Island Arc Setting of Andaman-Java Subduction Complex', *Journal of Asian Earth Sciences*, 39: 12–23, doi:10.1016/jseaes.2010.02.005.
16. D. Haldar (1992) 'Volcanic Eruption of the Barren Island Volcano, Andaman Sea', *Journal of the Geological Society of India*, 39: 411–9.

Chapter 14. The Making of the Himalaya, Karakoram, and Tibetan Plateau

1. R. D. Law, M. P. Searle, and L. Godin (eds) (2006) 'Channel Flow, Ductile Extrusion and Exhumation in Continental Collision Zones', *Geological Society of London*, Special Publication, Volume 268.
2. K. D. Nelson, W. Zhao, and project INDEPTH members (1996) 'Partially Molten Middle Crust Beneath Southern Tibet: Synthesis of Project INDEPTH Results. *Science*, 274: 1684–96.
3. P. Molnar, P. England, and J. Martinod (1993) 'Mantle Dynamics, Uplift of the Tibetan Plateau and the Indian Monsoon', *Reviews of Geophysics*, 31: 357–96.
4. J. F. Dewey and K. C. A. Burke (1973) 'Tibetan, Variscan and Precambrian Basement Reactivation: Products of a Continental Collision', *Journal of Geology*, 81: 683–92.
5. K. Priestley and D. McKenzie (2006) 'The Thermal Structure of the Lithosphere from Shear Wave Velocities', *Earth and Planetary Science Letters*, 244: 285–301.
6. P. Molnar and P. Tapponnier (1975) 'Cenozoic Tectonics of Asia: Effects of a Continental Collision', *Science*, 189: 419–26.
7. C. Beaumont, R. A. Jamieson, M. H. Nguyen, and B. Lee (2001) 'Himalayan Tectonics Explained by Extrusion of a Low-Viscosity Crustal Channel Coupled to Focussed Surface Denudation', *Nature*, 414: 738–42.
8. M. P. Searle, R. R. Parrish, A. V. Thow, S. R. Noble, R. J. Phillips, and D. J. Waters (2010) 'Anatomy, Age and Evolution of a Collisional Mountain Belt: The Baltoro Granite Batho-

lith and the Karakoram Metamorphic Complex, Pakistani Karakoram', *Journal of the Geological Society of London*, 167: 183–202.

9. M. P. Searle, J. R. Elliott, R. J. Phillips, and S-L Chung (2011) 'Crustal-Lithospheric Structure and Continental Extrusion of Tibet', *Journal of the Geological Society of London*, 168: 633–72, doi: 10.1144/0016-76492010-139.
10. E. B. Burov and A. B. Watts (2006) 'The Long-Term Strength of the Continental Lithosphere: "Jelly Sandwich" or "Crème Brûle"? *GSA Today*, 16, 4-10.
11. J. Jackson (2002) 'Strength of the Continental Lithosphere: Time to Abandon the Jelly Sandwich?' *GSA Today*, 12(9): 4–9.
12. M. R. St-Onge, N. Rayner, and M. P. Searle (2010) 'Zircon Age Determinations for the Ladakh Batholith at Chumathang (Northwest India): Implications for the Age of the India–Asia Collision in the Ladakh Himalaya', *Tectonophysics*, 495: 171–83.
13. L.R. Wager (1937) 'The Arun River drainage and the rise of the Himalaya', *Geographical Journal*, 89, 239–50.
14. B. Hallet and P. Molnar (2001) 'Distorted Drainage Basins as Markers of Crustal Strain East of the Himalaya', *Journal of Geophysical Research*, 196: 13697–709.
15. M. K. Clark et al. (2004) 'Surface Uplift, Tectonics and Erosion of Eastern Tibet from Large-Scale Drainage Patterns', *Tectonics*, 23, TC1006, doi:10.1029/2002TC001402.
16. M. E. Raymo and W. F. Ruddiman (1992) 'Tectonic Forcing of Late Cenozoic Climate', *Nature*, 359: 117–22.
17. F. M. Richter, D. B. Rowley, and D. J. DePaulo (1992) 'Sr Isotope Evolution of Seawater: The Role of Tectonics', *Earth and Planetary Science Letters*, 109: 11–23.

REFERENCES

Allégre, C. J. et al. (1984) 'Structure and Evolution of the Himalaya—Tibet Orogenic Belt, *Nature*, 307: 17–22.

Allen, C. (1982) *A Mountain in Tibet: The Search for Mount Kailas and the Sources of the Great Rivers of India*, London: Deutsch.

Ammon, C. J. et al. (2005) 'Rupture Process of the 2004 Sumatra–Andaman Earthquake', *Science*, 308: 1133–9.

Argand, E. (1924) *La tectonique de l'Asie*. Thirteenth International Geological Congress, Brussels 1924, 171–372.

Armijo, R., Tapponnier, P., and Han, T. (1989) 'Late Cenozoic Right-Lateral strike-Slip Faulting in Southern Tibet', *Journal of Geophysical Research*, 94: 2787–838.

Astill, T. (2005) Mount Everest—The Reconnaisance 1935: The Forgotten Adventure, privately published by the author.

Bailey, F.M. (1945) *China, Tibet, Assam*, London: Cape.

—— (1957) *No Passport to Tibet*, London: Rupert Hart-Davis.

Baker, I. (2004) *The Heart of the World: A Journey to the Last Secret Place*, New York: Souvenir Press.

—— (2006) *Heart of the Himalaya: A Journey to Tibet's Lost Paradise*, New York: Penguin.

Beaumont, C., Jamieson, R. A., Nguyen, M. H., and Lee, B. (2001) 'Himalayan Tectonics Explained by Extrusion of a Low-Viscosity Crustal Channel Coupled to Focused Surface Denudation', *Nature*, 414: 738–42.

Bedi, R., and Bedi, R. (1986) *Ladakh*, Delhi: Brijbasi Press.

Bilham, R. (2005) 'A Flying Start then a Slow Slip', *Science*, 308: 1126–7.

—— Engdahl, E. R., Feldl, N., Satyabala, S. P. (2005) 'Partial and Complete Rupture of the Indo-Andaman Plate Boundary 1847–2004, *Seismol. Res. Lett.* 76, 299–311 (2005).

Bonnington, C. (1989) *Mountaineer: Thirty Years of Climbing on the World's Great Peaks*, London: Diadem Books.

—— and Clarke, C. (1999) *Tibet's Secret Mountain: Sepu Kangri*, London: Trafalgar Square Publishing.

Booth, A. L., Chamberlain, C. P., Kidd, W. S. F., and Zeitler, P. K. (2009) 'Constraints on the Metamorphic Evolution of the Eastern Himalayan Syntaxis from Geochronologic and Petrologic Studies of Namche Barwa', *Geological Society of America Bulletin*, 121: 385–407.

Boukreev, A. and Weston DeWald, G. (1999) *The Climb: Tragic Ambitions on Everest*, New York: St. Martin's Griffin.

Breashers, D. (1999) *An Enduring Passion for Everest and Unforgiving Places*, New York: Simon & Schuster.

Briggs, R. W., Sieh, K., et al. (2006) 'Deformation and Slip along the Sunda Megathrust in the Great 2005 Nias-Simeulue Earthquake', *Science*, 311: 1897–901.

Bryson, B. (2003) *A Short History of Nearly Everything*, London: Doubleday Transworld Publishers.

Buhl, H. (1956) *Nanga Parbat Pilgrimage*, London: Hodder and Stoughton.

Burbank, D. and Anderson, R. (2005) *Tectonic Geomorphology*, Chichester: Wiley-Blackwell.

Burchfiel, B. C., Royden, L. H., van der Hilst, R. D., Hager, B. H., Chen, Z., King, R. W., Li, C., Lu, J., Yao, H., and Kirby, E. (2008) 'A Geological and Geophysical Context for the Wenchuan Earthquake of 12 May 2008, Sichuan, People's Republic of China', *GSA Today*, 18: 4–11.

Burchfiel, B. C., Zhiliang, C., Hodges, K. V., Yuping, L., Royden, L., Changrong, D., and Jiene, X. (1992) 'The South Tibetan Detachment System, Himalayan Orogen: Extension Contemporaneous with and parallel to Shortening in a Collisional Mountain Belt', *Geological Society of America Special Paper*, 269.

Burg, J-P. (1984) Geological map of South Tibet, CNRS, Paris.

Burrard, S. G. and Hayden, H. H. (1907) *A Sketch of the Geography and Geology of the Himalaya Mountains and Tibet*, Calcutta: Government of India.

Burov, E. B. and Watts, A. B. (2006) 'The Long-Term Strength of the Continental Lithosphere: "Jelly Sandwich" or "Crème Brûle"? *GSA Today*, 16: 4–10.

Butler, R. W. H. and Prior, D. J. (1988) 'Tectonic Controls on the Uplift of the Nanga Parbat Massif, Pakistan Himalaya', *Nature*, 333: 247–50.

Chan, G. H., Waters, D. J., Searle, M. P., Aitchison, J., Horstwood, M., Crowley, Q., Lo, C-H., and Chan, J. (2009) 'Probing the Basement of Southern Tibet: Evidence from Crustal Xenoliths Entrained in a Miocene Ultrapotassic Dyke', *Journal of the Geological Society of London*, 166: 45–52.

Chengfa, C., Shackleton, R. M., and Dewey, J. F. (1988) *Geological Evolution of Tibet*, Report on the 1985 Royal Society–Academia Sinica Geotraverse of the Qinghai–Xizang Plateau, The Royal Society, London, with map.

Chesner, C. A., Rose, W. I., Deino, A., Drake, R., and Westgate, J. A. (1991) 'Eruptive History of Earth's Largest Quaternary Caldera (Toba, Indonesia) Clarified', *Geology*, 19: 200–3.

Chung, S-L., et al. (2005) 'Tibetan Tectonic Evolution Inferred from Spatial and Temporal Variations in Post-Collisional Magmatism', *Earth Science Reviews*, 68: 173–96.

Clark, M. K. and Royden, L. H. (2000) 'Topographic Ooze: Building the Eastern Margin of Tibet by Lower Crustal Flow. *Geology*, 28: 703–6.

Clark, M. K., et al. (2004) 'Surface Uplift, Tectonics and Erosion of Eastern Tibet from Large-Scale Drainage Patterns', *Tectonics*, 23, TC1006, doi:10.1029/2002TC001402.

Colchen, M., Pêcher, A., and LeFort, P. (1986) *Annapurna—Manaslu—Ganesh Himalaya*, Paris: CNRS.

Coleman, R. G. and Wang, X. (eds) (1995) *Ultrahigh Pressure Metamorphism*, New York: Cambridge University Press.

Cottle, J. M., Jessup, M. J., Newell, D. L., Searle, M. P., Law, R. D., and Horstwood, M. S. A. (2007) 'Structural Insights into the Early Stages of Exhumation along an Orogen-Scale Detachment: The South Tibetan Detachment System, Dzakaa Chu section, Eastern Himalaya', *Journal of Structural Geology*, 29: 1781–97. doi:10.1016/j.jsg.2007.08.007.

Cottle, J. M., Searle, M. P., Horstwood, M. S. A., and Waters, D. J. (2009) 'Timing of Mid-Crustal Metamorphism, Melting and Deformation in the Mount Everest Region of South Tibet Revealed by U-Th-Pb Geochronology, *Journal of Geology*, 117: 643–64, doi: 10.1086/60599.

Cox, K., Storm, K., and Baker, I. (eds) (1999) *Frank Kingdon Ward's Riddle of the Tsangpo Gorges: Retracing the Epic Journey of 1924–25 in Southeast Tibet'*, Woodbridge, Suffolk: Antique Collectors Club.

Crowley, J. L., Waters, D. J., Searle, M. P., and Bowring, S. A. (2009) 'Pleistocene Melting and Rapid Exhumation of the Nanga Parbat Massif, Pakistan: Age and P-T Conditions of Accessory Mineral Growth in Migmatite and Leucogranite', *Earth and Planetary Science Letters*, 288: 408–20.

Curran, J. (1995) *K2: The Story of the Savage Mountain*, London: Hodder and Stoughton.

Curray, J. R. (2005) 'Tectonics and History of the Andaman Sea Region', *Journal of Asian Earth Sciences*, 25: 1–42.

Dalrymple, G. B. (1991) *The Age of the Earth*, Stanford, CA: Stanford University Press.

Desio, A. and Zanettin, B. (1974) *Geology of the Baltoro Basin: Italian Expeditions to the Karakoram and Hindu Kush*, Leiden: Brill.

Dewey, J. F. and Bird, J. M. (1970) 'Mountain Belts and the New Global Tectonics', *Journal of Geophysical Research*, 75, 2625–47.

Dewey, J. F. and Burke, K. C. A. (1973) 'Tibetan, Variscan and Precambrian Basement Reactivation: Products of a Continental Collision', *Journal of Geology*, 81: 683–92.

Dewey, J. F., Shackleton, R. M., Chang, C., and Sun, Y. (eds) (1990) *The Geological Evolution of Tibet*, Royal Society of London Memoir.

DeWit, M. J., Jeffery, M., Bergh, H., and Nicolaysen, L. (1988) *Geological Map of Sectors of Gondwana Reconstructed to their Disposition~150 Ma*, American Association of Petroleum Geologists and University of Witwatersrand.

Dickinson, L. (1993) *Ballooning over Everest*, Jonathan Cape.

Dilek, Y. and Newcomb, S. (2003) *Ophiolite Concept and the Evolution of Geological Thought*, Geological Society of America, Special Paper 373.

Drew, F. (1875) *Jummo and Kashmir*, London: Edward Stanford.

Dunai, T. (2010) *Cosmogenic Nuclides: Principles, Concepts and Applications in the Earth Surface Science*, Cambridge: Cambridge University Press.

England, P. and Houseman, G. (1988) 'The Mechanics of the Tibetan Plateau', *Philosophical Transactions of the Royal Society*, London, 326: 301–19.

Fellowes, P. F. M. (1934) *First Over Everest: The Houston Mount Everest Expedition, 1933*, New York: Robert M. McBride.

French, P. (1994) *Younghusband: The Last Great Imperial Adventurer*, London: HarperCollins.

Gan, W., Zhang, P., et al. (2007) 'Present-Day Crustal Motion within the Tibetan Plateau Inferred from GPS measurements', *Journal of Geophysical Research*, 112, B08846.

Gansser, A. (1964) *Geology of the Himalayas*, Chichester: John Wiley & Sons.

Glennie, K., Boeuf, M. G. A., Hughes-Clarke, M. W., Moody-Stuart, M., Pilaar W. F. H., and Reinhardt, B. M. (1974) *Geology of the Oman Mountains*, 2 volumes and geological map, Verhandelingen van het Koninklijk Nederlands geologisch mijnbouwkundig Genootschap.

Goodheart, A. (2000) 'The Lost Island of the Savages', *The American Scholar*, 69: 13–44.

Gruisen, J. Van, and Rao, N. (1989) *Ladakh: The Secret Land beyond the Himalaya*, New Delhi: Lustre Press.

Green, O. R., Searle, M. P., Corfield, R. I., and Corfield, R. M. (2008) 'Cretaceous-Tertiary Carbonate Platform Evolution and the Age of the India-Asia Collision along the Ladakh Himalaya (Northwest India)', *Journal of Geology*, 116, 331–53.

Gregson, J. *Massacre at the Palace—The Doomed Royal Dynasty of Nepal*, New York: Miramax.
Hagen, T. (1960) *Nepal*, Bern: Kümmerly and Frey.
Haldar, D., et al. (1992) 'Volcanic Eruption of the Barren Island Volcano, Andaman Sea', *Journal of the Geological Society of India*, 39: 411–19.
Hallet, B. and Molnar, P. (2001) 'Distorted Drainage Basins as Markers of Crustal Strain East of the Himalaya', *Journal of Geophysical Research*, 196: 13697–709.
Harrer, H. (1953) *Seven Years in Tibet*, London: Rupert Hart-Davis.
—— (1980) *Ladakh: Gods and Mortals Behind the Himalayas*, Innsbruck: Penguin-Verlag.
Hauck, M., Nelson, K. D., Brown, L. D., Zhao, W., and Ross, A. R. (1998) 'Crustal Structure of the Himalayan Orogen at 90° East Longitude from Project INDEPTH Deep Reflection Profiles', *Tectonics*, 17: 481–500.
Hedin, S. (1898) *Through Asia*, 2 vols, London: Methuen & Co.
—— (1910) *Trans-Himalaya Discoveries and Adventures in Tibet*, 4 vols, London: MacMillan.
Heim, A. and Gansser, A. (1939) *Central Himalaya: Geological Observations of the Swiss Expedition 1936*, Zurich: Gebrüder Fretz.
—— —— (1939) and Gansser, A. (1964) *Geology of the Himalaya*, Chichester: John Wiley & Sons.
Heron, A. M. (1922) 'Geological Results of the Mount Everest Reconnaissance Expedition', *Records of the Geological Survey of India*, LIV, 215–34.
Hildebrand, P. R., Noble, S. R., Searle, M. P., Parrish, R. R., and Shakirullah (1998) 'Tectonic significance of 24 Ma crustal melting in the eastern Hindu Kush, Pakistan', *Geology*, 26, 871–4.
Hildebrand, P. R., Noble, S. R., Searle, M. P., and Parrish, R. R. (2001) 'Old Origin for an Active Mountain Range: Geology and Geochronology of the Eastern Hindu Kush, Pakistan.' *Geological Society of America Bulletin*, 113: 625–39.
Hildebrand, P. R., Searle, M. P., Shakirullah, Z. K., and van Heijst, H. J. (2000) 'Geological Evolution of the Hindu Kush, NW Frontier Pakistan: Active Margin to Continent–Continent Collision Zone', in M. Asif Khan, P. J. Treloar, M. P. Searle, and M. Qasim Jan (eds) 'Tectonics of the Nanga Parbat Syntaxis and the Western Himalayas', *Geological Society of London Special Publication*, 170: 277–93.
Hillary, E. (1955) *High Adventure*. London: Hodder and Stoughton, reprinted by Oxford University Press.
Hodges, K. V., Parrish, R. R., and Searle, M. P. (1996) 'Tectonic Evolution of the Central Annapurna Range, Nepalese Himalayas', *Tectonics*, 15(6): 1264–91.
Hoke, L., Lamb, S., Hilton, D., and Pored, R. (2000) 'Southern Limit of Mantle-Derived Geothermal Helium Emissions in Tibet: Implications for Lithospheric Structure', *Earth and Planetary Science Letters*, 180: 297–308.
Holmes, A. (1944) *Principles of Physical Geology*, Edinburgh: Thomas Nelson & Sons.
Hopkirk, P. (1984a) *Foreign Devils on the Silk Road: The Search for the Lost Cities and Treasures of Chinese Central Asia*, London: John Murray.
—— (1984b) *Setting the East Ablaze: Lenin's Dream of an Empire in Asia*, London: Kodansha International.
—— (1995) *Trespassers on the Roof of the World: The Secret Exploration of Tibet*, London: John Murray.
—— (2006) *The Great Game*, London: John Murray.
Houston, C. and Bates, R. (1939) *Five Miles High*, New York: Dodd, Mead & Co.
—— —— (1954) *K2—The Savage Mountain*, New York: McGraw-Hill.
How The Earth Was Made, History Channel TV documentary.

Hunt, J. (1953) *The Ascent of Everest*, London: Hodder and Stoughton.

Iyer, L. A. N. (1953) 'The Geology and Gemstones of the Mogok Stone Tract, Burma', *Memoir of the Geological Survey of India*, 82: 1–100.

Jackson, J. (2002) 'Strength of the Continental Lithosphere: Time to Abandon the Jelly Sandwich?' *GSA Today*, 12(9): 4–9.

Jessup, M., Law, R. D., Searle, M. P., and Hubbard, M. S. (2006) 'Structural Evolution and Vorticity of Flow during Extrusion and Exhumation of the Greater Himalayan Slab, Mount Everest Massif, Tibet/Nepal: Implications for Orogen-Scale flow Partitioning', in R. D. Law, M. P. Searle, and L. Godin (eds) 'Channel Flow, Ductile Extrusion and Exhumation in Continental Collision Zones', *Geological Society of London Special Publication*, 268: 379–413.

Jessup, M. J., Searle, M. P., Cottle, J. M., Law, R. D., Newell, D. L., Tracy, R. J., and Waters, D. J. (2008) 'P-T-t-D Paths of Everest Series Schist, Nepal', *Journal of Metamorphic Petrology*, doi:10.1111/j.1525-1314.2008.00784.x.

Kane, R. E. and Kammerling, R. C. (1992) 'Status of Ruby and Sapphire Mining in the Mogok Stone Tract', *Gems and Gemology*, 28: 152–74.

Kapp, P., DeCelles, P. G., Gehrels, G., Heizler, M., and Ding, L. (2005) 'Geological Records of the Lhasa-Qiantang and Indo-Asian Collisions in the Nima Area Central Tibet', *Geological Society of America Bulletin*, 119: 917–32.

Kearey, P. and Vine, F. J. (1990) *Global Tectonics*, Oxford: Blackwell Science.

Keay, J. (1977) *When Men and Mountains Meet*, London: John Murray.

—— (1979) *The Gilgit Game*, London: John Murray.

—— (1991) *The Honourable Company: A History of the English East India Company*, London: Harper Collins.

—— (1999) *The Great Game*, Lodon: Jonathan Cape.

—— (2000) *The Great Arc*, London: HarperCollins.

Kingdon-Ward, F. (1913) *The Land of the Blue Poppy*, Cambridge: Cambridge University Press.

—— (1926) *The Riddle of the Tsangpo Gorges*, London: Edward Arnold & Co.

—— (1941) *A Plant-Hunter in Tibet*, London: Jonathan Cape.

—— (1949) *Assam Adventure*, London: Jonathan Cape.

—— (1949) *Burma's Icy Mountain*, London: Jonathan Cape.

Kirby, E., et al. (2000) 'Neotectonics of the Min Shan, China: Implications for Mechanisms driving Quaternary Deformation along the Eastern Margin of the Tibetan Plateau', *Geological Society of America Bulletin*, 112: 375–93.

Kirby, E., et al. (2002) 'Late Cenozoic Evolution of the Eastern Margin of the Tibetan Plateau: Inferences from $^{40}Ar/^{39}Ar$ and (U-Th)/He Thermochronology', *Tectonics*, 21, doi: 10.1029/2000TC001246.

Krakauer, J. (1998) *Into Thin Air: A Personal Account of the Mt. Everest Disaster*, New York: Anchor Books Doubleday.

Lacassin, R., Replumaz, A., and Leloup, P. H. (1998) 'Hairpin River Loops and Slip Sense Inversion on Southeast Asian Strike-Slip Faults', *Geology*, 26: 703–6.

Laccasin, R., Valli, F., et al. (2004) Large-Scale Geometry, Offset and Kinematic Evolution of the Karakoram Fault, Tibet', *Earth and Planetary Science Letters*, 219: 255–69.

Lacedelli, L. and Caenacchi, G. (2006) *K2 The Price of Conquest*, Ross-on-Wye: Carreg Ltd.

Lamb, S. and Sington, D. (2003) *Earth Story*, London: BBC Books.

Law, R. D., Searle, M. P., and Godin, L. (eds) (2006) 'Channel Flow, Ductile Extrusion and Exhumation in Continental Collision Zones', *Geological Society of London Special Publication*, Volume 268.

Law, R. D., Searle, M. P., and Simpson, R. L. (2004) 'Strain, Deformation Temperatures and Vorticity of Flow at the Top of the Greater Himalayan Slab, Everest Massif, Tibet', *Journal of the Geological Society of London*, 161: 305–20.

LeFort, P. (1981) 'Manaslu Leucogranite: A Collision Signature of the Himalaya, a Model for its Genesis and Emplacement', *Journal of Geophysical Research*, 86, 10545–68.

Leloup, P. H., Arnaud, N., Lacassin, R., Kienast, J. R., Harrison, T. M., Trong, T. P., Replumaz, A., and Tapponnier, P. (2001) 'New Constraints on the Structure, Thermochronology and Timing of the Ailao Shan-Red River Shear Zone, SE Asia', *Journal of Geophysical Research*, 106: 6657–71.

Leloup, P. H., Harrison, T. M., Ryerson, F. J., Chen Wenji, L. Q., Tapponnier, P., and Lacassin, R. (1993) 'Structural, Petrological and Thermal Evolution of a Tertiary Ductile Strike-Slip Shear Zone, Diancang Shan, Yunnan', *Journal of Geophysical Research*, 98: 6715–43.

Leloup, P. H., Lacassin, R., et al. (1995) 'The Ailao Shan–Red River Shear Xone (Yunnan, China): Tertiary Transform Boundary of Indochina', *Tectonophysics*, 251: 3–84.

Lewis, C. (2000) *The Dating Game*, Cambridge; New York; Melbourne: Cambridge University Press.

McCartney-Snape, T. (1991) *Everest: From Sea to Summit*, Sydney: Australian Geographic Society.

McPhee, J. (1982) *Basin and Range*, New York: Farrar, Straus and Giroux.

Mantovani, R. (1997) *Everest: The History of the Himalayan Giant*, New Delhi: Om Book Service.

—— and Diemberger, K. (1997) *K2, Challenging the Sky*, Seattle, WA: Mountaineers Books.

Margolis, E. S. (2000) *War at the Top of the World*, New York: Routledge.

Messner, R. (1980) *Solo, Nanga Parbat*, New York: Oxford University Press.

—— (1998) *The Crystal Horizon: Everest—The First Solo Ascent*, Seattle, WA: Mountaineers Books.

—— (1999) *Everest: Expedition to the Ultimate*, Seattle, WA: Mountaineers Books.

Misch, P. (1949) 'Metasomatic Granitization of Batholithic Dimensions', *American Journal of Science*, 247: 209–45.

Molnar, P., England, P., and J. Martinod (1993) 'Mantle Dynamics, Uplift of the Tibetan Plateau and the Indian Monsoon', *Reviews of Geophysics*, 31: 357–96.

Molnar, P. and Tapponnier P. (1975) 'Cenozoic Tectonics of Asia: Effects of a Continental Collision', *Science*, 189: 419–26.

—— —— (1978) 'Active Tectonics of Tibet', *Journal of Geophysical Research*, 83: 5361–75.

Monecke, K., et al. (2008) 'A 1,000 Year Sediment Record of Tsunami Recurrence in Northern Sumatra', *Nature*, 455(7217): 1232–4.

Morris, J. (2000) *Coronation Everest*, London: Faber & Faber.

Myrow, P. M., Hughes, N. C., Searle, M. P., Fanning, C. N., Peng, S-C., and Parcha, S. K. (2008) 'Stratigraphic Correlation of Cambrian-Ordovician Deposits Along the Himalaya: Implications for the Age and Nature of Rocks in the Mount Everest Region', *Bulletin of the Geological Society of America*, 120: 323–32.

National Geographic Map of Mount Everest, Boston Museum of Science.

NASA Jet Propulsion Laboratory. Website: <http://jpl.nasa.gov/>.

Nelson, K. D., Zhao, W., and project INDEPTH members (1996) 'Partially Molten Middle Crust beneath Southern Tibet: Synthesis of Project INDEPTH Results', *Science*, 274: 1684–96.

Noble, S. R. and Searle, M. P. (1995) 'Age of Crustal Melting and Leucogranite Formation from U-Pb Zircon and Monazite Dating in the Western Himalaya, Zanskar, India', *Geology*, 23: 1135–8.

Norgay, J. T. (2002). *Touching My Father's Soul: In the Footsteps of Tenzin Norgay*, London: Wisdom Books.

O'Brien, P., Zotov, N., Law, R., Khan, M. A. and Jan, M. Q. (2001) 'Coesite in Himalayan Eclogite and Implications for Models of India-Asia Collision', *Geology*, 29: 435–8.

Odell, N. E. (1925) 'Observations on the Rocks and Glaciers of Mount Everest', *Geographical Journal*, LXVI: 289–315.

Olschak, B., Gansser, A., and Bührer, E. (1987) *The Himalayas*, New York and Oxford: Facts on File.

Oppenheimer, C. (2011) *Eruptions that Shook the World*, Cambridge; New York: Cambridge University Press.

Pal, T., et al. (2010) 'The 2005–2006 Eruption of the Barren Volcano, Andaman Sea: Evolution of Basaltic Magmatism in Island Arc Setting of Andaman-Java Subduction Complex', *Journal of Asian Earth Sciences*, 39: 12–23, doi:10.1016/jseaes.2010.02.005.

Pandit, T. N. (1990) *The Sentinelese*, Kolkata: Seagull Books.

Parrish, R. R., Gough, S. J., Searle, M. P., and Waters, D. J. (2006) 'Plate Velocity Exhumation of Ultrahigh-Pressure Eclogites in the Pakistan Himalaya', *Geology*, 34: 989–92.

Parrish, R. and Noble, S. (2003) 'Zircon U-Th-Pb Geochronology by Isotope Dilution—Thermal Ionization Mass Spectrometry (ID-TIMS)', in J. Hanchar and P. Hoskin (eds) *Zircon: Reviews of Mineralogy and Geochemistry*, Mineralogical Society of America, 183–213.

Peach, B. N., Horne, J., Gunn, W., Clough, C. T., Hinxman, L. W., and Teall, J. J. H. (1907) *The Geological Structure of the North-West Highlands of Scotland*, Memoir of the Geological Survey of Great Britain, Glasgow: HMSO.

Pearce, J. A., Alabaster, T., Shelton, A. W., and Searle, M. P. (1981) 'The Oman Ophiolite as a Cretaceous Arc-Basin Complex: Evidence and Implications', *Philosophical Transactions of the Royal Society of London* Series A, 300, 299–317.

Pegler, G. and Das, S. (1998) 'An Enhanced Image of the Pamir–Hindu Kush Seismic Zone from Relocated Earthquake Hypocenters', *Geophysical Journal International*, 134: 573–95.

Peltzer, G. and Tapponnier, P. (1988) 'Formation and Evolution of Strike-Slip Faults, Rifts and Basins during the India–Asia collision: An Experimental Approach', *Journal of Geophysical Research*, 93: 15085–117.

Pettersen, M. G. (2010) 'A Review of the Geology and Tectonics of the Kohistan Island Arc', *Geological Society of London Special Publication*, 338: 287–327.

Phillips, R. J., Parrish, R. R., and Searle, M. P. (2004) 'Age Constraints on Ductile Deformation and Long-Term Slip Rates along the Karakoram Fault', *Earth and Planetary Science Letters*, 226: 305–19.

Phillips, R. J. and Searle, M. P. (2007) 'Macrostructural and Microstructural Architecture of the Karkoram Fault: Relationship between Magmatism and Strike-Slip Faulting', *Tectonics*, 26, TC3017, doi:10.029/2006TC001946.

Phillimore, R. H. (1950–1968) *Historical Records of the Survey of India*, Dehra Dun.

Prakash, A. R. (2004) *Maoists in the Land of Buddha*, Delhi and Kathmandu: Nirala Books.

Priestley, K. and McKenzie, D. (2006) 'The Thermal Structure of the Lithosphere from Shear Wave Velocities', *Earth and Planetary Science Letters*, 244: 285–301.

Rampino, M. R. and Self, S. (1992) 'Volcanic Winter and Accelerated Glaciation Following the Toba Super-Eruption', *Nature*, 359: 50–2.

Rampino, M. R. and Self, S. (1993) 'Climate-Volcanism Feedback and the Toba Eruption ~74,000 Years Ago', *Quaternary Research*, 40: 269–80.

Rashid, A. (2001) *Taliban, The Story of the Afghan Warlords*, London: Pan Books.

Raymo, M. E. and Ruddiman, W. F. (1992) 'Tectonic Forcing of Late Cenozoic Climate. *Nature*, 359: 117–22.

Rex, A. J., Searle, M. P., Tirrul, R., Crawford, M. B., Prior, D. J., Rex, D. C., and Barnicoat, A. (1988) 'The Geochemical and Tectonic Evolution of the Central Karakoram, N. Pakistan', *Philosophical Transactions of the Royal Society, London*, A326, 229–55.

Richter, F. M., Rowley, D. B., and DePaulo, D. J. (1992) 'Sr Isotope Evolution of Seawater: The Role of Tectonics', *Earth and Planetary Science Letters*, 109: 11–23.

Rider, M. (2005) 'Hutton's Arse: 3 Billion Years of Extraordinary Geology in Scotland's Northern Highlands', Sutherland: Rider-French.

Roger, F., Calasou, S., Lancelot, J., Malavielle, J., Mattauer, M., Zhiqin, X., Ziwen, H., and Liwei, H. (1995) 'Miocene Emplacement and Deformation of the Konga Shan Granite (Xianshui He Fault Zone, West Sichuan, China): Geodynamic Implications, *Earth and Planetary Science Letters*, 130: 201–16.

Rose, W. I., and Chesner, C. A. (1987) 'Dispersal of Ash in the Great Toba Eruption, 75 ka', *Geology*, 15: 913–7.

Rowell, G. (1977) *In the Throne Room of the Mountain Gods*, London: George Allen & Unwin.

Royden, L., Burchfield, C., King, R., Wang, E., Chen, Z., She, F., and Liu, Y. (1997) 'Surface Deformations and Lower Crustal Flow in Eastern Tibet', *Science*, 276: 788–90.

Ruttledge, H. (1934) *Everest 1933*, London: Hodder & Stouton.

St-Onge, M. R., Searle, M. P., and Wodicka, N. (2006) 'Trans-Hudson Orogen of North America and Himalaya-Karakoram-Tibetan Orogen of Asia: Structural and Thermal Characteristics of the Lower and Upper Plates', *Tectonics*, 25, doi:10.1029/2005TC001907.2006.

St-Onge, M. R., Rayner, N., and Searle, M. P. (2010) 'Zircon Age Determinations for the Ladakh Batholith at Chumathang (Northwest India): Implications for the Age of the India-Asia Collision in the Ladakh Himalaya', *Tectonophysics*, 495: 171–83.

Schofield, V. (2000) *Kashmir in Conflict*, London; New York: I.B. Tauris.

Scott, D. (1993) *Himalayan Climber: A Life Time's Quest to the World's Greater Ranges*, London: Bâton Wickes.

Searle, M. P. (1986) 'Structural Evolution and Sequence of Thrusting in the High Himalayan, Tibetan Tethys and Indus Suture Zones of Zanskar and Ladakh, Western Himalaya', *Journal of Structural Geology*, 8, 923–36.

—— (1991) *Geology and Tectonics of the Karakoram Mountains*, Chichester: John Wiley & Sons.

—— (1996) 'Geological Evidence Against Large-Scale Pre-Holocene Offsets along the Karakoram Fault: Implications for the Limited Extrusion of the Tibetan Plateau', *Tectonics*, 15, 171–86.

—— (1999a) 'Extensional and Compressional Faults in the Everest–Lhotse Massif, Khumbu Himalaya, Nepal', *Journal of the Geological Society of London*, 156, 227–40.

—— (1999b) 'Emplacement of Himalayan Leucogranites by Magma Injecion along Giant Sill Complexes: Examples from the Cho Oyu, Gyachung Kang and Everest Leucogranites (Nepal Himalaya)', *Journal of Asian Earth Sciences*, 17: 773–83.

—— (2003) Geological Map of the Mount Everest Region Nepal and South Tibet, scale 1:50,000, Oxford University. Second edition published 2007.
—— (2006a) 'Role of the Red River Fault, Yunnan and Vietnam, in the Continental Extrusion of Southeast Asia', *Journal of the Geological Society of London*, 163: 1–12.
—— (2006b) 'Co-seismic Uplift of Coral Reefs along the Western Andaman Islands during the December 26th 2004 Earthquake, *Coral Reefs*, 25(2), doi: 10.1007/s00338-005-0051-z.
—— (2010) 'Low-Angle Normal Faults in the Compressional Himalayan Orogen: Evidence from the Annapurna–Dhaulagiri Himalaya, Nepal', *Geosphere*, 6,(4): 296–315.
—— Cooper, D. J. W., and Rex, A. J. (1988) 'Collision Tectonics of the Ladakh-Zanskar Himalaya', London: Philosophical Transactions of the Royal Society, A326, 117–50.
—— Corfield, R., Stephenson, B., and McCarron, J. (1997) 'Structure of the North Indian Continental Margin in the Ladakh-Zanskar Himalaya: Implications for the Timing of Obduction of the Spontang Ophiolite, India-Asia Collision and Deformation Events in the Himalaya', *Geological Magazine*, 134: 297–316.
—— Cottle, J. M., Streule, M. J., and Waters, D. J. (2010) 'Crustal Melt Granites and Migmatites along the Himalaya: Melt Source, Segregation, Transport and Granite Emplacement Mechanisms', *Transactions of the Royal Society*, Edinburgh, 100: 219–33.
—— and Cox, J. S. (1999). 'Tectonic Setting, Origin and Obduction of the Oman Ophiolite', *Geological Society of America Bulletin*, 111, 104–22.
—— Elliott, J. R., Phillips, R. J., and Chung, Sun-Lin (2011) 'Crustal-Lithospheric Structure and Continental Extrusion of Tibet', *Journal of the Geological Society of London*, 168, 633–72, doi: 10.1144/0016-76492010-139.
—— and Godin, L. (2003) 'The South Tibetan Detachment and the Manaslu Leucogranite: A Structural Reinterpretation and Restoration of the Annapurna-Manaslu Himalaya, Nepal', *Journal of Geology*, 111: 505–23.
—— Hacker, B. R., and Bilham, R. (2001) 'The Hindu Kush Seismic Zone as a Paradigm or the Creation of Ultrahigh-Pressure Diamond- and Coesite-Bearing Continental Rocks', *Journal of Geology*, 109: 143–53.
—— Law, R. D., Godin, L., Larson, K. P., Streule, M. J., Cottle, J. M., and Jessup, M. J. (2008) 'Defining the Himalayan Main Central Thrust in Nepal', *Journal of the Geological Society of London*, 164: 523–34.
—— —— Jessup, M., and Simpson, R. L. (2006) 'Crustal Structure and Evolution of the Greater Himalaya in Nepal–South Tibet: Implications for Channel Flow and Ductile Extrusion of the Middle Crust', in R. D. Law, M. P. Searle, and L. Godin (eds) 'Channel Flow, Ductile Extrusion and Exhumation in Continental Collision Zones', *Geological Society of London Special Publication*, 268: 355–78.
—— and Malpas, J. (1980) 'Structure and Metamorphism of Rocks Beneath the Semail Ophiolite of Oman and their Significance in Ophiolite Obduction', *Philosophical Transactions of the Royal Society of Edinburgh*, 71: 213–28.
—— and Morley, C. K. (2010) 'Tectonics and Thermal Evolution of Thailand in the Regional Context of Southeast Asia', *Geological Society Memoir, Geology of Thailand*, Ch. 20: 540–71.
—— Noble, S. R., Hurford, A. J., and Rex, D. C. (1999) 'Age of Crustal Melting, Emplacement and Exhumation History of the Shivling Leucogranite', *Geological Magazine*, 136: 513–25.
—— —— Cottle, J. M., Waters, D. J., Mitchell, A. H. G., Hlaing, T., and Horstwood, M. S. A. (2007) 'Tectonic Evolution of the Mogok Metamorphic Belt, Burma (Myanmar) Con-

strained by U-Th-Pb Dating of Metamorphic and Magmatic Rocks', *Tectonics*, 26, TC3014, doi: 10.1029/2006TC002083.

—— Parrish, R. R., Thow, A. V., Noble, S. R., Phillips, R. J., and Waters, D. J. (2010) 'Anatomy, Age and Evolution of a Collisional Mountain Belt: The Baltoro Granite Batholith and Karakoram Metamorphic Complex, Pakistani Karakoram', *Journal of the Geological Society of London*, 167: 183–202, doi: 10.1144/0016-76492009-043.

—— —— Tirrul, R., Rex, D. C. (1990) 'Age of Crystallization and Cooling of the K2 Gneiss in the Baltoro Karakoram', *Journal of the Geological Society of London*, 147: 603–6.

—— —— Hodges, K. V., Hurford, A., Ayres, M. W., and Whitehouse, M. J. (1997) 'Shisha Pangma Leucogranite, South Tibet: Field Relations, Geochemistry, Age, Origin, and Emplacement, *Journal of Geology*, 105: 295–317.

—— and Phillips, R. J. (2007) 'Relationships between Right-Lateral Shear along the Karakoram Fault Metamorphism, Magmatism Exhumation and Uplift: Evidence from the K2-Gasherbrum-Pangong Ranges, North Pakistan and Ladakh', *Journal of the Geological Society of London*, 164: 439–50.

—— and Rex, A. J. (1989) 'Thermal Model for the Zanskar Himalaya,' *Journal of Metamorphic Geology*, 7: 127–34.

—— —— Tirrul, R., Rex, D. C., Barnicoat, A., and Windley, B. F. (1989) 'Metamorphic, Magmatic and Tectonic Evolution of the Central Karakoram in the Biafo-Baltoro-Hushe Regions of N. Pakistan', *Geological Society of America Special Paper* 232, 47–73.

—— Simpson, R. L., Law, R. D., Parrish, R. R., and Waters, D. J. (2003) 'The Structural Geometry, Metamorphic and Magmatic Evolution of the Everest Massif, High Himalaya of Nepal–South Tibet', *Journal of the Geological Society of London*, 160: 345–66.

—— and Szulc, A. G. (2005) 'Channel Low and Ductile Extrusion of the High Himalayan Slab—the Kangchenjunga–Darjeeling Profile, Sikkim Himalaya', *Journal of Asian Earth Sciences*, 25: 173–85.

—— Yeh, M-W., Lin, T-H., and Chung, S-L. (2010) 'Structural Constraints on the Timing of Left-Lateral Shear along the Red River Shear Zone in the Ailao Shan and Diancang Shan Ranges, Yunnan, SW China', *Geosphere*, 6(4): 1-23, doi: 10.1130/GES00580.1.

Shipton, E. (1938) *Blank on the Map*, London: Hodder and Stoughton.

—— (1969) *That Untravelled World*, London: Hodder and Stoughton.

Smith, A. G., Hurley, A. H., and Briden, J. C. (1981) *Phanerozoic Palaeocontinental World Maps*, Cambridge: Cambridge University Press.

Snellgrove, D. L. and Skorupsky, T. (1980) *Cultural Heritage of Ladakh*, Warminster: Aris & Philips.

Spicer, R. A., et al. (2003) 'Constant Elevation of South Tibet over the Past 15 million Years', *Nature*, 421: 622–4.

Stein, S. and Okal, E. A. (2005) 'Speed and Size of the Sumatra Earthquake', *Nature*, 434: 581–2.

Stephenson, B. J., Searle, M. P., Waters, D. J., and Rex, D. C. (2001) 'Structure of the Main Central Thrust Zone and Extrusion of the High Himalayan Deep Crustal Wedge, Kishtwar-Zanskar Himalaya,' *Journal of the Geological Society of London*, 158: 637–52.

Stephenson, B. J., Waters, D. J., and Searle, M. P. (2000) 'Inverted Metamorphism and the Main Central Thrust: Field Relations and Thermobarometric Constraints from the Kishtwar Window, NW Indian Himalaya,' *Journal of Metamorphic Geology*, 18, 571–90.

Stork, A. L., Selby, N. D., Heyburn, R., and Searle, M. P. (2008) 'Accurate Relative Earthquake Hypocenters Reveal Structure of the Burma Seismic Zone', *Bulletin of the Seismological Society of America*, 98: 2815–927.

Stothers, R. B. (1984) 'The Great Tambora Eruption in 1815 and its Aftermath', *Science*, 224: 1191–8.

Stow, D. (2010) 'Vanishing Ocean: How Tethys Reshaped the World', Oxford; New York: Oxford University Press.

Streule, M. J., Phillips, R. J., Searle, M. P., Waters, D. J., and Horstwood, M. (2009) 'Evolution and Chronology of the Pangong Metamorphic Complex Adjacent to the Karakoram Fault, Ladakh: Constraints from Thermobarometry, Metamorphic Modelling and U-Pb Geochronology', *Journal of the Geological Society of London*, 166: 919–32.

—— Searle, M. P., Waters, D. J., and Horstwood, M. (2010) 'Metamorphism, Melting and Channel Flow in the Greater Himalayan Sequence and Makalu Leucogranite: Constraints from Thermobarometry, Metamorphic Modeling and U-Pb Geochronology. *Tectonics*, 29, TC5011 doi:10.1029/2009TC002533.

Tobita, M., Suito, H., Imakire, T., Kato, M., Fujiwara, S., and Murakami, M. (2006) 'Outline of Vertical Displacement of the 2004 and 2005 Sumatra Earthquakes Revealed by Satellite Radar Imagery, *Earth Planets* Space, 58, e1–14.

Venables, S. (2000) *Everest: Alone at the Summit*, Bath: Odyssey Books.

Vine, F. J., and Matthews, D. H. (1963) 'Magnetic Anomalies over Oceanic Ridges', *Nature*, 199, 947–9.

Wadia, D. N. (1931) 'The Syntaxis of the Northwest Himalaya: Its Rocks, Tectonics and Orogeny, *Records of the Geological Survey of India*, 65: 189–220.

Wager, L. R. (1934) 'A Review of Geology and Some New Observations', in H. Ruttledge, *Everest 1933*, London: Hodder and Stoughton, 312–36.

—— (1937) 'The Arun River Drainage and the Rise of the Himalaya', *Geographical Journal*, 89: 239–50.

—— (1939) 'The Lachi Series of North Sikkim and the Age of the Rocks Forming Mount Everest', *Records of the Geological Survey of India*, LXXIV: 171–88.

—— (1965) 'Injected Granite Sheets of the Rongbuk Valley and the North Face of Mount Everest', D. N. Wadia Commemorative volume, Mining and Metallurgical Institute of India.

Wagner, G. A. and Van den Haute, P. (1992) *Fission-Track Dating*, Dordrecht: Kluwer Academic Publishers.

Wahid, S., and Storm, K. Jr. (1989) *Ladakh Between Earth and Sky*, Bombay: B.I. Publications.

Wake, C. and Searle, M. (1993) 'Rapid Advance of the Pumarikish Glacier, Hispar Glacier Basin, Karakoram Himalaya', *Journal of Glaciology*, 39(131): 204–6.

Walker, C. B., Searle, M. P., and Waters, D. J. (2001) 'An Integrated Tectono-Thermal Model for the Evolution of the High Himalayas in Western Zanskar, with Constraints from Thermobarometry Metamorphic Modelling', *Tectonics*, 20: 810–833.10.

Walker, J. D., Martin, M. W., Bowring, S., Searle, M. P., Waters, D. J., and Hodges, K. V. (1999) 'Metamorphism, Melting and Extension: Age Constraints from the High Himalayan Slab of SE Zanskar and NW Lahoul, Indian Himalaya', *Journal of Geology*, 107: 473–95.

Waller, D. (1988) *The Pundits: British Exploration of Tibet and Central Asia*, Lexington: University Press of Kentucky.

Waltham, T. (1999) 'The Ruby Mines of Mogok', *Geology Today*, 15: 143–9.

Ward, M. (2003) *Everest: A Thousand Years of Exploration*, Glasgow: The Ernest Press.

Watts, A. B. (2001) *Isostasy and Flexure of the Lithosphere*, Cambridge; New York; Melbourne: Cambridge University Press.

Wilson, C. J. L., Harrowfield, M. J., and Reid, A. J. (2006) 'Brittle Modification of Triassic Architecture in Eastern Tibet: Implications for the Construction of the Cenozoic Plateau', *Journal of Asian Earth Sciences*, 27: 341–57.

Winchester, S. (2001) *The Map that Changed the World*, London: Penguin, Viking.

—— (2003) *Krakatao: The Day the World Exploded: August 27, 1883*, Ontario: Harper.

Yardley, B. *Introduction to Metamorphic Petrology*, Harlow: Longman Earth Science Series.

Young, D. A. (2003) *Mind over Magma: The Story of Igneous Petrology*, Princeton, NJ: Princeton University Press.

Younghusband, F. (1896) *The Heart of a Continent*, London: John Murray, reprinted 1984 Oxford University Press.

Zeitler, P. K., et al. (2001a) 'Crustal Reworking at Nanga Parbat, Pakistan: Metamorphic Consequences of Thermal-Mechanical Coupling Facilitated by Erosion', *Tectonics*, 20: 712–28.

Zeitler, P. K., et al. (2001b) 'Erosion, Himalayan Geodynamics and the Geomorphology of Metamorphism', *GSA Today*, 11: 4–9.

COMMUNITY ACTION NEPAL (CAN)
WWW.CANEPAL.ORG

The mountains of Nepal are one of the world's most inspiring landscapes; however, life for remote mountain populations has many challenges and difficulties. Mountaineer Doug Scott CBE, the first Englishman to climb Mount Everest in 1975, has seen these daily challenges at first hand, when relying on the support of local mountain porters and Sherpas to help him achieve climbing success and relative fame and fortune. Doug explains it like this: 'In 1989 whilst walking up to Kangchenjunga, I made a decision that would eventually associate me more intimately with the local hill people of High Asia. Sitting around the kitchen fire, our Sherpas and porters were bemoaning the fact that the trekking industry offered little security or return for all their hard work. We therefore launched a trekking cooperative with guaranteed incomes and a chance for porters to determine their own future.'

In 1994 profits from this trekking company were first put into community work. This was at Ghunsa, where a school and health post were established at the request of the local community and Tej Tamang, a trek Sirdar. CAN still continues to work with this community today. Now Community Action Nepal supports over forty projects in the middle hill regions, where government welfare provision is limited. These range from running village health posts with trained nurses, where previously patients were faced with many days walk to obtain treatment, to building and staffing schools, which soon become the hub of village life. Doug has not forgotten the porters and, in partnership with the International Porter Protection Group (IPPG), CAN have built two Porter Shelters in the Khumbu and is currently raising funds to build a third at Gokyo Lakes. These shelters offer porters overnight accommodation, kitchen and sanitation facilities at very little cost, since the shelters are subsidized by fees charged to trekkers for treatment by the volunteer Western doctors who staff them.

CAN works very closely with local village communities and operates from a UK-based office and an office in Kathmandu. As Doug puts it: 'We work on the principle that we are all on our own journeys and no one is more important than the other. The local hill people have as much to teach us, as we, the outsiders, have to show them, if not more so.' To find out more about CAN's work or to make a donation visit their website at www.canepal.org.uk or call 01768 484842.

INDEX